Host Manipulation by Parasites

EDITED BY

David P. Hughes
Department of Entomology and Biology, Penn State University, USA

Jacques Brodeur
Department of Biological Sciences, University of Montreal, Canada

Frédéric Thomas
Centre National de la Recherche Scientifique (CNRS) and Institut de Recherche pour le Développement (IRD), France

OXFORD
UNIVERSITY PRESS

OXFORD
UNIVERSITY PRESS

Great Clarendon Street, Oxford OX2 6DP,
United Kingdom

Oxford University Press is a department of the University of Oxford.
It furthers the University's objective of excellence in research, scholarship,
and education by publishing worldwide. Oxford is a registered trade mark of
Oxford University Press in the UK and in certain other countries

First Edition published in 2012

Impression: 1

British Library Cataloguing in Publication Data

Data available Library of Congress Control Number: 2012934755

Library of Congress Cataloging in Publication Data

Library of Congress Control Number: 2012934755

ISBN 978–0–19–964223–6 (Hbk.)
      978–0–19–964224–3 (Pbk.)

Printed and bound by
CPI Group (UK) Ltd, Croydon, CR0 4YY

Contents

List of contributors

Shelley A. Adamo, Department of Psychology/ Neuroscience Institute, Dalhousie University, Halifax, NS, B3H 3X5, Canada
sadamo@dal.ca

John Alcock, School of Life Sciences, PO Box 874501, Arizona State University, Tempe, AZ 85287-4501, USA
j.alcock@asu.edu

Victoria Braithwaite, School of Forest Resources, University Park, Penn State University, PA 16802, USA
v.braithwaite1@mac.com

Jacques Brodeur, Département de Sciences Biologiques, Université de Montréal, 4101, rue Sherbrooke Est, Montréal (Québec), H1X 2B2, Canada
jacques.brodeur@umontreal.ca

Frank Cézilly, Université de Bourgogne, Equipe Ecologie Evolutive, UMR CNRS 5561 Biogéosciences, 6 blvd. Gabriel, 21000 Dijon, France
frank.cezilly@u-bourgogne.fr

Richard Dawkins, New College, University of Oxford, Oxford, UK.
Richard.Dawkins1@me.com

Frédérique Dubois, Université de Montréal, Département de Sciences Biologiques, CP 6128, Succursale Centre-ville, Montréal (Québec), H3C 3J7, Canada
frederique.dubois@umontreal.ca

Scott V. Edwards, Department of Organismic and Evolutionary Biology, Harvard University, 26 Oxford Street, Cambridge, MA 02138, USA
sedwards@fas.harvard.edu

Lee Ehrman, Natural Sciences, State University of New York, Purchase College, Purchase, New York, USA
lee.ehrman@purchase.edu

Bert Hölldobler, School of Life Sciences, Arizona State University, PO Box 874501, Tempe, AZ 85287, USA and University of Würzburg, Biozentrum, Zoologie II, Am Hubland, D-97074 Würzburg, Germany
Bert.Hoelldobler@asu.edu

David P. Hughes, Centre for Infectious Disease Dynamics, MSC, University Park, Penn State University, PA 16802, USA
dhughes@psu.edu

Pedro Jordano, Integrative Ecology Group, Estación Biológica de Doñana, CSIC, Sevilla, Spain
jordano@ebd.csic.es

Alex Kacelnik, Behavioural Ecology Research Group, Department of Zoology, Oxford University, Oxford, OX1 3PS, UK
alex.kacelnik@zoo.ox.ac.uk

Armand M. Kuris, Ecology, Evolution & Marine Biology, University of California, Santa Barbara, Santa Barbara, CA 93106-9620, USA
kuris@lifesci.ucsb.edu

Kevin D. Lafferty, United States Geological Survey, Western Ecological Research Center and Marine Science Institute, University of California, Santa Barbara, California 93106, USA
klafferty@usgs.gov

Naomi E. Langmore, Research School of Biology, Australian National University, Canberra 0200, Australia
naomi.langmore@anu.edu.au

Edward P. Levri, Department of Biology, Penn State Altoona, 3000 Ivyside Park, Altoona, PA 16601, USA
EPL1@psu.edu

Michel Loreau, McGill University, Department of Biology, 1205 ave Docteur Penfield, Montréal (Québec), H3A 1B1, Canada
michel.loreau@mcgill.ca

Mark C. Mescher, 539 ASI Bldg., Penn State University, University Park, PA 16802, USA
mcmescher@psu.edu

Wolfgang J. Miller, Laboratories of Genome Dynamics, Centre of Anatomy and Cell Biology, Medical University of Vienna, Waehringerstrasse 10, A-1090 Vienna, Austria
wolfgang.miller@meduniwien.ac.at

Janice Moore, Biology Department, Colorado State University, Fort Collins, CO 80523, USA
Janice.Moore@ColoState.EDU

Robert Poulin, Department of Zoology, University of Otago, PO Box 56, Dunedin, New Zealand
robert.poulin@stonebow.otago.ac.nz

Andrew Read, Centre for Infectious Disease Dynamics, MSC, University Park, Penn State University, PA 16802, USA
a.read@psu.edu

Thierry Rigaud, Equipe Ecologie Evolutive, Laboratoire Biogéosciences, UMR CNRS 5561, Université de Bourgogne, 6 boulevard Gabriel, 21000 Dijon, France
thierry.rigaud@u-bourgogne.fr

Gene Robinson, Department of Entomology and Institute for Genomic Biology, University of Illinois at Urbana-Champaign, 320 Morrill Hall, 505 South Goodwin Avenue, Urbana, IL 61801, USA
generobi@illinois.edu

Bernard D. Roitberg, Evolutionary and Behavioural Ecology Research Group, BioSciences, Simon Fraser University, Burnaby, BC V5A 1S6, Canada
roitberg@sfu.ca

Daniela Schneider, Laboratories of Genome Dynamics, Centre of Anatomy and Cell Biology, Medical University of Vienna, Waehringerstrasse 10, A-1090 Vienna, Austria
daniela.schneider@meduniwien.ac.at

Claire N. Spottiswoode, Department of Zoology, University of Cambridge, CB2 3EJ, UK and DST/ NRF Centre of Excellence at the Percy FitzPatrick Institute, University of Cape Town, Rondebosch 7701, South Africa
cns26@cam.ac.uk

Stephen C. Stearns, Department of Ecology and Evolutionary Biology, Yale University, Box 208106, New Haven, CT 06520-8106, USA
stephen.stearns@yale.edu

Frédéric Thomas, MIVEGEC, UMR CNRS/IRD/ UM2/UM1, 911 av Agropolis, BP 64501, 34394 Montpellier Cedex 5, France
frederic.thomas2@ird.fr

Foreword

Richard Dawkins

If I were asked to nominate my personal epitome of Darwinian adaptation, the *ne plus ultra* of natural selection in all its merciless glory, I might hesitate between the spectacle of a cheetah outsprinting a jinking Tommie in a flurry of African dust, or the effortless streamlining of a dolphin, or the sculptured invisibility of a stick caterpillar, or a pitcher plant silently and insensibly drowning flies. But I think I'd finally come down on the side of a parasite manipulating the behaviour of its host—subverting it to the benefit of the parasite in ways that arouse admiration for the subtlety, and horror at the ruthlessness, in equal measure. No need to single out any particular example, they abound on every page of this splendid book and they are thrilling in their dark ingenuity.

Hughes, Brodeur, and Thomas have assembled a team of experts to fill eleven chapters with different aspects of this engrossing topic, skilfully edited to achieve a rare lightness of style and with the added bonus—which I have not seen before—of an afterword by yet another expert, reflecting on each chapter and giving a fresh perspective. My own perspective is scarcely fresh—it is exactly thirty years since *The Extended Phenotype* was published (Dawkins, 1982)—but forgive me for hoping it is worth briefly restating. I shall refrain from detailed citations of that book in what follows. Parasitic manipulations of host behaviour are all extended phenotypes of parasite genes.

We take the *organism* for granted as the level in the hierarchy of life at which we attribute *agency*. It is the organism—not the gene or the cell or the species—that chases, pounces or flees; the organism that patiently rasps its way through the shell of another organism in order to devour it; the organism that inserts its proboscis (or its penis) inside the body of another organism in order to suck blood (or inject sperm). But my thesis is that the organism is a phenomenon whose very existence needs explaining, no less than the phenomenon of sex or the phenomenon of multicellularity.

One could imagine a form of life—on another planet, say—where living matter is not bunched up into discrete bodies—organisms. Just as life didn't have to become multicellular, and just as reproduction didn't have to be sexual, so living material didn't have to become packaged into discrete, individual organisms, bounded by a skin, distinct and separate from other individual organisms, behaving as unitary, purposeful agents. The only thing that is really fundamental to Darwinian life is self-replicating, coded information—genes, in the terminology of life on this planet. For reasons that should be positively explained not just accepted, genes cooperated to develop the organism as their primary vehicle of propagation. Thinking about parasites, and especially parasitic manipulation of hosts, leads us directly to an understanding of the nature and significance of the organism—indeed to what I shall offer as the very definition of an organism.

"An organism is an entity, all of whose genes share the same stochastic expectations of the distant future" (Dawkins, 1990). What does this mean? A typical organism, such as a pheasant or a wombat, is a collection of trillions of cells, each containing near-identical copies of the same genome, which is itself a collection of tens of thousands of genes. Why do all the genes "work together" for the good of the organism rather than going their separate ways as a pure "selfish gene" view of life might expect? It is because they are all constrained to share the same exit route from the present organism into the future: the sperms or eggs or propagules of the present organism. To be more accurate, all have arrived in the present organism by travelling through a large succession of ancestral organisms, leaving each one via shared propagules. All have therefore been selected to "agree" on the same "interests": the organism must do whatever is necessary, given the species way of life, to survive long enough to reproduce; it must successfully woo a member of the opposite sex; it must be good at caring for its young, if that is what the species happens to do. That is why the genes within an organism work harmoniously as a cooperative, rather than as an anarchistic rabble with every gene out for its own selfish survival.

If there is a parasite within the organism, say a fluke, the genes of the parasite potentially can affect the host phenotype—the extended phenotype. As abundantly demonstrated in this book, they normally bend the host phenotype in a direction hostile to that favoured by the host's "own" genes. We take this for granted, but it is true only because they usually do not share a common route into the genetic future and therefore do not stand to gain from the same outcomes. But, to the extent that parasite genes and host genes do share future expectations, to that extent will the parasite become benign. In extreme cases, the distinction between parasite and host will be blurred, and eventually the two merge to form a single organism—as with lichens, for example, or mitochondria in eucaryotic cells.

Consider two kinds of parasite, extremes at the ends of a continuum. To clarify the distinction, we can make them both bacteria and both feed inside the cells of the host. They differ crucially by virtue of their route into their next host. *Verticobacter* can only infect the progeny of its host, and it does so by travelling inside the eggs or sperms of the host. *Horizontobacter* can infect any members of the host species, with no special preference for the individual host's own children.

The point of the difference is this. The genes of *Horizontobacter* have no particular interest in the survival of the individual host, and especially no

interest in the host's successful reproduction. The genes of *Horizontobacter* might happily evolve towards causing its bacterial phenotypes to eat the host to death from within and finally cause the host to explode, releasing a cloud of bacterial spores to the four winds, to be breathed in by new hosts. The genes of *Verticobacter*, on the other hand, have identical interests to the genes of the host. Both "want" the host to survive long enough to reproduce, both "want" the host to be attractive to the opposite sex, both "want" the host to be a good parent, and so on. There is no difference between the interests of host genes and parasite genes, because both stand to gain from exactly the same outcome: the successful passing of gametes in which both sets of genes travel together into the future. The extended phenotypic effects of the genes of *Verticobacter* will tend to improve the survival, sexual attractiveness and parental effectiveness of the host, in exactly the same way as the host's "own" genes, and they will work in cooperation with the host's "own" genes in exactly the same way as the host's "own" genes cooperate with each other. My repetition of quotation marks may seem like overkill, but it is very deliberate. Remember, my thesis is that the only reason an organism's own genes cooperate with each other is that they share their exit route from the present organism into the future.

So closely bound are the genes of host and *Verticobacter*, it is likely that the bacteria will cease to be clearly distinguishable. They will eventually merge into the corporate identity of the organism. That, indeed, is exactly what happened long ago with the bacterial ancestors of mitochondria and chloroplasts, which is why these organelles so long escaped recognition as bacteria at all. The same process of merging identities has been re-enacted more recently a number of times in evolution, for example in some of the large protozoans that inhabit the guts of termites (see "The Mixotrich's Tale" in Dawkins, 2004).

What does it mean to say that genes "want" something? It means that, as a consequence of natural selection, they can be expected to influence phenotypes in the direction of that something, whether or not the phenotype in question can be regarded as belonging to their "own" individual organism.

Indeed it is part of the logic of the extended phenotype argument that the very idea of an "own" individual organism has no necessary validity. The genes of a "parasite" at the *Verticobacter* end of the continuum have just as much right to claim the phenotype of the "host" as the host's "own" genes.

These were the thirty-year-old thoughts that were revived in me as I read this book, and this was the thread with which I bound its provocative wonders together in my mind. But other readers may find other threads and this Foreword has already been self-indulgent enough. The book itself is fascinating. It should be read not just by parasitologists but by ethologists, ecologists, evolutionary biologists, poets and—importantly—doctors, vets and all involved in the health professions. As Nesse and Williams (1994) explain in their pathbreaking book, many of the symptoms of disease are manifestations of parasitic manipulation of patients, and many are counter-manipulations by patients. This book should be on the syllabus of medical schools the world over.

Poets? Yes, poets. And philosophers and theologians too. Samuel Taylor Coleridge attended Humphry Davy's science lectures at the Royal Institution in London "to improve my stock of metaphors." The distinguished philosopher Daniel Dennett begins his magisterial account (2006) of religion as a natural phenomenon with a description of the behavior of an ant, possessed by the brainworm *Dicrocoelium dendriticum*. It is such a powerful metaphor for a human brain, possessed by a powerful, parasitic idea and foregoing all the normal benefits and comforts of life in its service. What might the author of "Yea, slimy things did crawl with legs upon the slimy sea" not have done with the rich materials furnished by this book?

Dawkins, R. (1982). *The Extended Phenotype*. Oxford: Oxford University Press.

Dawkins, R. (1990). Parasites, desiderata lists, and the paradox of the organism. In: *The Evolutionary Biology of Parasitism* Supplement to *Parasitology* (eds Keymer A. E. and Read A. F.) **100**, 63–73.

Dawkins, R. (2004). *The Ancestor's Tale*. New York: Houghton Mifflin.

Nesse, R. and Williams, G. C. (1994). *The New Science of Darwinian Medicine*. New York: Random House.

Richard Dawkins

CHAPTER 1

A history of parasites and hosts, science and fashion

Janice Moore

1.1 Introduction

Not so long ago, the chasm between the study of disease and the study of ecology was so great that no one seemed able to cross it. Then again, few people tried. This is hard to understand now, but it is the way it was. Ecologists saw parasitology as something fit only for medical schools—parasites were for people in white laboratory coats, not for guys like ecologists who attended urban academic conferences kitted out by L. L. Bean, not for guys who viewed themselves as the heirs of John Audubon, if not Ernest Shackleton. Some of the folks alleged to wear white lab coats disagreed, but it was difficult for them to get a hearing, simply because everybody knew that ecology was mainly about birds and mammals, perhaps insects and plants, but certainly not about worms, much less things invisible without a microscope.

This began to change in the late 1970s, with profound effects on the study of evolution and behavioral ecology, and this book will illuminate some of those effects. My task here is to review some of the literature published prior to that sea change in what everybody knew, and I will suggest some things that might have contributed to that shift. For decades, there had been ample evidence supporting the idea that infectious organisms exert powerful evolutionary and ecological effects, but it was evidence that was largely ignored by ecologists and behavioral biologists. I will emphasize the literature around parasite-induced behavioral change; the ecology of

parasites themselves also deserves something of a mention (e.g. Holmes 1961; Schad 1963; Paperna 1964; Simberloff and Moore 1997), in part because that topic has yet to attract the kind of attention that parasite–host interactions have attracted.

That said, it is difficult to understand what contributes to fashion in science. This difficulty must be compounded by the fact that, as scientists, we want to believe our own hype, which states that we are preternaturally rational and open-minded truth-seekers. Scientists are indeed those things in our better moments, but we are also social creatures, and so we tend to travel, intellectually speaking, in herds, a tendency that has been encouraged by the equally marvelous tendency of grant funding to form large clots around a small percentage of research programs. These attributes have played a role in parasite ecology, broadly defined—that is, behavioral, ecological, and evolutionary aspects of parasitism. Much like country music, there was a time when parasites were not cool, and much like country music, they are cool now.[1] This is one view of how parasites came to be cool.

1.2 The days before cool

All manner of people who would not know a parasite life cycle if one bit them nonetheless know about two parasites: the worm that gets into the snail (intermediate host) and transforms its tentacles so that final host birds eat it (i.e., the trematode, *Leucochloridium paradoxum* and congeners) and the

---

[1] "I Was Country when Country Wasn't Cool" –a song by Kye Fleming and Dennis Morgan, made popular by American country singer Barbara Mandrell in 1981—which was about the time parasites began to be seen as cool.

worm that gets into the second intermediate host ant brain and takes control of the ant (i.e., the trematode, *Dicrocoelium dendriticum* and relatives). These have starred in most television documentaries and popular magazine articles about parasites and behavior, and until they recently began to share the spotlight with the even more amazing *Toxoplasma gondii* (and relatives), were among the invertebrates most likely to give bower birds and whales a run for their money in terms of charisma or, more likely, morbid fascination.

I will begin with *L. paradoxum*; unless otherwise indicated, the history I report for this trematode will be taken from Wesenberg-Lund's delightful 1931 account of this organism, supplemented by general aspects of parasitology history as reported in Cheng (1973) and Foster (1965). Adult *Leucochloridium* live in songbirds; eggs pass out with feces and await consumption by snails (e.g., *Succinea*), where they develop into colorful, striped sporocysts that invade the snails' tentacles. In addition to their striped appearance, the sporocysts have been observed to pulsate, especially in the presence of light.

Our knowledge of *Leucochloridium* begins over 200 years ago, and holds several surprises. Even the events leading to the description of the genus have a mysterious air. For instance, Kagan (1952, p. 29) tells us that in 1831, the polymath C. G. Carus reported seeing a copper engraving "of unknown origin and age" in which an infected snail was shown with eight broodsacs. (Carus went on to describe *L. paradoxum* in 1835.) For a while, there was some suggestion that these remarkable sporocysts might be the result of spontaneous generation. Perhaps this is not surprising, because the first trematode life cycle, that of *Fasciola hepatica*, was not worked out until 1881; that trematode itself had been recognized as early as 1379. It was probably much easier to believe that parasites appeared spontaneously in various tissues than to comprehend the fabulous journey that they really do make from host to host.

The notion that life could come from inanimate objects apparently held great appeal; though challenged with clever experiments beginning in the seventeenth century, it was not firmly put to rest until Pasteur's work approximately 150 years ago.

Ironically, a demonstration that fly eggs develop into maggots by Francesco Redi (he of "redia" fame) was among the first to take issue with the notion of spontaneous generation; this physician-entomologist-poet was also the first person that we know of to search through animals with the goal of finding parasites. Because of this, he is considered the founder of parasitology and the redia stage in the trematode life cycle bears his name. None of that saved *Leucochloridium* from spending some time, along with tapeworms, pinworms, and maggots, on the spontaneous generation list. Yet all through the history of parasitology—well before the elucidation of life cycles and Koch's postulates—keen observers were asking if flies could transmit yaws, if mosquitoes might transmit malaria. I doubt that any of us in this post-Enlightenment world could muster the mental agility required to entertain ideas about pathogenic miasma while simultaneously considering the possibility that flies might transmit disease.

Meanwhile, the thought that parasites could increase their appeal or the appeal of their host to predators is older than we might think. For instance, C.T. Siebold (as cited in Kagan 1951) is credited with suggesting in 1853 that *Leucochloridium* in snails might attract avian predators, a suggestion made all the more remarkable because the first trematode life cycle awaited discovery decades in the future (see also Ahrens 1810, cited in Lewis 1977). Wesenberg-Lund (1931) reports that when he finally found an area that contained parasitized snails, he could see their swollen, pulsating tentacles from a distance of two meters. Wesenberg-Lund becomes the first, and perhaps only, person to express compassion for a trematode sporocyst: "…I confess I have been sorry for all the wasted efforts of the sacks pumping and pumping at a rate of about 70 strokes a minute" (p. 98; see Berg 1948 for more about Professor Wesenberg-Lund).

Wesenberg-Lund then presented a puzzle: "For my own part, I could have snails with *Leucochloridia* before my eyes for hours. The birds were singing only a few metres from them, but nevertheless I never had the good fortune to see the birds take the parasites" (p. 98). In this, he is like most of us. Despite the notoriety of *Leucochloridium*, few people have seen birds eat them. In part, this may be because *Leucochloridium* and its gastropod host are both difficult to maintain in the laboratory and patchily distributed in nature. In part, it may be because few people today have the patience of Wesenberg-Lund.

But what of *Dicrocoelium*? *Dicrocoelium dendriticum* is nowhere nearly as conspicuous as *Leucochloridium*; it certainly does not visibly pulsate, nor is it striped. The adult lives in the liver of ruminants; terrestrial snails get infected with larval stages when they eat the shed eggs. The cercariae escape from the snail in masses of mucous called "slime balls," which seem to be considered good eating by ants. It is possible to feel slightly smug that the ant—that paragon of industry and unfailing cooperation—is tempted like the rest of us by unhealthy snacks. (*Dicrocoelium hospes* Looss, 1907 has a similar life cycle with similar effects on the hosts; see Anokhin 1966; Romig et al. 1980 and references therein.)

The snack in question, however, is not merely unhealthy, but potentially deadly. Upon ingestion, one of the cercariae in the slime ball leaves its compatriots and encysts in the subesophageal ganglion of the ant. In the case of *D. hospes*, two metacercariae tend to encyst in the antennal lobes. Infected ants behave normally during the day, but at day's end, when temperatures cool, they crawl up on blades of grass and enter a kind of torpor, often firmly seizing grass blades with mandibles, thus securing their elevated positions—positions that likely increase their vulnerability to unintentional predation by grazing ruminants. Carney (1969) described a similar phenomenon in *Brachylecithum mosquensis*, a dicrocoeliid parasite of perching birds, snails, and once again, ants.

The role of the ant in the life cycle of *Dicrocoelium* is a fairly recent discovery (Krull and Mapes 1953), and the parasite does not have the *frisson* associated with *Leucochloridium*. ("The first sight of the 'pulsating brood-sacs' in the tentacles of the snails is a thrilling sight to a parasitologist." Woodhead 1935, p. 337). Nonetheless, both trematodes have captured public imagination and been grist for the mills of science writers. It is notable that neither *Dicrocoelium* nor *Leucochloridium*—both emblems of parasite manipulation—have been demonstrated to enhance their own transmission to the final host. Neither of these trematodes is well suited to experi-

mental work, and tracking the fate of parasitized ants on a pasture is a daunting prospect.

Nematodes also entered the altered behavior/ transmission story at an early stage, ushered in by Eloise Cram. She was a remarkable woman: in more than 85 years of existence, the American Society of Parasitologists has had almost as many presidents. Five of those presidents have been women, and of those, Cram was the first, in 1956 (the second, Marietta Voge, would be elected 20 years later, and the third, Lillian Mayberry, in another 20 years). Eloise Cram was, indeed, a pioneer, rising to the top of her profession, becoming a world expert first on parasites of poultry and later, on control of schistosomiasis, and leaving over 160 publications in her wake (National Agricultural Library). Early in her career, in one of these papers (Cram 1931), she observed that the larvae of the nematode *Tetrameres americana* cause grasshopper intermediate hosts to be "droopy and inactive, a condition which would make them easy prey for food-seeking fowls in nature." (p. 4). The thought that parasites alter behavior in ways that could affect transmission by influencing predator–prey interactions was in the air, if one cared to notice.

Dame Miriam Rothschild did notice, as she seemed to notice most living things. Described by the *New York Times* (25 January 2005) as "the heiress who discovered how fleas jump," she was fascinated by the natural world, and trematodes in snails were among the many things that caught her attention. In this respect, she was particularly interested in snail gigantism (e.g., Rothschild 1941). She noted in 1940 that there were behavioral differences in the population of *Pseudosuccinea columella* that she studied; snails infected with xiphidiocercariae were found on top of waterlily leaves, in direct sunlight, whereas uninfected snails were not found in such a location (Rothschild 1940, 1941a). In response to Lewis and Wright (1962), who suggested that trematodes in black flies might inhibit biting, Rothschild (1962, p. 1312) began with a statement that reflects the state of parasites and behavior at the time: "Dr. Lewis and Dr. Wright…are to be congratulated on emphasizing an extremely interesting point which has hitherto attracted little, if any, attention—namely, that infestation with larval trematodes can alter the behaviour of the intermediate hosts". After giving a few examples of this phenomenon, she makes one of the earliest suggestions that the behavioral alterations induced by parasites can lead to broader misperceptions about what is going on in nature, including a fair amount of sampling error: "A 'random sample' of specimens may prove to be only a random sample of the infested portion of a population." (p. 1312). In this same publication, Rothschild also reports Jean Baer's observation that the movement of sticklebacks infected with an immature stage of the cestode *Schistocephalus solidus* was impaired to the point of rendering them subject to capture by hand. She clearly saw parasites as players in the behavior of animals, players that had not received the attention they probably deserved.

Sticklebacks are remarkably amenable to laboratory investigation, and beginning in the 1980s, *S. solidus* and sticklebacks were the subject of many elegant studies aimed at elucidating all the ways that the tapeworm influenced stickleback behavior (see Moore 2002; Barber and Scharsack 2010 for reviews). Long before that, however, a related cestode (*Ligula intestinalis*) and its intermediate host, the roach (*Leuciscus rutilus*), were the subjects of the first field study to demonstrate that altered behaviors associated with parasites could result in increased predation on intermediate hosts. While examining the diet of cormorants in The Netherlands, W. H. van Dobben (1952) made some interesting parasitological observations, including the fact that while 30% of the roach taken by cormorants were infected with *L. intestinalis*, only 6.5% of the roach taken by fishermen were infected. Van Dobben suggested that the increased vulnerability of infected roach was the result of impaired movement.

Rothschild (1969) also noticed parasites in her beloved fleas. The way that she came to this observation was serendipitous: specifically, she found a copulating female flea on a pregnant rabbit. At first glance, without consideration of Dame Rothschild's other work on fleas, this might not seem all that notable. Recall, however, that Rothschild and Ford (1964) reported the first case of hormonal synchronization of reproduction in parasite (flea) and host (rabbit), and so this particular flea was not supposed to be mating yet—not until it had transferred to the new-

born rabbits. As Rothschild (1969, p. 10) put it, "…much to our dismay, we observed a pair of rabbit fleas copulating on the ears of a pregnant doe rabbit. We at first assumed that, somehow, we had missed this type of behaviour in the past, and that, on the quiet, some fleas were pairing on the pregnant doe….But on less panicky second thoughts this seemed most improbable, so we fixed the two fleas and sectioned them". The sectioning revealed that the female was host to a mermithid nematode—"coil upon coil of the parasite." (p. 10). Rothschild was left mystified by how the female flea's receptivity was induced, and no wonder. In over 4,000 sectioned fleas, Rothschild found only one mermithid, and that nematode was in the only prematurely copulating flea she observed. This presages what would become a rich literature on compensation for parasite-reduced fecundity, including Minchella and Loverde's (1981) suggestion that parasitized animals facing fecundity reduction might invest more heavily in earlier reproduction.

The fact that mermithids could exert such a powerful effect on an insect host was not as surprising to Rothschild as it might have been, for she was well aware of William Morton Wheeler's early work on mermithids in ants (Wheeler 1907, 1928). As early as 1901, Wheeler had collected highly unusual specimens of *Pheidole dentatus*. These were twice as long as normal individuals, and because their gasters were greatly enlarged, Wheeler estimated their volume to be 10–12x that of normal ants. These mermithergates, as he called them, each contained a mermithid nematode which, when uncoiled, was ten times the length of its enlarged host. These parasitized hosts were not only larger than their uninfected conspecifics, but they also behaved differently. They did not participate in colony tasks such as foraging or caring for brood, but instead begged for food, incessantly and voraciously.

Returning to fleas, these insects were the first hosts of parasites to reveal to investigators that altered host behavior can result in increased parasite transmission, albeit by blood-feeding vectors, not by intermediate hosts encountering predators. Bacot and Martin (1914) were remarkably understated when they said (p. 431), "…whereas certain of our fleas sucked energetically and persistently, no blood entered their stomachs, but the oesophagus became unusually distinct". They went on to report that dissection of these fleas revealed blocked proventriculi, and "It occurred to us that fleas whose proventriculi were obstructed with plague-culture were likely to be responsible for the conveyance of infection. " (p. 432). By following the feeding of individual blocked fleas on rats, Bacot and Martin showed that such exposure led to a high probability of plague transmission.

Finally, in addition to their effects on host behavior, parasites were shown to alter host ecological interactions themselves. Thomas Park, who was cited by the *New York Times* (4 April 1992) as being "instrumental in transforming the field of ecology into a science with quantification and controlled experiments," found that the outcome of competition between two species of flour beetle was reversed by the presence or absence of a protistan parasite (Park 1948). Typically, *Tribolium castaneum* competitively excludes *Tribolium confusum*. If the coccidian parasite *Adelina tribolii* is present, the outcome is reversed.

This review of our knowledge up through, say, the 1960s, is nowhere near exhaustive (see Moore 2002, for a more comprehensive review). It does not have to be exhaustive in order to support the contention that early in the development of modern biology, at least some scientists recognized that parasites could have major effects on host behavior and ecology. This recognition began with hypotheses about *Leucochloridium* in the first half of the nineteenth century and continued with examples from other trematodes, cestodes, nematodes, protists, and even bacteria, living in snails, fish, grasshoppers, fleas, beetles, and ants. The scientists who wondered about parasites and host behavior included some of the bright lights of their era, Fellows of the Royal Society, members of the United States National Academy of Sciences, presidents of learned societies. Casting a wider net to encompass not only behavior and what was becoming behavioral ecology, but also evolution, we see that such luminaries as J. B. S. Haldane (1949) had remarked on the potential significance of disease in evolution. This idea about parasites and how they might affect hosts was not a secret.

Nonetheless, the topic of parasitism, be it in the broad areas of ecology or evolution, or the developing study of animal behavior, was largely ignored. As Haldane (1949, p. 68) put it, "When however an attempt is made to show how natural selection acts, the structure or function considered is almost always one concerned either with protection against natural forces such as cold or against predators, or one which helps the organism to obtain food or mates". In other words, pathogens and parasites did not enter the picture.

This was certainly the world that I found as a graduate student. Because my personal experience encompassed the shift from the world that rejected the behavioral/ecological/evolutionary study of parasite–host interactions to the world that accepts it, I am going to take a brief detour here and describe my personal experience—not because I am somehow special, but because the scientists who didn't live through this era might find it otherwise hard to believe. I had first heard of parasites changing behavior as an undergraduate, when I had the good fortune of studying parasitology at Rice University under the tutelage of Clark Read, acknowledged as one of the more creative thinkers and teachers the field has seen. Read allocated the critical nuts and bolts of parasitology—the identification, diagnosis, and life cycles—to the laboratory, and spent the lecture time pacing the floor, sharing the mysteries of the field and posing thought experiments. One of the mysteries that made its way into Read's classroom was Holmes and Bethel's now classic work on acanthocephalan-induced behavioral alterations (Holmes and Bethel 1972; Bethel and Holmes 1977 and references therein). I had never heard of such a thing, and I was intrigued—by the magic of the life cycles, by the highly derived structures (all those hooks and suckers!), and now by the ability of these animals to seemingly reach out and transform the lives and activities of other animals. I informed Read that I wanted to study parasites and how they affected behavior. He told me that no one was doing much of that kind of research, but that if I got a strong foundation in parasitology and animal behavior, I could certainly put it together myself. In retrospect, I am amazed by his optimism and grateful for my own response to it, however naïve.

I followed Read's advice, and completed a Master's degree focused on insect behavior. I then decided to add parasites to the mix, and looked for a doctoral program where I could do that. The struggle to get a hearing as a graduate student who was interested in parasites and host behavior was an education in itself. Both ecologists and behavioral biologists informed me that parasites had very little to do with their own disciplines and that I should study what they were studying, not parasites. I decided to take some time away from degree programs and found employment as a technician with a truly remarkable scientist, Lynn Riddiford; during that time, I probably learned more about how to conduct science and how to conduct myself as a scientist than I did in any other two years of my professional life. When I returned to school, I understood how to develop an independent research program, and I found a department populated by scientists who were perfectly willing to question what everybody knew, and to believe that if we could study behavior and ecology with birds, we could do it with worms! The University of New Mexico was not rich, but it offered graduate students small research grants to pursue independent projects. It was the perfect setting to study parasites and host behavior.

Of course, there was the nagging thought that this was a temporary respite, that I would finish my degree and the great world of biology would respond with yet another collective yawn when I trotted out my dissertation about parasites and host behavior (see Moore 1983, 1984). Luckily for me, about the time that I finished that dissertation, the scientific world began to change.

This change didn't happen overnight, of course. It had been building for a while. For instance, clever ecologists and evolutionary biologists had been eyeing parasitic castration, offering evolutionary explanations for this widespread phenomenon, a trait of parasites that were possibly having their cake (a living host) and eating it, or at least some of it, too (e.g., Kuris 1974; Baudoin 1975). In addition, stories of altered host behavior had continued to surface. William Bethel and John Holmes (1973, 1974, 1977) had matched van Dobben's field demonstration of altered host behavior resulting in

enhanced predation on intermediate hosts, producing their own clear and elegant laboratory confirmation of the same phenomenon, this time using gammarids infected with acanthocephalans and exposed to duck predation. They categorized types of alterations and made ecological predictions about predator/final host traits that would favor each type. Rothschild's observations of gastropod behavior altered by trematodes were confirmed by other scientists (e.g., Sinderman and Farrin 1962; Lambert and Farley 1968). Questions arose about the possibility of host suicide as an adaptation to parasitism (Shapiro 1976; Smith Trail 1980) and about the evolutionary history of altered behaviors (Szidat 1969). Following the work of Bacot decades earlier, medical entomologists began to find that feeding and flight in biting flies were dramatically affected by all manner of parasites (reviewed in Moore 1993). And off in the world of insect pathology, ignored by not only ecologists and behavioral biologists but also parasitologists, reports of summit disease caused by viruses and fungi continued to pile up. Indeed, when we consider the vast number of ecological studies carried out on organisms like ants and gastropods, animals whose abundance and distribution can be profoundly affected by whether or not they have certain parasites, and when we consider how many of those studies do not acknowledge the possible existence of parasites, much less estimate their levels, the idea that some truths are inconvenient takes on additional meaning.

1.3 Becoming cool

There was indeed an accretion of evidence supporting the fact that parasites have ecological, evolutionary and behavioral significance, but that might have gone unnoticed for quite a while longer were it not for three endeavors that I view as primarily responsible for shifting the attention of ecologists and evolutionary and behavioral biologists toward parasites. They address different areas of parasite evolutionary biology and ecology; their combined effect was a sea change in how other organismal biologists viewed parasites.

Traditionally, parasites were thought to evolve toward a benign coexistence with their hosts; this was an idea widely accepted and taught in the medical community. It had been questioned on occasion (e.g., Ball 1943), but the doubt had not penetrated the seemingly common-sense certainty that supported the idea: killing one's dwelling and food source (host) did not appear to be a particularly adaptive thing to do. Beginning in the 1970s, Roy Anderson and Robert May (1991 and references therein), began to question the conventional wisdom about how parasites evolved, especially when it came to their effects on their hosts. They did so by returning to a fundamental concept in evolution and population ecology—that of fitness, one measure of which can be expressed as reproductive rate (R_0). They examined microparasites and macroparasites, that is, parasites that are relatively small and that have short life spans, during which they reproduce within the host, and parasites that are somewhat larger, live longer, and whose offspring exit the host and colonize new hosts. Microparasites tend to induce stronger immune responses than macroparasites do, and this in turn affects the likelihood of reinfection.

Of course, May and Anderson knew that the biological reality of parasites did not segregate so neatly, but the distinction worked well for the purposes of their models, which in their most streamlined forms addressed reproductive rate in terms of host population density, host recovery rate, and host death rate from parasites (virulence) and from other causes. One of the many outcomes of their work, and perhaps one of the most relevant for the study of manipulation, is that traits like life history, resource use, and transmission can greatly influence virulence. For instance, consider transmission: if virulence is linked to that trait in such a way that transmission increases with virulent infections, then if all else is equal, virulence will be favored by natural selection, that is, reproductive rate will increase. This means that given the complexity and diversity of host–parasite associations, there is no single roadway to virulence or avirulence, there is no one-size-fits-all statement about how parasite–host associations should evolve; instead, the diversity of host–parasite interactions produces countless possibilities. Anderson and May's work planted the study of how hosts and parasites interact squarely in the middle of evolutionary theory, causing science and medicine to see the evolution of sickness and health in a different light.

Peter Price entered the world of parasite ecology and evolution from the vantage point of a forest entomologist. Having done extensive research on sawflies and parasitoids, he was impressed with the vast array of organisms that might be considered parasitic, what they had in common, and the extent to which all of this was underappreciated. Some of this he attributed to historical anthropocentrism that saw "mammals and birds, and particularly man…[as] the glories of the evolutionary process". (Price 1980, p. 13). This same vision shuttled parasites down "blind alleys".

In contrast to this view, Price wrote a book for the Princeton Monographs in Population Biology (Price 1980), introducing it with the following sentence (p. 3): "Parasites form a large proportion of the diversity of life on earth". Price explored definitions of the word "parasite" (always a slippery prospect), and expanded its traditional application, noting that most of an individual parasite's food comes from a single organism. As a result, his parasite could be any one of a number of organisms taken from an array not unlike the passenger manifest for a Noah's Ark of Very Small Things—from leafhoppers to beetles to bugs (and beyond), as well as the usual culprits that populate parasitology books. Much like a colonial explorer stepping onto an uncharted continent—but with a much greater understanding of what he might be getting into—Price courageously laid claim to well over half of the world's living creatures as appropriate objects of study for those interested in parasites (see also Price 1975, 1977 for the earliest explorations of this theme). From these, he extracted some ecological and evolutionary commonalities (e.g., small size relative to host, high rates of evolution and speciation, high potential for adaptive radiation and sympatric speciation). This, in turn, led most notably to explorations of phylogeny and community ecology (Holmes and Price 1980) and parasite-mediated host interactions (Price et al. 1986).

Bill Hamilton had been interested in the relationship between disease and sexual reproduction for some time (e.g., Hamilton 1980). In 1982, he and

Marlene Zuk presented what became their eponymous hypothesis to explain the fact that in many species, males have traits associated with courtship that are conspicuous in behavior and/or appearance (Hamilton and Zuk 1982). They hypothesized that such showiness can only be sustained by a healthy male, one that is resistant to parasites, and that in preferring males with such traits, females are selecting males with resistant genes. The hypothesis predicts that within a species, females should prefer males with lower parasitemia; across species, those species that are subject to the highest levels of parasite attack should demonstrate the most conspicuous sexually selected traits.

Behavioral ecologists sprang into action; the eagerness to test this hypothesis produced something akin to a stampede in some quarters, and the resulting literature is still being plumbed. Behavioral ecologists and evolutionary biologists began to discover some things that immunologists and parasitologists had known for quite some time, such as linkages among hormones, stress, and the immune system.

Of course, memory is notoriously fickle, so I decided to "test" my sea change hypothesis—that parasites and host behavior became fashionable in eco-evo-behavioral circles around 1980, when May

and Anderson, Price, and Hamilton and Zuk published their ideas. I used the bibliography of Moore (2002) as a source of publications in the field. This bibliography is fairly comprehensive and was compiled for purposes other than examining historical trends; it ranges widely and deeply, to say the least, for in my reading and writing I honestly attempted to leave no stone unturned. In the analysis for this chapter, after dismissing a number of references that were from collections, edited volumes, and the like, along with some papers that were irrelevant to this question, I allocated each of the remaining publications cited in Moore (2002) to one of three groups based on journal type as reflected in journal title: parasitological (including medical, pathology, N = 417), ecological/evolutionary/behavioral (N = 277), and general (e.g., physiology, entomology (and other "-ologies,"), *Nature*, *Science*, PNAS; N = 441). I did not use papers published after 1999, as these were newer at the time the bibliography was generated, and therefore more subject to unintentional drift into one area or another that might bias the current analysis. I then asked how articles about parasite–host interactions sorted within these groups across decades: What proportion of articles in each of the three groups was written in each decade? The results, shown in Table 1.1, support the

Table 1.1 Percentage of scientific papers containing information about parasite–host interactions that pertain to host behavior in each of three categories of journals, sorted by decade of publication.

Decade of publication	Parasitology*	General**	Behavior/Ecology***
	N = 417	N = 441	N = 277
1990s	42.9	32.9	67.1
1980s	32.6	31.3	23.1
1970s	11.7	17.5	6.1
1960s	7.7	9.5	2.2
1950s	3.1	4.5	1.1
1940s	<0.5	1.4	<0.5
1930s	0.7	1.4	0
1920s	<0.5	0.7	0
1910s	<0.5	0.9	0

*"Parasitology" category journals were those with some references to parasites, pathology, veterinary or human medicine, and the like in their titles.

**"General" category journals were largely those devoted to broad categories of science or to organismal studies.

***"Behavior/Ecology" category journals were largely those that are devoted to behavioral, ecological, or evolutionary studies.

observation that there was a decided increase in parasitology-related interests among ecologists and evolutionary and behavioral biologists later in the twentieth century. Of course, there are many sources of potential bias in such an analysis, ranging from the varying birth rates of different types of journals to my own sorting rules, however straightforward I tried to make them. Nonetheless, the sudden and belated recognition of parasites in ecological/ behavioral/evolutionary circles shows in the fact that of all the papers with parasitological content published in that category of journals, 67% of them were published in the 1990s (compared to 43% for parasitology-oriented journals and 33% for general journals during that time period). Put another way, out of 143 cited papers published in the 1970s, 12% were published in the "ecology" group journals; that increased to 18.9% of 338 in the 1980s and doubled to 36.5% of 510 in the 1990s.

These three influences, then—the work of May and Anderson, Price, and Hamilton and Zuk— seemed to be the starting point for a new look at parasites in the world of behavioral ecology and evolution. In introducing his book, Peter Price said (p. 3), "Indeed, I think there is much room for recasting the image of parasites held among biologists, and in so doing, certain views in ecology, evolutionary theory, and biology must also be recast". That process has begun.

1.4 Beyond manipulation

Finally, there is the fact that not all altered behaviors benefit the parasite. Parasite benefit was the focus of much early work in parasites and host behavior, perhaps a lingering influence of the visual drama around *Leucochloridium*. It continues to command attention, as investigators address possible mechanisms and seek ways to explain its evolution. Although parasite benefit is implicit in the word "manipulation," as if the parasite is a puppeteer, there are times that behavior is altered by parasites in ways that benefit the host. The first examples of this were in the area of parasite avoidance—especially the avoidance of and defense against biting flies—and because these studies were often sequestered in the medical entomology literature, they

received little attention from ecologists and other non-entomologists. Edman's work in particular ranged across the 1970s and 1980s (Edman and Spielman 1988 and references therein). As time went on, other ways to avoid parasites, especially those having to do with mate choice, captured the attention of a wide range of biologists.

Even if an animal gets infected, there are behavioral changes that occur that can benefit the host instead of the parasite. These behaviors range from self-medication to grooming, but some of the earliest observations were around homeostasis, and in particular, behavioral ways to regulate body temperature in animals that could not do so metabolically. For instance, in 1974, Vaughn and co-workers showed that desert iguanas responded behaviorally to an injection of a pyrogen (in this case, killed pathogenic bacteria), increasing their body temperature by 2 °C. Matthew Kluger did much to raise awareness of fever as an adaptive response (e.g., Kluger 1986; 1991). Some of the behaviors displayed by parasitized animals that seem to be a result of manipulation leading to increased predation risk (e.g., conspicuous behavior, elevation seeking, hyperactivity, positive phototaxis, etc.) might have originated with fever responses (Horton and Moore 1993; Moore 2002).

Given this background, it remained for Benjamin Hart, a professor of veterinary medicine, to introduce the idea that the constellation of behaviors associated with febrile sickness was not evidence of debilitation, as traditionally thought by the medical community, but was instead adaptive (Hart 1988). These behaviors are consistent across host taxa, type of pathogen, and disease severity, and are typified by anorexia, sleep, "depressed" behavior—in other words, behaviors that redirect resources toward combating the pathogen. At the time Hart suggested this, behavioral biologists had become more interested in the effects that parasites had on animals, but they were still not terribly interested in the behavior of animals that were simply "sick". Instead, they assumed that sick animals were abnormal, and therefore not suitable for study. Hart's background in veterinary medicine may well have been crucial to his ability to see the behavior of sick animals in a way that had escaped other behavioral scientists. In so doing, he, along with others, helped create a conceptual counterweight to the purely manipulative (parasite benefit) explanation for altered host behavior. Hart has also synthesized much of the literature on grooming behavior as an adaptive measure (see also Clayton 1990), and has worked extensively on the antibacterial activity of saliva, that is, the wisdom of licking one's wounds (e.g., Hart and Powell 1990).

There is at least one more aspect of the study of parasite–host interactions that intrigues me. When parasites were first becoming cool, women were still rare in scientific circles. Because of this, I have found it pleasantly puzzling that so many women seemed to gravitate toward the study of parasite–host interactions, specifically those with a greater or lesser behavioral twist. In addition to myself, there are at least eight women that I can think of who

were beginning their careers during this 1980-ish era. That is a lot of women during a time when many entire departments had no women faculty members, or perhaps one token female; that is a lot of women in a field that itself was not so large. I have sometimes speculated about why this happened, but I lack testable hypotheses. What I can do is briefly introduce these women; they are creative scientists and good people, well met (see Box 1.1). (In addition, the field seemed to gain more traction among UK and UK-trained scientists than in the United States, but that may be a topic for another chapter!)

1.5 Conclusion

I am thankful that this history has no real conclusion, at least not in the sense of "ending". I am

---

Box 1.1 Rothschild and Cram may have led the way…In the early 1980s, the study of parasite–host interactions attracted an unusually large number of women, many of whom were just beginning their scientific careers.

Nancy Beckage (University of California, Riverside) began her career studying parasite–host interactions at the endocrine level and then moved into studies of pathogens of mosquitoes and symbionts of parasitic wasps. (See Beckage 1985, 1997, 2008.) Shelley Adamo, a contributor to this volume, worked with Nancy as a post-doctoral associate.

Hilary Hurd at Keele University has studied parasite effects on host fecundity in a variety of systems, including tapeworm–beetle as well as malaria–mosquito associations, and has written several papers about manipulation (e.g., Hurd and Arme 1984; Hurd 2005, 2003; Webb and Hurd 1999).

Marijke de Jong-Brink, now retired from Vrije Universiteit Amsterdam, investigated the ways that trematode parasites can interfere with snail host endocrine systems and gene expression, thus altering snail physiology and behavior (e.g., de Jong-Brink 1995; de Jong-Brink et al. 1999).

Anne Keymer of University of Oxford worked at the intersection of parasite population ecology, immunology, and host nutrition. She forayed into behavioral research with a study of parasite-induced taste aversion (Keymer

et al. 1983). A remarkable scientist, she died of cancer in 1993 at the age of 36.

Marilyn Scott of McGill University is interested in the interaction of parasites, host nutrition, and immunology, and explores this from a variety of perspectives, ranging from controlled laboratory experiments and experimental populations to assessment of parasite control in developing countries (e.g., Payne et al. 2007).

Nancy Stamp (Binghamton University) studies plant–insect interactions. Using tethered caterpillars at different elevations, her early work elegantly tested the host-suicide hypothesis in gregarious caterpillars with parasitoids (Stamp 1981).

Deborah Smith Trail (Deborah Smith at Kansas University) suggested host suicide as a possible beginning point for complex life cycles involving transmission to predators (Smith Trail 1980). She now works on social arthropods, especially bees and spiders.

Marlene Zuk (University of California, Riverside) continues to be interested in the way that parasites influence mate choice, and the conflicts between natural and sexual selection that can occur as a result (e.g., Sheridan et al. 2000).

fairly certain that parasites, including manipulating parasites, will be cool for the foreseeable future, and that they provide remarkably fertile subjects for study. We still have much to understand, including fundamental questions about genetic bases of manipulation, geographic variation in host–parasite interactions, and evolutionary antecedents of manipulation. Indeed, the list seems without end. Moreover, having studied parasites when they were not cool, and then when they were, I am fairly convinced that cool is better.

I am not sure, however, that this leap into coolness—the trend that became apparent in the 1990s—is cause for unfettered celebration. Some reflection is in order: here in this book, we have a collection of chapters about phenomena that might have been dismissed as irrelevant three decades ago. Given the increasing costs of research and diminishing resources, how likely is it today for the next uncool field to emerge into the spotlight? What do we all lose when fashion—what "everybody knows"—closes the door on what might be discovered? I suspect we lose more than information, as precious as that is. We lose some of the creative impulse, the angle that no one else has noticed, the patient, delighted watchfulness of Wesenberg-Lund.

So in the midst of celebrating what we have learned about manipulative parasites, I suggest that we also remember that just recently we were intellectual refugees. With such a heritage, we should of course savor what we are doing—and we should keep watching, peering into dark, neglected corners for the next thing that everyone knows is not very cool.

Acknowledgements

I am grateful to the editors of this volume who invited me to participate, who gave me permission to write this sort of piece and even encouraged it, and who were remarkably patient when other writing commitments played havoc with my schedule. I am grateful to David Hughes in particular for suggestions, comments, and encouragement. I thank John Holmes, the doyen of this field, for taking time out from an active retirement in order to give me valuable feedback on an earlier draft.

References

Anderson, R. M. and May, R. M. (1991). *Infectious Diseases of Humans*. Oxford University Press, Oxford.

Anokhin, I. A. (1966). Daily rhythm in ants infected with metacercariae of *Dicrocoelium lanceceatum*. *Doklady Akademi Nauk SSSR* **166**, 757–759.

Bacot, A. W. and Martin, C. J. (1914). Observations on the mechanism of the transmission of plague by fleas. *Journal of Hygiene Plague Supplement* **3**, 423–439 + plates.

Ball, G. H. (1943). Parasitism and evolution. *American Naturalist* **77**, 345–364.

Barber, I. and Scharsack, J. P. (2010). The three-spined stickleback-Schistocephalus solidus system: an experimental model for investigating host-parasite interactions in fish. *Parasitology* **137**, 411–422.

Baudoin, M. (1975). Host castration as a parasitic strategy. *Evolution* **29**, 335–352.

Beckage, N. (ed.) (1997). *Parasites and Pathogens: Effects on Host Hormones and Behavior*. Chapman and Hall. New York

Beckage, N. (ed.) (2008). *Insect Immunology*. Academic Press/Elsevier, San Diego.

Beckage, N. E. (1985). Endocrine interactions between endoparasitic insects and their host. *Annual Review of Entomology* **30**, 371–413.

Berg, K. (1948). Personalia. C. Wesenberg-Lund. *Hydrobiologia* **1**, 322–324.

Bethel, W. M. and Holmes, J. C. (1973). Altered evasive behavior and responses to light in amphipods harboring acanthocephalan cystacanths. *Journal of Parasitology* **59**, 945–956.

Bethel, W. M. and Holmes, J. C. (1974). Correlation of development of altered evasive behavior in *Gammarus lacustris* (Amphipoda) harboring cystacanths of *Polymorphus paradoxus* (Acanthocephala) with the infectivity to the definitive host. *Journal of Parasitology* **60**, 272–274.

Bethel, W. M. and Holmes, J. C. (1977). Increased vulnerability of amphipods to predation owing to altered behavior induced by larval acanthocephalans. *Canadian Journal of Zoology* **55**, 110–115.

Carney, W. P. (1969). Behavioral and morphological changes in carpenter ants harboring dicrocoeliid metacercariae. *American Midland Naturalist* **82**, 605–611.

Cheng, T. C. (1973) *General Parasitology*. Academic Press, New York.

Clayton, D. H. (1990). Mate choice in experimentally parasitized rock doves: lousy males lose. *American Zoologist*, **30**, 251–262.

Cram, E. B. (1931). Developmental stages of some nematodes of the spiruroidea parasitic in poultry and game birds. *USDA Technical Bulletin* **227**, 27 pp.

de Jong-Brink, M. (1995). How schistosomes profit from the stress responses they elicit in their hosts. *Advances in Parasitology* **35**, 178–256.

de Jong-Brink, M., Reid, C. N., Tensen, C. P., and Ter Maat, A. (1999). Parasites flicking the NPY gene on the host's switchboard: why NPY? *The FASEB Journal* **13**, 1972–1984.

Edman, J. D. and Spielman, A. (1988). Blood-feeding by vectors: physiology, ecology, behavior and vertebrate defense. In: *The Arboviruses: Epidemiology and Ecology* (ed. Monath T. P.) vol. 1, pp. 153–189. CRC Press, Boca Raton, Florida.

Foster, W. D. (1965). *A History of Parasitology*. E & S Livingstone, Edinburgh.

Haldane, J. B. S. (1949). Disease and evolution. *La Ricerca Scientifica Supplementa* **19**, 68–76.

Hamilton, W. D. (1980). Sex versus non-sex versus parasite. *Oikos* **35**, 282–290.

Hamilton, W. D. and Zuk, M. (1982). Heritable true fitness and bright birds: a role for parasites? *Science* **218**, 384–386.

Hart, B. L. (1988). Biological basis of the behavior of sick animals. *Neuroscience and Biobehavioral Reviews* **12**, 123–137.

Hart, B. L. and Powell, K. L. (1990). Antibacterial properties of saliva: role in maternal periparturient grooming and in licking wounds. *Physiology and Behavior* **48**, 383–386.

Holmes, J. C. (1961). Effects of concurrent infections on *Hymenolepis diminuta* (Cestoda) and *Moniliformis dubius* (Acanthocephala). I. General effects and comparison with crowding. *The Journal of Parasitology* **47**, 209–216.

Holmes, J. C. and Bethel, W. M. (1972). Modification of intermediate host behavior by parasites. In: *Behavioral Aspects of Parasite Transmission* (eds Canning E. U. and Wright C. A.), pp. 123–149. Academic Press, New York.

Holmes J. C. and Price, P. W. (1980). Parasite communities: the roles of phylogeny and ecology. *Systematic Zoology* **29**, 203–213.

Horton, D. R. and Moore, J. (1993). Behavioral effects of parasites and pathogens in insect hosts. In: *Parasites and Pathogens of Insects* (eds Beckage N. E., Thompson S. N., B. A. Federici), vol. 1, pp. 107–124. Academic Press, New York.

Hurd, H.(2003). Manipulation of medically important insect vectors by their parasites. *Annual Review of Entomology* **48**, 141–161.

Hurd, H. (2005). Parasite manipulation: stretching the concept. *Behavioural Processes* **68**, 235–236.

Hurd, H. and Arme, C. (1984). Pathophysiology of *Hymenolepis diminuta* infections in *Tenebrio molitor*: effect of parasitism on haemolymph proteins. *Parasitology* **89**, 253–262.

Kagan, I. G. (1951). Aspects in the life history of *Neoleucochloridium problematicum* (Maagath, 1920) new comb. and *Leucochloridium cyanocittae* McIntosh, 1932 (Trematoda: Brachylaemidae). *Transactions of the American Microscopical Society* **70**, 281–318.

Kagan, I. G. (1952). Further contributions to the life history of *Neoleucochloridium problematicum* (Magath, 1920) new comb. (Trematoda: Brachylaemidae). *Transactions of the American Microscopical Society* **71**, 20–44.

Keymer, A., Crompton, D. W. T., and Sahakian, B. J. (1983). Parasite-induced learned taste aversion involving *Nippostrongylus* in rats. *Parasitology* **86**, 455–460.

Kluger, M. J. (1986). Is fever beneficial? *Yale Journal of Biology and Medicine* **59**, 89–95.

Kluger, M. J. (1991). Fever: role of pyrogens and cryogens. *Physiological Reviews* **71**, 93–127.

Krull, W. H., and Mapes, C. R. (1953). Studies on the biology of Dicrocoelium dendriticum (Rudolphi, 1819) Looss, 1899 (Trematoda: Dicrocoeliidae), including its relation to the intermediate host Cionella lubrica (Müller). IX. Notes on the cyst, metacercaria, and infection in the ant, Formica fusca. *Cornell Veterinarian* **43**, 389–410.

Kuris, A. M. (1974). Trophic interactions: similarity of parasitic castrators to parasitoids. *Quarterly Review of Biology* **49**, 129–148.

Lambert, T. C. and Farley, J. (1968). The effect of parasitism by the trematode *Cryptocotyle lingua* (Creplin) on zonation and winter migration of the common periwinkle, *Littorina littorea* (L.). *Canadian Journal of Zoology* **46**, 1139–1147.

Lewis, D. J. and Wright, C. A. (1962). A trematode parasite of *Simulium*. *Nature* **193**, 1311–1312.

Lewis, P. D., Jr. (1977). Adaptations for the transmission of species of *Leucochloridium* from molluscan to avian hosts. *Proceedings of the Montana Academy of Sciences* **37**, 70–81.

Minchella, D. J. and Loverde, P. T. (1981). A cost of increased early reproductive effort in the snail *Biomphalaria glabrata*. *American Naturalist* **118**, 876–881.

Moore, J. (1983). Responses of an avian predator and its isopod prey to an acanthocephalan parasite. *Ecology* **64**, 1000–1015.

Moore, J. (1984). Parasites and altered host behavior. *Scientific American* **250**, 108–115.

Moore, J. (1993). Parasites and the behavior of biting flies. *Journal of Parasitology* **79**, 1–16.

Moore, J. (2002). *Parasites and the Behavior of Animals.* Oxford University Press. Oxford Series in Ecology and Evolution. (eds R. M. May and P. H. Harvey). New York

National Agricultural Library (USDA), Eloise Cram Papers. http://riley.nal.usda.gov/nal_display/index.php?info_center=8&tax_level=4&tax_subject=158&topic_id=1982&level3_id=6419&level4_id=10857&level5_id=0&placement_default=0 [accessed 1 December 2011].

Paperna, I. (1964). Competitive exclusion of *Dactylogyrus extensus* by *Dactylogyrus vastator* (Trematoda, Monogenea) on the gills of reared carp. *The Journal of Parasitology* **50**, 94–98.

Park, T. (1948). Experimental studies of interspecific competition. I. Competition between populations of the flour beetles, *Tribolium confusum* Duval and *Tribolium castaneum* Herbst. *Ecological Monographs* **18**, 265–308.

Payne, L. G., K. G. Koski, E. Ortega-Barria, and M. E. Scott. (2007). Benefit of vitamin A supplementation on *Ascaris* reinfection is less evident in stunted children. *Journal of Nutrition* **137**, 1455–1459.

Price, P. W. (1975). *Evolutionary Strategies of Parasitic Insects and Mites*. Plenum, New York.

Price, P. W. (1977). General concepts on the evolutionary biology of parasites. *Evolution* **31**, 405–420.

Price, P. W. (1980). *Evolutionary Biology of Parasites*. Monographs in Population Biology (ed. R. M. May) Princeton University Press, New Jersey.

Price, P. W., Westoby, M., Rice, B., Atsatt, P. R., Fritz, R. S., Thompson, J. N., and Mobley, K. (1986). Parasite mediation in ecological interactions. *Annual Review of Ecology and Systematics* **17**, 487–505.

Romig, T., Lucius, R., and Frank, W. (1980). Cerebral larvae in the second intermediate host of *Dicrocoelium dendriticum* (Rudolphi, 1819) and *Dicrocoelium hospes* Looss, 1907 (Trematoda, Docrocoeliidae). *Zeitschrift fur Parasitenkunde* **63**, 277–286.

Rothschild, M. (1940). *Cercaria pricei*, a new trematode, with remarks on the specific characters of the "Prima" group of Xiphidiocercariae. *Journal of the Washington Academy of Sciences* **30**, 437–448.

Rothschild, M. (1941). Observations on the growth and trematode infections of *Peringia ulvae* (Pennant) 1777 in pool in the Tamar saltings, Plymouth. *Parasitology* **33**, 406–415.

Rothschild, M. (1962). Changes in behaviour in the intermediate hosts of trematodes. *Nature* **193**, 1312–1313.

Rothschild, M. (1969). Notes on fleas, with the first record of a mermithid nematode from the order. *Proceedings and Transactions of the British Entomological and Natural History Society* **1**, 9–16.

Rothschild, M. and Ford, B. (1964). Breeding of the rabbit flea (*Spilopsyllus cuniculi* (Dale)) controlled by the reproductive hormones of the host. *Nature* **201**, 103–104.

Schad, G. A. (1963). Niche diversification in a parasitic species flock. *Nature* **198**, 404–406.

Shapiro, A. M. (1976) Beau geste! *American Naturalist* **110**, 900–902.

Sheridan, L. A. D., Poulin, R., Ward, D. F., and Zuk, M. (2000). Sex differences in parasitic infections among arthropod hosts: is there a male-bias? *Oikos* **88**, 327–334.

Simberloff, D. and Moore, J. (1997). Communities of parasites and free-living animals. In *Host-Parasite Evolution: General Principles and Avian Models* (eds D. H. Clayton and J. Moore), pp. 174–197. Oxford University Press, Oxford.

Sindermann, C. J. and Farrin, A. E. (1962). Ecological studies of *Cryptocotyle lingua* (Trematoda: Heterophyidae) whose larvae cause "pigment spots" of marine fish. *Ecology* **43**, 69–75.

Smith Trail, D. R. (1980). Behavioral interactions between parasites and hosts: host suicide and the evolution of complex life cycles. *American Naturalist* **116**, 77–91.

Stamp, N. E. (1981). Behavior of parasitized aposematic caterpillars: advantageous to the parasitoid or the host? *American Naturalist* **118**, 715–725.

Szidat, L. (1969). Structure, development, and behaviour of new strigeatoid metacercariae from subtropical fishes of South America. *Journal of the Fisheries Research Board of Canada* **26**, 753–786.

Van Dobben, W. H. (1952). The food of the cormorant in the Netherlands. *Ardea* **40**, 1–63.

Webb, T. J. and Hurd, H. (1999). Direct manipulation of insect reproduction by agents of parasite origin. *Proceedings of the Royal Society of London B* **266**, 1537–1541.

Wesenberg-Lund, C. (1931). Contributions to the development of the trematoda digenea. Part I. The biology of *Leucochloridium paradoxum*. *Memoires de l''Academie Royale des Sciences et des Lettres de Danemark, Copenhague, section des sciences* **4**, 90–142.

Wheeler, W. M. (1907). The polymorphism of ants, with an account of some singular abnormalities due to parasitism. *Bulletin of the American Museum of Natural History* **23**, 1–93.

Wheeler, W. M. (1928). *Mermis* parasitism and intercastes among ants. *Journal of Experimental Zoology* **50**, 165–237.

Woodhead, A. E. (1935). The mother sporocysts of *Leucochloridium*. *Journal of Parasitology* **21**, 337–346.

Afterword

John Alcock

Janice Moore asks many intriguing questions about parasites and behavioral research in her chapter on the prehistory of the modern field of parasite-influenced animal behavior. The two that I wish to comment on here are: (1) why were researchers so slow to catch on to the possibility that a great deal of animal behavior is profoundly influenced by parasite pressure? (2) Is the current interest in the field of behavioral parasitology a result of a bandwagon effect, with modern researchers hurrying to cash in on the latest fashionable topic before the funding moves to a new hot topic?

At the heart of these questions is the reality that behavioral biologists have only relatively recently become interested in the evolutionary consequences of interactions between microscopic parasites and their hosts. I am representative of this group of latecomers to the topic. In the first edition of my textbook, *Animal Behavior, An Evolutionary Approach*, published in 1975, I did describe how avian brood parasites afflicted their victims, but microscopic parasites received nary a mention. Not until the fifth edition, published in 1989, did the textbook deal briefly with the issue of the possible effects of debilitating microparasites on the evolution of animal behavior.

And yet, as Moore points out, stunning examples of parasite-induced changes in animal behavior were well known long before the more recent surge of interest in parasitology. For example, she describes how, ever since the 1930s, biologists were familiar with the amazing case of a trematode parasite that induces a snail to make sporocyst-infected tentacles resemble worms, the better to lure worm-eating birds to eat the tentacles and ingest the sporocysts. Yet this and other similar

accounts had little effect on the field of animal behavior. It was not until the 1980s that parasites suddenly became a hot topic for behavioral types, a phenomenon that Moore traces to the ecological work of Peter Price and a paper by William Hamilton and Marlene Zuk. In their paper, Hamilton and Zuk argued that blood parasites and the like had affected the evolution of female mate choice in birds with colorful plumage. Hamilton and Zuk proposed that the feathers of some male birds had properties that permitted females to evaluate the parasite load carried by a potential partner, the better to pick a healthy, non-infectious partner or perhaps even parasite-resistant male able to pass on his good (anti-parasite) genes to the female's offspring.

These studies marked the beginning of the era when it became "cool" to study microscopic parasites. Now it could be that others stampeded after Price and Hamilton because these persons were well-established, highly successful researchers from whom others took their cues about how to succeed in science. (Zuk was just beginning her fine career in 1982.) There is little doubt that Hamilton and Zuk's paper did catch the imagination of many, myself included. Their paper was featured in my fifth edition and in every subsequent edition to the present. But if the enthusiasm that I, as a textbook writer, and others, as behavioral researchers, had and have for this paper happened because of a bandwagon effect, we still have to explain why Hamilton and Zuk were able to get the bandwagon rolling whereas earlier studies of parasitic life cycles did not do so.

Let me suggest why parasitological studies in the earlier part of the twentieth century failed to generate much interest in microparasites by behavioral biologists at the time. This work, astonishing though the complex parasite–host interactions were, was nonetheless primarily descriptive with little or no theoretical foundation, ecological or evolutionary. In this, parasitologists at the time were not fundamentally different from the ornithologists, mammalogists, entomologists, herpetologists, ichthyologists, and other -ologists who dominated biology departments then. As George C. Williams demonstrated, by the mid-1900s, almost all biologists had forgotten the essence of Darwinian logic and were prone to explain adaptations in terms of their supposed benefits to the species as a whole. Williams's book, *Adaptation and Natural Selection*, published in 1966, helped biologists rediscover the point that evolutionary change by natural and sexual selection occurs when individuals, not entire species or groups within species, differ in their hereditary ability to produce surviving descendants. Williams, along with W.D. Hamilton, Richard Alexander, and Robert Trivers, demonstrated just how productive this fundamental idea could be for studies of animal behavior. Hamilton and Zuk's paper was one more powerful demonstration that you might be able to explain the adaptive value of certain traits in novel, testable ways—if you really understood evolutionary theory.

The development of an "adaptationist programme" by Williams and its popularization by Richard Dawkins (in *The Selfish Gene*, published in 1975), encouraged all sorts of biologists to understand Darwinian evolutionary theory. This in turn made possible the explosive development of an entire discipline, behavioral ecology, of which behavioral parasitology is now one component. At the time this was happening, some researchers complained that the new breed of behavioral biologists were giving way too much attention to evolutionary studies at the expense of those focused on the immediate, proximate causes of behavior. But this complaint ignored the fact that, prior to the evolutionary revolution initiated by Williams, researchers had largely failed to properly investigate the adaptive basis of behavior, which had nothing to do with preventing the extinction of species or groups. The shift to modern behavioral ecology was largely compensatory for this earlier failure. Thanks to Williams, and later Hamilton and Zuk, we know now that evolutionary theory is immensely helpful in moving beyond natural history, important though descriptive natural history is. The complex life cycles of certain minute parasites are wonderful to know about, but it is also valuable to consider how a female tanager's criteria for mate choice may have evolved because of variation among the males of her species in their susceptibility to invisible microparasites. What a cool idea.

CHAPTER 2

Evolutionary routes leading to host manipulation by parasites

Frédéric Thomas, Thierry Rigaud, and Jacques Brodeur

2.1 Introduction

Determining why and how host manipulation by parasites evolves is a fascinating but challenging question for evolutionary biologists. Pioneer authors addressing this question (e.g., Smith-Trail 1980; Moore 1984) proposed that host manipulations probably occurred after the establishment of complex life cycles involving more than one host species. Natural selection would favor parasites able to alter the behavior of their host to increase their transmission success (Poulin 1994; Combes 1998). Incidental side-effects of parasitism would then be progressively shaped by selection into refined manipulation mechanisms. In a 1995 paper, Robert Poulin emphasized that the evolution of host manipulation must be considered within a phylogenetic context since the capacity to manipulate host behavior could potentially be inherited from an ancestor of the parasite rather than being a product of natural selection (Poulin 1995). Some parasite phyla, such as the acanthocephalans, are indeed entirely composed of manipulative members, suggesting that this capacity was inherited from a common ancestor and then evolved into subsequent behavioral modifications.

Following this ground-breaking work, hypotheses and studies on the evolution of manipulation have flourished, especially over the last ten years. In this chapter, we will focus on several aspects emerging from recent studies on host manipulation within an evolutionary perspective: (1) the acknowledgement that different evolutionary routes can lead to host manipulation, (2) the determinants of intraspecific variation in host manipulation, and (3) the evolution of multidimensional manipula-

tion. In recent studies exploring hypotheses on the evolution of host manipulation, the original causes of phenotypic changes (i.e., answering the macroevolutionary question "why did this phenomenon appear?") and the possibility of an ongoing evolutionary phenomenon (i.e., answering the microevolutionary question "after its appearance, how is manipulation able to evolve?") are often difficult to distinguish. While these two questions can be intimately linked, we treat them separately below.

2.2 The origins of host manipulation

According to Lefèvre et al. (2009), three main evolutionary routes can lead to host manipulation: (1) the manipulation *sensu stricto*, (2) the mafia-like strategy, and (3) the exploitation of compensatory responses. Route 1 is a decidedly parasite-oriented view (albeit not excluding some host responses), while routes 2 and 3 correspond to scenarios where the host genotype is more involved in the evolution of manipulation. In addition, another recent hypothesis has proposed that parasite manipulation did not emerge as an adaptation of parasites increasing their transmission, but rather as a consequence of the evolution of other parasite traits (Cézilly et al. 2010). This suggests that behavioral changes are the cause rather than the consequence of the evolution of complex life cycles in some parasites.

2.2.1 Manipulation *sensu stricto*

Most studies still consider this scenario as the main process used by parasites to control their host's behavior. Under this hypothesis, host changes that lead to parasite transmission are considered to be illustrations of the "extended phenotype" concept proposed by Dawkins (1982): the aberrant behavior displayed by parasitized hosts results from the expression of the parasite's genes. In this view, parasite genes are selected for their effect on host behavior. This hypothesis is not really in contradiction with the phylogenetic context described by Poulin (1995) if we consider that this capacity could have been the product of natural selection in an ancestral parasite with the trait then being inherited from this common ancestor after radiation. Some parasite phyla, such as acanthocephalans, are entirely composed of manipulative members. In addition to inheritance, true convergences also exist between parasite taxa evolving under similar ecological pressures for their transmission (e.g., Moore and Gotelli 1990; Thomas and Poulin 1998; Ponton et al. 2006a; Franceschi et al. 2007); these examples provide powerful evidence that natural selection can lead to similar manipulations independently in different parasite lineages (see Moore 2002; Thomas et al. 2005; and Poulin 2010 for reviews).

2.2.2 Complex parasitic cycles: the cause or the consequence of parasite manipulation?

As noted in the introduction, pioneering works suggested that host manipulations followed the establishment of complex life cycles. In this context, and under the hypothesis of manipulation *sensu stricto*, phenotypic changes in the intermediate hosts could have been selected to optimize parasite transmission to the next host. Recently, Cézilly et al. (2010) suggested that the question of the adaptiveness of host manipulation could also result in a chicken and egg problem: given the close interactions between the host's nervous and immune systems, ancestor species of contemporary manipulative parasites could have induced various byproducts affecting the host's behavioral phenotype by fighting against the host's immune system. Similarly, it has also been hypothesized that certain parasites (e.g., trematodes) manipulate the behavior of their hosts by becoming encysted in the host's brain, although this action came about initially to avoid an immune response from the host (Poulin 1994). The cerebral location would then have allowed the parasite to interfere with the host's behavior (Poulin 1994). Under these hypotheses, the resulting behavioral changes might then have led to an increased vulnerability to predators, thus creating a strong selective pressure for the parasite to include some predators in its life cycle. In this scenario, the inclusion of a final host in the cycle would be the true adaptive phenomenon, not the phenotypic alterations that arose as byproducts or pathological effects.

Could this hypothesis be generalized to behavioral manipulation? There are many parasites that change the behavioral characteristics of their hosts that have only one host in their life cycle, thus improving their chances of completing their development (e.g., fungi, hairworms, parasitoids). In these parasites, we need to discover—at the phylogenetic level—if manipulation preceded key characteristics of the parasite's life cycles or, conversely, emerged after these characteristics. For example, some hairworm species infecting Orthoptera change the behavior of their hosts in a way that forces them to jump into water, where the worms can emerge from the host and begin to search for a sexual partner. In this example, to test the hypothesis of manipulation as a cause of the appearance of key characteristics of parasite life cycles, the question to answer would be whether the hairworms adapted to water ancestrally or only after the appearance of this behavioral change?

2.2.3 Host-driven scenarios of manipulation

Few studies have explored the degree to which parasite-manipulated behaviors could be a compromise between host and parasite strategies (Thomas et al. 2005). However, "making the host do *something*" can sometimes be achieved when the *something* is better than nothing for the host. In this context, behavioral changes in infected hosts can be the direct products of natural selection acting on the host genome, even when they result in significant fitness benefits for the parasite. There are at least two scenarios in which parasitic manipulation can enhance host fitness as well.

2.2.3.1 *Facultative virulence: the mafia-like strategy*
Zahavi (1979) proposed that parasites could select for collaborative behavior in their hosts by imposing extra fitness costs in the absence of compliance. Parasites could increase their virulence when the host does not behave as expected, thereby rendering non-collaborative behaviors a more expensive option for the host than collaborative ones. In this view, parasite genes are selected for their ability to detect non-collaborative behaviors and their ability to produce retaliatory behaviors. In particular systems (like birds; see the example below), parasites can "teach" the host that it should comply rather than resist. In other cases, selection is expected to produce population-specific phenotypic plasticity through time and to act on the pattern of condition-dependent expression of behavior, causing individuals to behave differently when infected (Ponton et al. 2006b; Lefèvre et al. 2009).

The first and by far the most popular example of the mafia strategy is the association between the great spotted cuckoo (*Clamator glandarius*) and the magpie (*Pica pica*). In this association, the host can raise at least some of its own young along with those of the cuckoo (see Chapter 6). Soler et al. (1995) found that magpies that eject the cuckoo's egg from the nest suffered from considerably higher nest predation by cuckoos than those that accept parasite eggs (see Hoover and Robinson 2007 for another example). In addition, the probability of being revisited and parasitized a second time by a cuckoo is also higher for a rejecter magpie nest (Soler et al. 1998). By modelling the evolution of retaliation by brood parasites, Pagel et al. (1998) showed that retaliation can also evolve even when hosts rear only the parasite's young (e.g., nests parasitized by *Cuculus canorus*). This can arise if brood parasites have a good memory for the location and status of nests in their territory. This implies that during the breeding season, non-ejectors benefit from lower rates of parasitism in later clutches than ejectors, that is, while ejectors are likely to be re-parasitized, non-ejectors are able to rear a clutch of their own following the rearing of a cuckoo nestling.

The mafia-like strategy of manipulation works from a theoretical point of view, but concrete examples in nature are still very scarce (see Ponton et al. 2006b). A possible reason is that most studies do not even consider this scenario despite its relevance. However, except for cuckoos, the few available studies testing this scenario did not find conclusive results (Biron et al. 2005; Ponton et al. 2009). Nevertheless, when studying host behavior manipulation, one should keep in mind the need to demonstrate that there is no fitness compensation to infected hosts whose behavior benefits the parasite. If this is not done, the mafia scenario could be an alternative to the manipulation *sensu stricto* hypothesis.

2.2.3.2 *Exploiting host compensatory responses*
A hypothesis recently proposed by Lefèvre et al. (2008) suggests that parasites could affect fitness-related traits in their hosts (e.g., survival, growth, fecundity, competitiveness) in order to induce host compensatory responses since these responses can at least partially match with the transmission routes of parasites. Compensation is a widespread strategy among living organisms, and parasitism can be a decisive variable initiating compensatory responses in hosts. Because of their detrimental effects on host fitness, parasites can themselves be the causal agents of the compensatory response displayed by the host. Parasites can also directly exploit the compensatory responses that have been selected in other ecological contexts by mimicking the causes that induce them.

An examination of several published papers on host manipulation supports the idea that parasites could indeed exploit host compensatory responses instead of manipulating *sensu stricto* their host (Lefèvre et al. 2008, 2009). For example, the host compensation hypothesis predicts that parasites with direct transmission could benefit from decreasing the reproductive output of their host since this should promote compensatory sexual activities that increase social interactions (see Polak and Starmer 1998; Adamo 1999) and hence parasite transmission (Lefèvre et al. 2008, 2009). This scenario apparently occurs with the sexually transmitted ectoparasite *Chrysomelobia labidomerae* parasitizing the leaf beetle host *Labidomera clivicollis*: infected males, which have a reduced survival compared to uninfected ones, exhibit increased sexual behavior and, as a result, enhanced inter- and intra-sexual contacts (i.e., copulation and competition) that provide more opportunities for parasite transmission (Abbot and Dill 2001; Drummand et al. 1989).

Similarly, it is well known that parasitized hosts are often more vulnerable to predators because they have to forage more to compensate for the negative effects of infection. In this context, inducing some form of energy depletion could be an efficient strategy for trophically transmitted parasites to be transmitted to the definitive predatory hosts. Interestingly, such situations were traditionally considered as perfect examples of infection byproducts that have fortuitous benefits for

transmission and thus are not true adaptations. However, recent review papers now acknowledge that this scenario can indeed be adaptive (e.g., Poulin 2010).

These few examples illustrate well that parasites could theoretically achieve transmission by triggering host compensatory responses when these responses fit, at least partially, with the transmission route. From an evolutionary perspective, we believe that selection is likely to favor parasites that exploit these responses instead of manipulating (*sensu stricto*), not only because this meets their objectives, but also because it requires no manipulative effort: the host is doing the job. The main task for the parasite is to induce a fitness-related

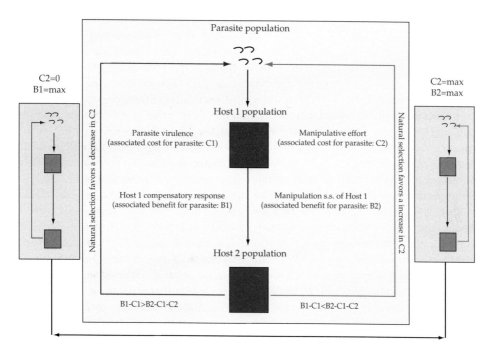

Figure 2.1 Schematic representation by Lefèvre et al. (2008) of the evolutionary dynamics of two manipulative strategies: (i) the manipulation *sensu stricto* and (ii) the exploitation of host compensatory responses. The two strategies can be represented by a continuum (arrow in bold) along which a parasite can both induce a host's compensatory response (via its virulence, e.g., fecundity reduction) and invest energy in manipulation *sensu stricto* (i.e., a mixed strategy). Such a parasite has fitness costs associated with the manipulative effort *sensu stricto* **(C2)** and with the virulence incurred to the hosts **(C1)**, but gains transmission benefits from the *sensu stricto* manipulation **(B2)** and from the host's compensatory responses **(B1)**. When B1−C1 > B2−C1−C2, natural selection favors a decrease in manipulative effort (i.e., decrease in C2) and thus favors the strategy based on the exploitation of host's compensatory response. Conversely, when B1−C1 < B2−C1−C2, natural selection favors a higher investment in the manipulation *sensu stricto* strategy. The case shown at the extreme left is where the parasite induces a host compensatory response that completely matches the parasite's objectives. At the extreme right, no compensatory response exists in the host's phenotypic repertoire; hence the host behaves in such a way that benefit to the parasite can only be achieved by manipulation *sensu stricto*. This simplistic view has the advantage of emphasizing the fact that when the compatibility between the type of host compensatory response and the parasite's objective is strong enough, exploitation of host's compensatory response is the best strategy of manipulation.

cost to the host, something that, by definition, parasites normally do. It is also interesting that non resistance from the host is less likely to evolve because it is better for the host to behave in a way that alleviates the costs of parasitism even if it benefits the parasite. Manipulation *sensu stricto* and exploitation of host compensatory responses are not mutually exclusive; rather, these two processes occur along a continuum (Lefèvre et al. 2008, Fig. 2.1). Further studies are clearly needed to determine if the exploitation of host compensatory responses is a common manipulative strategy. One way this could be done would be to decouple the host response from the parasite transmission strategy, for example, by inducing a stress comparable to the parasite's presence but in the absence of the parasite, and seeing if similar compensatory responses are obtained.

2.2.4   Exaptation?

Some of the patterns described above can be considered as exaptations rather than as adaptations *sensu stricto* (Gould and Vrba 1982). Exaptations are traits that "are fit for their current role ... but were not designed for it" (Gould and Vrba 1982). They could have been selected originally either as adaptations to other pressures or linked to other selected traits (Andrews et al. 2002). Both the immune system hypothesis and the host compensatory response hypothesis rely on the appearance of traits that are either a byproduct of the infection or a consequence of the parasite's reaction to the host's defenses. Exaptations can then be followed by secondary adaptations related to the new role. In the case of manipulation, an *a posteriori* selection of some phenotypic changes is possible if parasite transmission is increased, thus improving parasite fitness. This would result in rerouting these traits at a pure parasite benefit. This could occur independently of the source of the trait's appearance, leading to a pure rerouting of the trait, but could also coexist along with the original cause for which it was originally selected (possibly leading to multidimensionality in manipulation; see Section III). This last possibility also corresponds to recent definitions of exaptation (e.g., Ayala 2010).

2.3   The evolution of manipulation after its emergence

There are many arguments suggesting that phenotypic changes do not remain fixed after their emergence. In parasite clades that include only manipulator parasites, it is rare that all species of a given clade induce the same manipulations. A good example of this is found in acanthocephalans: two different species (*Polymorphus minutus* and *Pomphorynchus laevis*) induce distinct behavioral changes in the same intermediate host species (the crustacean *Gammarus pulex*): *Pomphorynchus laevis* induces changes in phototaxis while *Polymorphus minutus* causes changes in geotaxis (Cézilly et al. 2000). These differences fit the "purposive design" criterion proposed by Poulin (1995): the geotactic changes induced by *P. minutus* increase the probability of encountering a bird (the definitive host for this parasite), while the phototactic changes increase the probability of encountering fish (the definitive hosts for *P. laevis*). In addition, the mechanisms to achieve these changes are probably distinct (Tain et al. 2006): in this case for example, the disruption of serotoninergic neurons is involved in behavioral changes induced by *P. laevis* but not by *P. minutus*, see Chapter 3. This example illustrates that behavioral manipulation has evolved in acanthocephalans, even if the original trait may have been inherited from a common ancestor.

At a finer level, that is, within species, manipulation should also evolve. A primary example could be the evolution in manipulation intensity. Most studies assume that costs are inevitably associated with manipulation *sensu stricto* (Poulin et al. 2005). Thus, natural selection could only optimize, not maximize, the influence of parasites on host behavior because the energy spent by parasites on host manipulation would not be available for other functions (e.g., fecundity, growth, survival) (Poulin 1994). A number of key predictions can be made on how the optimal manipulative effort should increase—or conversely decrease—in relation to various factors (Fig. 2.2). However, despite considerable research on manipulative parasites, the lack of knowledge on manipulative costs currently limits our understanding of the evolution of manipulative processes based on this scenario.

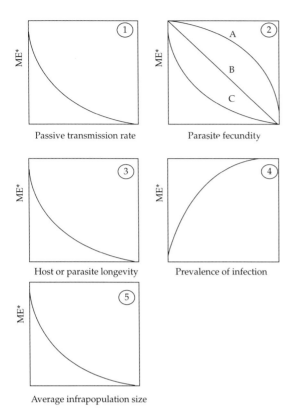

Figure 2.2   Predicted variation of optimal manipulation effort (ME*) according to ecological and life history variables (from Poulin 1994).

Evolution requires heritable variation within populations. Most, if not all, researchers studying host-manipulative parasite systems observe strong variations in the intensity and/or the frequency of the phenotypic changes displayed by infected hosts, even when they are collected in the same environment and at the same time (Perrot-Minnot 2004; Thomas et al. 2005). Three main categories of factors (parasite, host, and environmental) are responsible for this variability (Fig. 2.3). It is also important to distinguish between genetic and non-genetic components (Thomas et al. 2011).

2.3.1   Genetically based variation in phenotypic alterations

Because host manipulation generally reduces (often dramatically, i.e., the case of trophically transmit-

ted parasites) host fitness, natural selection is expected to favor host individuals that are able to resist infection by manipulative parasites and/or to tolerate manipulative effects. A potentially significant portion of the intraspecific variation in host manipulation may result from host individuals having different abilities in this respect. Indeed, Hamilton et al. (2006) showed that two inbred strains of laboratory mice parasitized by the roundworm *Toxocara canis* were not equally sensitive to the behavioral changes induced by the parasite. Similarly, within the association between *G. pulex* and the acanthocephalan *P. laevi*, Franceschi et al. (2010a) recently showed that naïve host individuals are more intensively manipulated than those coevolving with parasites, suggesting that host resistance to manipulation can evolve locally. There are more clear examples of genetically based host effects in the literature. In particular systems, such as mermithid nematodes infecting mayflies or arthropods infected by *Wolbachia* bacteria (both of which improve their transmission by manipulating the host's reproduction or sexuality), parasitic transmission relies on a sex-specific characteristic of the host (see Vance 1996 and Engelstädter and Hurst 2009). In some cases, manipulative strategies exerted by parasites are sex dependent, for example feminization or male-killing is induced only when the parasite finds itself in a male host. In addition, host genotype has been reported to influence the manipulation of host reproduction by identical *Wolbachia* strains (see the review by Engelstädter and Hurst 2009).

Furthermore, there is evidence that genetically based variations in parasites are also important. For example, by comparing the levels of behavioral alteration induced in *G. pulex* by different *P. laevis* families (sibships), Franceschi et al. (2010b) found significant among-family differences in the intensity of early behavioral manipulation. Similarly, Leung et al. (2010) provided evidence that more or less manipulative genotypes exist in the trematode *Curtuteria australis* that infects the cockle *Austrovenus stutchburyi* (intermediate host). In the latter case, Poulin (2010) argued that the selection of multiple genotypes with different strategies could result from trade-offs among different parasite strategies. In

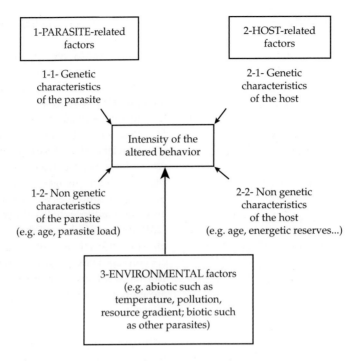

Figure 2.3 Factors influencing manipulation intensity in hosts parasitized by manipulative parasites (from Thomas et al. 2011).

cases of co-infections by *C. australis* (which is a common situation in the field), certain genotypes would be true manipulators while others would be hitch-hikers—benefiting from the manipulation of others without paying the direct and indirect costs of the manipulation. Indeed, there are (at least) indirect costs in this system: encysting at the top of the foot is risky because those metacercariae are more likely to be eaten by unsuitable definitive hosts like benthic fishes (Mouritsen and Poulin 2003).

In the gonochoric parasite, the intensity of phenotypic alterations in the host can be influenced by the parasite's sex. For example, Benesh et al. (2009) found that isopods (*Asellus aquaticus*) infected by a male acanthocephalan (*Acanthocephalus lucii*) tend to have darker abdominal pigmentation compared to those infected by females. The reasons behind these differences are unclear, but it could illustrate divergent manipulative strategies between sexes. For instance, males could have more resources available to allocate toward modifying host pigmentation because they invest less in growth than females do (fecundity constraint). In addition, males also have

less to gain by remaining in and growing mutually with the host than females do (Benesh et al. 2009).

In a recent study using a full reciprocal design of infection under controlled conditions, Franceschi et al. (2010a) showed that only some *P. laevis* populations induce a strong manipulation in gammarid behavior. However, while a mosaic of local parasite adaptations was found for the capacity of infecting hosts, no comprehensive pattern of local adaptation was identified for behavioral manipulation. A genetic basis similar to the "matching alleles" or "gene for gene" interactions (generally invoked to account for variation in infectivity) is therefore not likely to explain variability in manipulation intensity in the *G. pulex–P. laevis* system.

2.3.2 Other sources of variation

2.3.2.1 *Environmental heterogeneity and state-dependent responses*

The intensity of host manipulation (as well as multidimensional manipulation; see Section 2.4) is likely to evolve in response to spatially heterogeneous factors influencing transmission probabilities. Indeed, a given manipulation intensity may be manifested differently among sites because of various abiotic differences among microhabitats (e.g., sediment color, wind exposure). Manipulation efficiency is also expected to vary depending on various biotic factors, for example when suitable definitive hosts correspond to different species with different foraging strategies and/or when suitable hosts are locally outnumbered by non-host predators. In a metapopulation context (probably the case of most manipulative parasites), it is predicted that migration from habitats with contrasting selection pressures might generate sink populations that are permanently maladapted. Because of the resulting selective pressures, state-dependent responses (the ability to recognize fitness-related environmental cues and adjust strategic decisions) should play an important role for both the host and the parasite. For example, when the life expectancy of the host decreases, it is predicted that parasites should secure transmission by increasing their manipulative effort (Thomas et al. 2002a). Hosts of different ages might also plastically adjust their levels of resistance to manipulation (Poulin 2010). Manipulative parasites are expected to be sensitive to density-dependent phenomena; this is illustrated by the arrival of new infective *C. australis* larvae that migrate toward the distal part of the cockle's foot only when a threshold number of encysted parasites has been achieved (Leung et al. 2010).

Manipulative parasites can adjust manipulation intensity in response to both internal and external variables. Thomas et al. (2002a) argued that a potentially large number of direct and indirect cues can theoretically provide accurate information about the host's external environment. For instance, certain trophically transmitted parasites can perceive cues emanating from the definitive host when developing inside intermediate hosts (e.g., Poulin 2003; Lagrue and Poulin 2007; Loot et al. 2008). This suggests that manipulation intensity could be plastic and conditional upon the opportunities for transmission evaluated by the parasite itself from internal (the host) and external (the host's habitat) information.

Although displaying an aberrant behavior likely engenders an energetic cost (see Fig. 2.7A for an example), the hypothesis that heterogeneity in energetic reserves among hosts induces variation in manipulation intensity has received no attention until now. Further studies should test if hosts in poor condition are less subject to intensive manipulation even if they are genetically sensitive to manipulation and infected with a parasite having strong manipulative abilities. In several associations, especially crustaceans parasitized by various helminths (Amat et al. 1991; Plaistow et al. 2001), an increased lipid content is observed in manipulated individuals. Although speculative at the moment, this could correspond to a physiological manipulation aimed at reducing the problem associated with energetic costs related to manipulated behaviors.

2.3.2.2 *Sharing the host: parasite load and multiple infections*

In many systems, host manipulation is caused by parasites impairing the normal function of a specific host organ. Parasite load (i.e., the number of conspecific parasites sharing the same individual host) is thus likely to influence the intensity of host manipulation: in general, when more individual parasites are encysted, impairment is usually higher (e.g., Thomas and Poulin 1998). However, this relationship can be modulated by other processes. Indeed, such situations are likely to lead to a "conflict of interest" between conspecific parasites that are not mature/transmissible at the same time. As recently demonstrated, this phenomenon significantly influences manipulation intensities: in the acanthocephalan *P. laevis*, immature/non-transmissible stages decrease the intensity of the manipulation of *G. pulex* induced by mature/transmissible parasites (see Dianne et al. 2010).

In addition to hosting several conspecific parasites, hosts are usually parasitized by several parasite species (multiple infections; Rigaud et al. 2010) that may have either shared or opposing interests for their transmission (Brown 1999; Lafferty et al. 2000). As is the case for intraspecific conflicts (between parasites of different ages interspecific conflicts may explain a significant portion of the variation generally observed in behavioral responses. Several studies have demonstrated that manipulation intensity can be reduced in cases of conflicts of interest because certain parasites sabotage the manipulation initiated by other parasites (Cézilly et al. 2000; Thomas et al. 2002b). For instance, the behavioral manipulation induced by the trophically transmitted acanthocephala *P. minutus* is weaker in hosts also infected by the microsporidia *Dictyocoela* compared to hosts infected by *P. minutus* alone (Haine et al. 2005) (see also Fig. 2.4). Although often undetected, vertically transmitted parasites are ubiquitous in invertebrates, and conflict via transmission may be of great importance in the study of host–parasite relationships.

2.3.2.3 *Age and size of parasites*
An important part of the natural variation in host manipulation is undoubtedly due to the fact that all parasitized hosts in nature do not harbor parasites synchronized to their age. Manipulative parasites

do not necessarily exert the same level of manipulation on their host through time. For instance, in orthopterans parasitized by hairworms, a first behavioral change is the induction of an erratic behavior coupled with a photophilic behavior (Ponton et al. 2011). The changes induced by sub-adult worms presumably allow crickets to reach aquatic areas. Crickets displaying the erratic behavior do not yet enter the water (Sanchez et al. 2008). Later, crickets harboring fully mature worms display the typical water-seeking behavior that permits the worms to physically enter their reproductive habitat (Fig. 2.5) (see also Chapter 3 and 9). Finally, in the hours and days following worm emergence, surviving ex-parasitized crickets progressively recover a normal behavior (Ponton et al. 2011).

An increasing number of studies also show that trophically transmitted parasites and immature par-

Figure 2.4 The nematode *Gammarinema gammari*. This nematode uses *Gammarus insensibilis* as a habitat and source of nutrition, not as a vector for transmission to birds. There is evidence that this nematode prefers non-manipulated gammarids and seems also able to reverse the manipulation induced by *Microphallus papillorobustus*, changing the gammarid's altered behavior back to a normal behavior (Thomas et al. 2002b) © Pascal Goetgheluck. See also Plate 1.

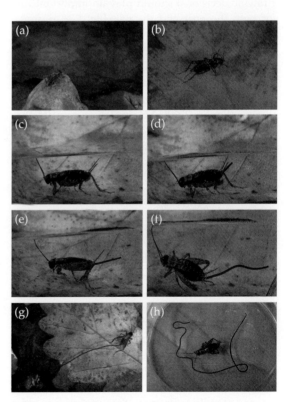

Figure 2.5 Water-seeking behavior in crickets (*Nemobius sylvestris*) harboring the hairworm *Paragordius tricuspidatus*. This behavioral sequence is only the second step within the host manipulation process. © Pascal Goetgheluck. See also Plate 2.

asitoids first impose some "protective" behavior on their host that is thought to reduce predation risk on the host to avoid premature transfer or mortality until parasites mature. It is only when parasite larvae become infective for the definitive host that the reverse pattern is observed (e.g., see Hammerschmidt et al. 2009; Brodeur and Boivin 2004). Dianne et al. (2011) recently showed in an acanthocephalan that the behavioral changes are indeed initially a protection of the host against predation (before the parasite reaches the transmission stage) and then an increased exposure of the hosts to predation (after the parasite has reached the transmission stage). Similarly, immature and mature stages of vector-borne parasites are not expected to induce the same manipulative changes. Vector organisms parasitized with non-transmissible stages have improved survival, which increases the parasite's chance of becoming mature. Mature stages may be confronted with a trade-off between increasing the vector's life span and thus survival on the one hand, or increasing the vector organism's biting frequency at the expense of survival on the other, since biting is risky in nature. Experimental evidence exists for these behavioral changes (e.g., see Koella et al. 2002).

Can age-dependent manipulation persist once infectivity is reached? According to Parker et al. (2009), the optimal time for switching manipulation is a trade-off between increasing the probability of establishment in the next host and reducing mortality in the present host. For instance, in the case of cestodes infecting copepods (intermediate hosts), as larvae become older and usually larger, the relative benefits of remaining in the intermediate host decrease while the potential costs increase (e.g., resource depletion, within-host space constraints, probability of natural host mortality) (Benesh et al. 2008a). Among the few empirical studies on this topic, Franceschi et al. (2008) found that older cystacanths in acanthocephalans induce a higher and less variable degree of manipulation compared with young ones. Benesh et al. (2009) showed that the alteration of isopod colouration following infection by *A. lucii* still increases after the cystacanth stage is reached. The idea that senescence effects (i.e., old parasites would less intensively manipulate their hosts) exist in manipulative parasites has not yet been explored, but would deserve some

consideration. There are undoubtedly many instances in which not all manipulative parasites achieve successful transmission; for example, not all intermediate hosts manipulated by trophically transmitted parasites are captured by predators.

2.3.2.4 *Seasonality*
Host manipulation by parasites may be a short seasonal phenomenon (e.g., the "suicide" of crickets in southern France occurs only during summer months; Thomas et al. 2002), but it can also concern parasites that are active for long periods of time in ecosystems. For example, crazy gammarids harboring the trematode *Microphallus papillorobustus* in brackish lagoons as well as freshwater crustaceans parasitized by acanthocephalans are present in all seasons. In these long-term associations, seasonal variation in manipulation intensity nevertheless exists. For instance, when *Acanthocephalus lucii* are maintained under continuous conditions, they exhibit constant increases in the behavioral alteration they induce in their isopod host, suggesting that their higher manipulation in spring is because this is when they need to reach maturity (Benesh et al. 2008a). However, seasonal behavioral differences induced by *P. laevis* are due to differences in growth rapidity and a subsequent trade-off between growth and behavioral manipulation (parasites rapidly reach maturity but induce little behavioral alteration at that stage in winter while the observed pattern is opposite in spring; Franceschi et al. 2010b). The seasonal cues for these changes are unknown in this case, but it is possible that seasonal variations in host immunocompetence or physiological condition are present.

2.3.2.5 *Parasite trade-offs*
Phenotypic traits are usually selected to optimize fitness in response not only to the environmental variation discussed above, but also to internal trade-offs. As predicted by Poulin (1994) and recently shown experimentally by Franceschi et al. (2010a), manipulative parasites cannot optimize both their growth rate and manipulative intensity: individual *P. laevis* growing slowly manipulate their host as soon as they reach the infective stage. Conversely, individuals that rapidly reach the infective stage do not induce behavioral changes at that stage. Another compelling illustration

concerns parasite fecundity as described by Maure et al. (2011) in a ladybird–parasitoid system. *Dinocampus coccinellae* (Braconidae) is a solitary endoparasitoid of the ladybird *Coleomegilla maculata*. Following larval development, which occurs inside the host body cavity, a parasitoid larva egresses from the ladybird and spins a cocoon between the ladybird's legs (Fig. 2.6) and initiates pupation. Remarkably, parasitoid cocoons tended by a living ladybird suffered significantly less predation than others. However, pupae that benefited from this protection for longer paid a cost in terms of their fecundity. Parasitoids develop using host resources, but these same resources are also needed to keep the ladybird alive following parasitoid egression since the host does not eat during the manipulation. Because host manipulation and reproduction are competing demands, parasitoids cannot simultaneously maximize these two traits. This example also supports the idea expressed earlier—that energetic reserves are indeed important to accomplish manipulated behaviors. These types of trade-offs are probably common in parasites, especially those with complex life cycles since they have to cope with multiple hosts. In addition to growth rate in the intermediate host and fecundity, other traits such as the ability to infect the final host and the growth rate in the final host could be constrained by the intensity of behavioral manipulation.

Figure 2.6   (a) Parasitoid wasp *Dinocampus coccinellae* egressing from the ladybird *Coleomegilla maculata* (photograph by Mathieu B. Morin). (b) Ladybird grasping a parasitoid cocoon between its legs © Pascal Goetgheluck. (c) Incidence of predation (%) by *Chrysoperla carnea* larvae on *D. coccinellae* cocoons attached or not attached to the legs of adult coccinellids *C. maculata*. Fisher exact test, *** $P < 0.0001$. Sample sizes are given within bars. (d) Relationship between the number of days during which *C. maculata* coccinellid remained alive on the *D. coccinellae* cocoon and the potential parasitoid fecundity measured at emergence. Residuals correspond to fecundity data corrected for size and pupal development time of the parasitoid as well as sex and size of the ladybird. $R^2 = 0.219$ and $P = 0.0137$ (from Maure et al. 2011).

2.4   Multidimensional manipulations: evidence of evolution or a syndrome?

It is increasingly acknowledged that manipulated hosts are not merely normal hosts with one aberrant trait (e.g., color or behavior); instead they are deeply altered organisms with a range of modifications occurring simultaneously and/or successively. Exploring this multidimensionality is an emerging and promising field in the study of host manipulation by parasites. According to Thomas et al. (2010a), a first condition before a manipulation can be considered multidimensional is that at least two changes in different phenotypic traits (behavior, morphology, and/or physiology) or in the same phenotypic traits must be observed in the manipulated host. In addition, because the label *manipulation* usually indicates phenotypic changes that are involved in parasite transmission processes (see Poulin 1995), Thomas et al. (2010b) proposed that the label *multidimensionality* be restricted to these specific changes as well. For instance, gammarids (*G. insensibilis*) harboring the trematode *M. papillorobustus* display—in addition to several changes that favor parasite transmission—a longer intermoult duration. This latter change, although typical of crazy gammarids, should not be included in the multidimensionality because its value for transmission is probably nil.

Cezilly and Perrot Minnot (2010) provided a slightly different view, proposing that all traits within infection syndromes induced by manipulative parasites be included under the umbrella of multidimensional manipulation, that is even those that have no function for parasite transmission or survival. This is because symptoms might actually be a consequence of the dysregulation of some key neuromodulators that arise as byproducts of the subversion of the host's immune system by the parasite. In this respect, it might be inadequate, from a functional point of view, to separate phenotypic effects that appear to increase trophic transmission from those that do not. This debate remains open (see Thomas et al. 2010b). Multidimensionality in host manipulation can also have mixed origins (i.e., both the parasite and the host) because, as seen before (see Section 2.2), manipulation *sensu stricto*

and exploitation of host's compensatory responses are not mutually exclusive scenarios.

2.4.1   Why do multidimensional manipulations evolve?

Natural selection would presumably favor individual parasites able to alter several traits in their intermediate hosts because such a phenomenon could greatly enhance the probability of transmission in many systems. For instance, in the context of trophically transmitted parasites, prey displaying both color and behavioral alterations (e.g., artemia parasitized by bird cestodes; Sanchez et al. 2009) probably also have both an enhanced detectability and vulnerability, and therefore have more chance to be captured by predators than those displaying only one of the two changes. A challenging question is then to determine the nature and outcome of each alteration on parasite transmission. This may require elegant protocols that are specific to each system (e.g., see Kaldonski et al. 2009).

2.4.2   Simultaneous versus sequential multidimensional manipulations

While many multidimensional manipulations occur simultaneously, phenotypic alterations within the same transmission strategy can clearly occur one after the other. For instance, as mentioned in the section on age and size of parasites, orthopterans parasitized by sub-adult and adult hairworms do not display the same behavioral changes: sub-adult worms induce erratic and photophilic behaviors (Ponton et al. 2011) while adults induce the water-seeking behavior per se (Sanchez et al. 2008). In other systems, such as the gammarid *G. insensibilis* parasitized by *M. papillorobustus*, certain phenotypic changes induced by parasites (i.e., photophilic behavior, negative geotactism, and aberrant escape behavior) occur simultaneously, but others (increased lipid content) start before the worm is infective for the bird and alters the behavior (Ponton et al. 2005). Mosquitoes parasitized with non-transmissible stages of malaria parasites like oocysts are less likely to bite while individuals infected with sporozoites (transmissible stages) have a higher motivation to bite vertebrates (Lefèvre and Thomas 2008).

2.4.3 How did multidimensional manipulations evolve?

Thomas et al. (2010a) argued that ancestral manipulations probably involved only one dimension and that fitness benefits were subsequently gained by adding dimensions to a simple manipulation. In other words, multidimensional manipulations would arise from single ones. However, disentangling the mix of adaptive forces that shaped the transition from a simple manipulation to a multidimensional one is complex. A first possibility is that the addition of a novel dimension(s) to a simple manipulation is favored by selection when the transmission benefits compensate for the extra costs of any new dimension. For instance, it is realistic in many systems, especially those involving intermediate hosts of trophically transmitted parasites, that adding a second dimension (e.g., color in addition to behavior or the reverse) can boost transmission in a synergistic fashion. Alternatively, once a dimension has been selected, it is also possible that selection favors adjustments (i.e., other dimensions) that reduce the associated costs to improve its efficiency. The case of *G. insensibilis* parasitized by *M. papillorobustus* may be useful to illustrate these statements. The aberrant

escape behavior of crazy gammarids (i.e., swimming toward the surface and turning at the air–water interface) following a mechanical disturbance in the water is likely to be energetically costly (Fig. 2.7A). Therefore, positive phototaxis could have evolved because it decreased the energetic costs linked to the first alteration (i.e., the aberrant escape behavior) just by making the gammarid stay at the water surface (Fig. 2.7B). However, other evolutionary scenarios can also be proposed. The first change that enhanced the transmission to birds could be positive phototaxis, after which natural selection would have favored individuals able to induce an aberrant escape behavior in the host. In the first scenario, phototaxis would have evolved as a secondary adjustment without direct value for transmission while each dimension has in itself a positive effect on transmission in the second scenario.

An interesting aspect of multidimensional manipulation concerns a dimension that evolved in order to protect the parasite from unfavorable situations. There are an increasing number of studies suggesting that some features of parasite-induced behavioral changes serve more to limit the risk of predation by the wrong (non-host) predator than to increase transmission to appropriate hosts (e.g., Levri 1998). For example, the bird acanthocephalan *P. minutus* not only increases the probability of *G. pulex* being eaten by birds, it also enhances its escape

(a) (b)

Figure 2.7 (a) Crazy gammarids actively swimming toward the surface following a mechanical disturbance in the water. (b) Photophilic behavior of crazy gammarids in the absence of a mechanical disturbance. © Pascal Goetgheluck. See also Plate 3.

performance towards a crustacean predator like *Dikerogammarus villosus* (Medoc and Beisel 2008). In other systems, especially those involving braconid wasps, bodyguard manipulation (protection of parasitoid pupae by the host) is the main behavioral manipulation occurring, and several features could be viewed as different protective dimensions (Brodeur and Vet 1994; Maure et al. 2011).

In accordance with their infection syndrome hypothesis, Cezilly and Perrot Minnot (2010) provided a different perspective on how different dimensions can appear in multidimensional manipulations. According to them, it is more parsimonious to assume that parasites secrete only one single compound resulting in a cascade of effects that include the dimensions that increased the transmission (see also Chapter 3). Genes that code for the ability to secrete that compound will then evolve in direct relation to the balance between the overall consequences of all induced phenotypic effects in the host on the parasite's fitness—whether each one decreases it, increases it, or is neutral—and the cost of producing the compound. Under this scenario, only one trait is under selection: the ability of the parasite to secrete the particular compound.

We believe that the scenarios proposed by Thomas et al. (2010) and Cezilly and Perrot-Minnot (2010) are not mutually exclusive: within a given multidimensional manipulation, it could be possible that certain dimensions result from an infection syndrome effect while others do not.

2.4.4 Proximate aspects of multidimensionality

Very little is known about the mechanisms by which parasites alter the phenotype of their host (Thomas et al. 2005; see Chapter 5); accordingly, mechanisms involved in multidimensionality are poorly understood. Within the context of their infection syndrome hypothesis, Cezilly and Perrot Minnot (2010) provided different examples of neuromodulatory effects triggering a cascade of biochemical reactions with consequences for the organism that go beyond the simple excitation or inhibition of classical neurotransmitters. However, no specific studies have addressed the link between the phenomena and

concrete cases of multidimensional manipulation. One way to explore how different dimensions could be mechanistically related would then be to study correlations between the extent of modification in different altered traits (e.g., see Benesh et al. 2008b). The detection of positive correlations might indicate that parasites alter one or a few components of the host's physiology, which results in a cascade of effects.

2.5 Concluding remarks

Despite considerable progress, our understanding of the evolution of host manipulation by parasites is still in its infancy. As exemplified in this chapter, there are different evolutionary routes that allow parasites to make their hosts behave in a way that is favorable to parasite transmission and survival. These scenarios are at the root of the evolution of behavioral manipulation, but we must keep in mind that parasites and their hosts are in constant coevolution: manipulating host behavior is an ever-evolving interaction rather than a static process, and parasites altering host behavior are not the definitive winners, they are merely typical parasites coevolving with their hosts. Issues of manipulation *sensu stricto* vs. interactive scenarios remain a hot topic and have much to gain from an attention to mechanisms (see Chapter 3).

It also appears that intraspecific variability in host manipulation arises from complex interactions between genomes, the environment, and the age of hosts and parasites. Because of the huge number of non-genetic variables involved, the question arises as to whether evolution of host manipulation should be considered within the framework of phenotypic plasticity rather than in a purely genetic framework. More common garden experiments, as well as findings on local adaptation patterns or genetic and plastic variations in the parasite's ability to manipulate host behavior (or the host's ability to resist or tolerate manipulation), would provide helpful information on how this strategy of host exploitation evolves. In particular, we believe that if manipulation is undergoing adaptive evolution, the purposive design criterion proposed by Poulin (1995) should be fulfilled at the ecological scale as well as at the phylogenetic scale (as originally proposed by Poulin). For example, in a single trophically transmitted parasite species that is a generalist for its definitive host, there could be variation in the intensity (or even in the pattern) of manipulation fitting the variety of feeding habits of the different hosts. If the different hosts live in allopatry, we can predict a geographic variation in manipulation that is not driven by abiotic conditions but rather through fitting the local predator's feeding habit. If hosts live in sympatry, we can predict the coexistence of several manipulation strategies leading at the extreme to genetic differentiation between hosts; this pattern is compatible with the "host-race" observed in some parasites. Such patterns are not predicted if the evolution of manipulation is not adaptive.

To understand the evolution of host manipulation by parasites, one must consider manipulated hosts within their ecological context since only this approach permits a true assessment of the selective pressures experienced by both the host and the parasite. We believe that why and how the multidimensionality of host manipulation evolves is a research direction requiring collaboration among parasitologists and researchers from other disciplines, especially physiology, morphology, behavioral ecology, and developmental biology.

References

Abbot, P. and Dill, L. M. (2001). Sexually transmitted parasites and sexual selection in the milkweed leaf beetle, *Labidomera clivicollis*. *Oikos* **92**, 91–100.

Adamo, S. A. (1999). Evidence for adaptive changes in egg-laying in crickets exposed to bacteria and parasites. *Animal Behaviour* **57**, 117–124.

Amat, F., Gozalbo, A., Navarro, J., Hontoria, F., and Varó, I. (1991). Some aspects of Artemia biology affected by cestode parasitism. *Hydrobiologia* **212**, 39–44.

Andrews, P. W., Gangestad, S. W., and Matthews, D. (2002). Adaptationism, exaptationism, and evolutionary behavioral science. *Behavioral and Brain Sciences* **25**, 534–553.

Ayala, F. J. (2010). *Am I A Monkey? Six Big Questions about Evolution.* Johns Hopkins University Press, Baltimore.

Benesh, D., Hasu, T., Seppälä, O., and Valtonen, E. (2008a). Seasonal changes in host phenotype manipulation by

an acanthocephalan: time to be transmitted? *Parasitology* **136**, 219–230.

Benesh, D. P., Valtonen, E. T., and Seppälä, O. (2008b). Multidimensionality and intra-individual variation in host manipulation by an acanthocephalan. *Parasitology* **135**, 617–626.

Benesh, D., Seppälä, O. and Valtonen, E. (2009). Acanthocephalan size and sex affect the modification of intermediate host colouration. *Parasitology* **136**, 847–854.

Biron, D. G., Ponton, F., Joly, C., Menigoz, A., Hanelt, B., and Thomas, F. (2005). Water seeking behavior in insects harboring hairworms: should the host collaborate? *Behavioral Ecology* **16**, 656–660.

Brodeur, J. and Boivin, G. (2004). Functional ecology of immature parasitoids. *Annual Review of Entomology* **49**, 27–49.

Brodeur, J. and Vet, L. E. M. (1994). Usurpation of host behaviour by a parasitic wasp. *Animal Behaviour* **48**, 187–192.

Brown, S. (1999). Cooperation and conflict in host-manipulating parasites. *Proceeding of the Royal Society of London. B, Biological Sciences* **266**, 1899–1904.

Cézilly, F., Grégoire, A., and Bertin, A. (2000). Conflict between co-occurring manipulative parasites? An experimental study of the joint influence of two acanthocephalan parasites on the behaviour of *Gammarus pulex*. *Parasitology* **120**, 625–630.

Cézilly, F. and Perrot-Minnot, M.-J. (2010). Interpreting multidimensionality in parasite-induced phenotypic alterations: panselectionism vs parsimony. *Oikos* **119**, 1224–1229.

Cézilly, F., Thomas, F., Médoc, V., and Perrot-Minnot, M. J. (2010). Host manipulation by parasites with complex life cycle: adaptive or not ? *Trends in Parasitology* **26**, 311–317.

Combes, C. (1998). *Parasitism, The Ecology and Evolution of intimate interactions.* The University of Chicago Press, Ltd., London.

Dawkins, R. (1982). *The Extended Phenotype.* Oxford University Press, Oxford.

Dianne, L., Rigaud, T., Léger, E., Motreuil, S., Bauer, A., and Perrot-Minnot, M. J. (2010). Intraspecific conflict over host manipulation between different larval stages of an acanthocephalan parasite. *Journal of Evolutionary Biology* **23**, 2648–2655.

Dianne, L., Perrot-Minnot, M. J., Bauer, A., Gaillard, M., Leger, E., Rigaud, T. (2011). Protection first then facilitation: a manipulative parasite modulates the vulnerability to predation of its intermediate host according to its own developmental stage. *Evolution* **65**, 2692–2698.

Drummand, F. A., Cassagrande, R. A., and Logan, P. A. (1989). Population dynamics of *Chrysomelobia labidomerae* Eickwort, a parasite of the Colorado potato beetle. *International Journal of Acarology* **15**, 31–45

Engelstädter, J., and Hurst, G. (2009). The ecology and evolution of microbes that manipulate host reproduction. *Annual Review of Ecology and Systematics* **40**, 127–149.

Franceschi, N., Bauer, A., Bollache, L., and Rigaud, T. (2008). The effects of parasite age and intensity on variability in acanthocephalan-induced behavioural manipulation. *International Journal of Parasitology* **38**, 1161–1170.

Franceschi, N., Bollache, L., Cornet, S., Bauer, A., Motreuil, S., and Rigaud, T. (2010b). Co-variation between the intensity of behavioural manipulation and parasite development time in an acanthocephalan-amphipod system. *Journal of Evolutionary Biology* **23**, 2143–2150.

Franceschi, N., Bollache, L., Dechaume-Moncharmont, F., Cornet, S., Bauer, A., Motreuil, S., and Rigaud, T. (2010a). Variation between populations and local adaptation in acanthocephalan-induced parasite manipulation. *Evolution* **64**, 2417–2430.

Franceschi, N., Rigaud, T., Moret, Y., Hervant, F., and Bollache, L. (2007). Behavioural and physiological effects of the trophically transmitted cestode parasite *Cyathocephalus truncatus* on its intermediate host, *Gammarus pulex*. *Parasitology* **134**, 1839–1847.

Gould, S. J. and Vrba, E. S. (1982). Exaptation: a missing term in the science form. *Paleobiology* **8**, 4–15.

Haine, E., Boucansaud, K., and Rigaud, T. (2005). Conflict between parasites with different transmission strategies infecting an amphipod host. *Proceeding of the Royal Society of London B* **272**, 2505–2510.

Hamilton, C. M., Stafford, P., Pinelli, E., and Holland, C. V. (2006). A murine model for cerebral toxocariasis: characterization of host susceptibility and behaviour. *Parasitology* **132**, 791–801.

Hammerschmidt, K., Koch, K., Milinski, M., Chubb, J., and Parker, G. A. (2009). When to go: optimization of host switching in parasites with complex life cycles. *Evolution* **63**, 1976–1986.

Hoover, J. P. and Robinson, S. K. (2007). Retaliatory mafia behavior by a parasitic cowbird favors host acceptance of parasitic eggs. *Proceeding of the National Academy of the United State of America* **104**, 4479–4483.

Kaldonski, N., Perrot-Minnot, M. J., Dodet, R., Martinaud, G., Cezilly, F. (2009). Carotenoid-based colour of acanthocephalan cystacanths plays no role in host manipulation. *Proceedings of the Royal Society B-Biological Sciences* **276**, 169–176.

Koella, J., Rieu, L., and Paul, R. (2002). Stage-specific manipulation of a mosquito's host-seeking behavior by the malaria parasite *Plasmodium gallinaceum*. *Behavioural Ecology* **13**, 816–820.

Lafferty, K., Thomas, F., and Poulin, R. (2000). Evolution of host phenotype manipulation by parasites and its consequences. In: *Evolutionary Biology of Host-Parasite Relationships: Theory meets Reality* (eds Poulin, R., Morand, S., and Skorping, A.) pp. 117–127. Elsevier, Amsterdam,

Lagrue, C. and Poulin, R. (2007). Life cycle abbreviation in the trematode *Coitocaecum parvum*: can parasites adjust to variable conditions? *Journal of Evolutionary Biology* **20**, 1189–1195.

Lefèvre, T., Adamo, S. A., Biron, D. G., Missé, D., Hughes, D., and Thomas, F. (2009). Invasion of the body snatchers: the diversity and evolution of manipulative strategies in host-parasite interactions. *Advances in Parasitology* **68,** 45–83.

Lefèvre, T., Roche, B., Poulin, R., Hurd, H., Renaud, F., and Thomas, F. (2008) Exploitation of host compensatory responses: the "must" of manipulation? *Trends in Parasitology* **24**, 435–439.

Lefèvre, T. and Thomas, F. (2008). Behind the scene, something else is pulling the strings: emphasizing parasitic manipulation in vector-borne diseases. *Infection, Genetics and Evolution* **8**, 504–519.

Leung, T., Keeney, D., and Poulin, R. (2010). Genetics, intensity-dependence, and host manipulation in the trematode *Curtuteria australis*: following the strategies of others? *Oikos* **119**, 393–400.

Levri, E. P. (1998). The influence of non-host predators on parasite-induced behavioural changes in a freshwater snail. *Oikos* **81,** 531–537.

Loot, G., Blanchet, S., Aldana, M., and Navarrete, S. (2008). Effect of human exclusion on parasite reproductive strategies. *Journal of Parasitology* **94**, 23–27.

Maure, F., Brodeur, J., Ponlet, N., Doyon, J., Firlej, A., Elguero, E., and Thomas, F. (2011). The cost of a bodyguard. *Biology Letters* **7**, 843–846.

Medoc, V. and Beisel, J. N. (2008). An acanthocephalan parasite boosts the escape performance of its intermediate host facing non host predators. *Parasitology* **135**, 977–984.

Moore, J. (1984). Altered behavioral responses in intermediate hosts- an acanthocephalan parasite strategy. *American Naturalist* **123**, 572–577.

Moore, J. (2002). *Parasites and the Behavior of Animals.* Oxford Series in Ecology and Evolution. Oxford University Press , USA.

Moore, J. and Gotelli, N. J. (1990). A phylogenetic perspective on the evolution of altered host behaviours: a critical look at the manipulation hypothesis. In: *Parasitism and host Behaviour* (eds Barnard, C. J. and Behnke, J. M.) pp. 193–233. Taylor and Francis, London.

Mouritsen, K. and Poulin, R. (2003). Parasite-induced trophic facilitation exploited by a non-host predator: a manipulator's nightmare. *International Journal for Parasitology* **33**, 1043–1050.

Pagel, M, Møller, A. P., and Pomiankowski, A. (1998). Reduced parasitism by retaliatory cuckoos select for hosts that rear cuckoo nestlings. *Behavioural Ecology* **9**, 566–572.

Parker, G. A., Ball, M. A., Chubb, J. C., Hammerschmidt, K., and Milinski, M. (2009). When should a trophically transmitted parasite manipulate its host? *Evolution* **63**, 448–458.

Perrot-Minnot, M. J. (2004). Larval morphology, genetic divergence, and contrasting levels of host manipulation between forms of *Pomphorhynchus laevis* (Acanthocephala). *International Journal for Parasitology* **34**, 45–54.

Plaistow, S., Troussard, J., and Cézilly, F. (2001). The effect of the acanthocephalan parasite *Pomphorhynchus laevis* on the lipid and glycogen content of its intermediate host *Gammarus pulex*. *International journal for parasitology* **31**, 346–351.

Polak, M. and Starmer, W. T. (1998). Parasite-induced risk of mortality elevates reproductive effort in male Drosophila. *Proceeding of the Royal Society of London B* **265**, 2197–2201.

Ponton, F., Biron, D. G., Joly C., Duneau, D., and Thomas, F. (2005). Ecology of populations parasitically modified: a case study from a gammarid (*Gammarus insensibilis*)-trematode (*Microphallus papillorobustus*) system. *Marine Ecology Progress Series* **299**, 205–215.

Ponton, F., Biron, D. G., Moore, J., Moller, A. P., and Thomas, F. (2006b). Facultative virulence as a strategy to manipulate hosts. *Behavioural Processes* **72**, 1–5.

Ponton, F., Duneau, D., Sanchez, M., Courtiol, A., Terekhin, A. T., Budilova, V., Renaud, T., and Thomas, F. (2009). Effect of parasite-induced behavioral alterations on juvenile development. *Behavioral Ecology* **20**, 1020–1026.

Ponton, F., Lefèvre, T., Lebarbenchon, C., Thomas, F., Loxdale, H. D., Marché, L., Renault, L., Perrot-Minnot, M. J., and Biron, D. G. (2006b). Do distantly parasites rely on the same proximate factors to alter the behaviour of their hosts? *Proceedings of Royal Society of London B* **273**, 2869–2877.

Ponton, F., Otálora-Luna, F., Lefèvre, T., Guerin, P., Lebarbenchon, C., Duneau, D., Biron, D. G., and Thomas, F. (2011). Water-seeking behavior in worm-infected crickets and reversibility of parasitic manipulation. *Behavioral Ecology* **22**, 392–400.

Poulin, R. (1994). The evolution of parasite manipulation of host behaviour: a theoretical analysis. *Parasitology* **109**, S109–S118.

Poulin, R. (1995). "Adaptive" change in the behaviour of parasitized animals: a critical review. *International Journal for Parasitology* **25**, 1371–1383.

Poulin, R. (2003). Information about transmission opportunities triggers a life-history switch in a parasite. *Evolution* **57**, 2899–2903.

Poulin, R. (2010). Parasite manipulation of host behavior: an update and frequently asked questions. *Advances in the Study of Behaviour* **41**, 151–186.

Poulin, R., Fredensborg, B. L., Hansen, E., and Leung, T. L.F. (2005). The true cost of host manipulation by parasites. *Behavioural Processes* **68**, 241–244

Rigaud, T., Perrot-Minnot, M. J., and Brown M. J. F. (2010). Parasite and host assemblages: embracing the reality will improve our knowledge of parasite transmission and virulence. *Proceedings of Royal Society of London B* **277**, 3693–3702.

Sanchez, M. I., Hortas, F., Figuerola, J., and Green, A. J. (2009). Sandpipers select red brine shrimps rich in both carotenoids and parasites. *Ethology* **115**, 196–200.

Sanchez, M., Ponton, F., Schmidt-Rhaesa, A., Hughes, D., Misse, D., and Thomas, F. (2008). Two steps to suicide in crickets harbouring hairworms. *Animal Behaviour* **76**, 1621–1624.

Smith-Trail, D. R. (1980). Behavioural interactions between parasites and host: host suicide and evolution of complex life cycles. *American Naturalist* **116**, 77–91.

Soler, M., Soler, J. J., Martinez, J. G., and Møller, A. P. (1995). Magpie host manipulation by great spotted cuckoos: evidence for an avian mafia? *Evolution* **49**, 770–775.

Soler, M., Soler, J. J., Martinez, J. G., Perez-Contreras and T., Møller, A. P. (1998). Micro-evolutionary change and population dynamics of a brood parasite and its primary host: the intermittent arms race hypothesis. *Oecologia* **117**, 381–390.

Tain, L., Perrot-Minnot, M .J., and Cezilly, F. (2006). Altered host behaviour and brain serotonergic activity caused by acanthocephalans: evidence for specificity. *Proceedings of Royal Society of London B* **273**, 3039–3045.

Thomas, F., Adamo, S. A., and Moore, J. (2005). Parasitic manipulation: where are we and where should we go? *Behavioural Processes* **68**, 185–199.

Thomas, F., Brodeur, J., Maure, F., Franceschi, N., Blanchet, S., and Rigaud, T. (2011). Intraspecific variability in host manipulation by parasites. *Infection, Genetics and Evolution* **11**: 262–269.

Thomas, F., Brown, S., Sukhdeo, M., and Renaud, F. (2002a). Understanding parasite strategies: a state-dependent approach? *Trends in Parasitology* **18**, 387–390.

Thomas, F., Fauchier, J., and Lafferty, K. (2002b). Conflict of interest between a nematode and a trematode in an amphipod host: test of the "sabotage" hypothesis. *Behavioural Ecology and Sociobiology* **51**, 296–301.

Thomas, F., Schmidt-Rhaesa, A., Martin, G., Manu, C., Durand P., and Renaud F. (2002). Do hairworms (Nematomorpha) manipulate their terrestrial host to seek water ? *Journal of Evolutionary Biology* **15**, 356–361.

Thomas, F. and Poulin, R. (1998). Manipulation of a mollusc by a trophically transmitted parasite: convergent evolution or phylogenetic inheritance? *Parasitology* **116**, 431–436.

Thomas, F., Poulin, R., and Brodeur, J. (2010a). Host manipulation by parasites: a multidimensional problem. *Oikos* **119**, 1217–1223.

Thomas, F, Poulin, R., and Brodeur, J. (2010b). Infection syndrome and multidimensionality two terms for two different issues. *Oikos* **119**, 1230.

Vance, S. A. (1996). Morphological and behavioural sex reversal in mermithid-infected mayflies. *Proceedings of Royal Society of London B* **263**, 907–912.

Zahavi, A. (1979). Parasitism and nest predation in parasitic cuckoos. *The American Naturalist* **113**, 157–159.

Afterword

Stephen C. Stearns

The theory of life history evolution developed at a time in the history of evolutionary thought when kin selection, genetic conflict, and hierarchical selection had not yet been well assimilated into general evolutionary theory. They were not included in the classical theory, and their implications for life histories are even now not yet completely understood. The situation is similar, and the lack of understanding more impressive, for the impact of extended phenotypes on life histories. The manipulated behavior of the host is part of the extended phenotype of the parasite, and the concrete intimacy of the host–parasite interaction is a good model system in which to explore the extent to which one species can influence the behavior and ecology of another. If we think not of interacting species but of interacting agents of any sort—the two sexes interacting in sexual selection, the participants in a group interacting in social selection—then perhaps we can see a broader range of potential application for insights won in the study of parasite manipulation.

One such insight is already available; it concerns a classic and as yet unanswered question about life histories. How do complex life cycles evolve? Part of the solution may lie in genetic assimilation that starts with the first exposure of a dispersal stage to a new habitat. If the phenotypic plasticity of that stage allows it to survive and reproduce in the new habitat, subsequent genetic modification of the reaction norm could eventually transform it into a stage in the life cycle specifically adapted to the new habitat. Support for this idea can be found in the chapter above, where it is suggested that because of the well-documented connections between the immune system and the nervous system, the ancestors of parasites that are now manipulative could have,

through their direct interactions with the hosts's immune system, generated behavioral byproducts. Those byproducts could then have been acted on by further selection of parasite variants to build up, step-by-step, a manipulated host phenotype that significantly increased the probability of successful transmission back to the primary host. The original host responses were variable, some benefiting and others penalizing the parasite, and through selection operating on genetic variants affecting the parasite interaction with the host immune system, the host behavior, viewed as a reaction norm of the parasite, was shaped and stabilized into a manipulated behavior that reliably benefited the parasite. In so doing, selection may have changed the characteristics of the parasite in the primary host as well as in the newly acquired secondary host. If the reader thinks this line of thought worth pursuing, they might want to consult West-Eberhard (2003) for background and significance and Lande (2009) for a useful model of genetic assimilation as reaction norm evolution in a novel environment.

We can also contemplate the host–parasite relationship in search of general principles for the shaping of phenotypes that extend beyond the epidermis. One such principle must surely be the regularity and reliability with which a species encounters a feature of its extra-epidermal environment that influences its fitness. A relevant example comes from the literature on the Hygiene Hypothesis in evolutionary medicine. An axenic infant mouse (a mouse completely lacking bacteria) does not develop gut-associated lymphoid tissue, the first line of immune defense that surrounds the gut. If we infect such a mouse during its first week of life with a gut microbe, such as *Salmonella*, the microbe induces the formation of gut associated lymphoid tissue (Rhee et al. 2004). What are the implications? I see at least two. First, the genome of the mouse has delegated to the genome of the bacterium the responsibility for emitting the signal that induces the development of part of the mouse's immune system. That could only have evolved if the presence of such bacteria in the gut of infant mice was completely reliable, for lacking such tissue is quite dangerous. Second, in making that coevolutionary move, the mouse genome extended its phenotype to include a gut commensal and demonstrated that extension through the integration of the signaling pathway initiated by the bacterium into its development. The message is that the acquisition of a new element of an extended phenotype depends on the reliability of the presence the environmental element being assimilated in every generation.

Going back to the evolution of complex life cycles and manipulation of hosts by parasites, I suggest that such manipulation can only evolve after the new stage has been reliably established as an una- voidable part of the life cycle of the host. Once that has happened, delegations of responsibility for fit- ness-altering events can be handed over to the pre- viously independent host genome, but not until then. Subsequent coevolution could then well wipe out or cover up any traces of the initial dynamic, leaving us to puzzle over whether some interesting pattern is a cause or a consequence of the host– parasite relationship.

References

Lande, R. (2009). Adaptation to an extraordinary environ- ment by evolution of phenotypic plasticity and genetic assimilation. *Journal of Evolutionary Biology* **2**, 1435–1446.

Rhee, K.-J., Sethupathi, P., Driks, A., Lanning, D. K., and Knight, K. L. (2004). Role of commensal bacteria in development of gut-associated lymphoid tissues and preimmune antibody repertoire. *Journal of Immunology* **172**, 1118–1124.

West-Eberhard, M.J. (2003). *Developmental Plasticity and Evolution*. Oxford University Press, New York.

CHAPTER 3

The strings of the puppet master: how parasites change host behavior

Shelley A. Adamo

3.1 Introduction

The concept of parasites manipulating behavior fascinates both scientists and non-scientists. The idea that an invading organism can control a person's mind remains a popular theme in film, literature and other media (e.g., "28 Days Later", "The Crazies"). One of the reasons this topic resonates with many inside and outside of science is that it touches on core philosophical issues such as the existence of free will. If the mind is a machine, then anyone can control it—anyone, that is, who knows the code and has access to the machinery. Is there evidence that parasites can control minds? While many of the behavioral changes induced by parasites are quite general (e.g., increased or decreased activity, Moore 2002), some seem spookily specific. For example, hairworms induce crickets to leap into water (Thomas et al. 2002), and a brain-dwelling protozoan causes rodents to be attracted to cat urine (Berdoy et al. 2000). Although some parasites alter host behavior by destroying sensory receptors or by disturbing connections to the motor system (see Moore 2002), other parasites alter host behavior by influencing central nervous system function (Thomas et al. 2005; Lefèvre et al. 2009). Therefore, some parasites do appear to hijack the brain.

This chapter reviews some of the best-understood examples of parasite-induced changes in host brain and behavior. I compare the methods parasites are thought to use to alter host behavior with how a neurobiologist might induce the same behavioral changes. Understanding how parasites manipulate behavior, as well as how their methods may differ from a neurobiologist's, will be of interest to neuro-scientists as well as parasitologists and behavioral ecologists.

Parasite-induced changes in host behavior are often thought to increase parasite fitness (Lefèvre et al. 2009; see Chapter 2). If this is true, it suggests that evolution has selected for the parasitic mechanisms that produce these changes. Demonstrating whether host behavioral change does, in fact, increase parasite fitness is difficult (Cézilly et al. 2010) and has not been conclusively shown for some of the systems described below. Regardless of the fitness effects, the systems discussed in this chapter are all examples of parasites producing robust and striking behavioral changes.

In this chapter I use the term parasite to mean any organism that lives in or on its host and extracts resources from it, leading to a loss of host fitness. Therefore, in this chapter, parasites include microorganisms and viruses.

3.2 How do parasites alter host behavior? Vertebrate examples

3.2.1 *Toxoplasma gondii*

One of the best-studied cases of manipulation of a vertebrate host is that of the protozoan intracellular parasite, *Toxoplasma gondii*, and its effects on its rodent intermediate host (see also Chapter 10). *T. gondii* requires a feline host for sexual reproduction, but it typically passes through a rodent first (Roberts and Janovy 1996). When a rodent ingests feline feces that are contaminated by *T. gondii* oocysts, the rodent becomes infected. After an acute infection, *T. gondii* forms cysts within various organs, including the brain (Dubey et al. 1998). Once a cat consumes an infected rodent, the life cycle is complete and *T. gondii* undergoes sexual reproduction (Roberts and Janovy 1996). Because *T. gondii* requires a specific definitive host, changes in rodent behavior that would increase the odds of being eaten by the correct host would greatly enhance transmission. Given that *T. gondii* tends to infect the brain, it may have required only a small number of evolutionary steps for this parasite to manipulate its host behavior.

The discussion below focuses on the behavioral changes that occur after cyst formation (latent phase). It is the bradyzoites inside the cysts that are infectious (Roberts and Janovy 1996); therefore behavioral changes that occur prior to bradyzoite formation would not promote parasite transmission. This discussion also focuses on studies that use infected rats instead of mice, because the effects on mice are more difficult to interpret due to a long-lasting acute phase (Hrdá et al. 2000). The acute phase involves substantial immune activity by the host. This activity can stimulate immune-neural connections, leading to changes in host behavior (Dantzer 2004). These host-induced behavioral changes make the effect of the parasite on host behavior harder to discern (Hrdá et al. 2000).

T. gondii induces a range of behavioral changes in its host. Some are likely to increase the odds of being eaten by any predator. For example, infected rats show: increased activity (Webster 1994), increased grooming in exposed areas—that is increased conspicuousness (Webster et al. 2006), and decreased neophobia (Webster et al. 1994). Infected rats also show decreased anxiety in some tests (e.g., increased exploration in an elevated maze, Gonzalez et al. 2007) but not others (e.g., time spent in the center of an open-field arena, Vyas et al. 2007a), suggesting that the effect is not simply a decrease in overall anxiety, but will result in increased exploration of novel areas. Infected rodents are more likely to be caught by traps in the wild than are uninfected rodents (Webster et al. 1994), suggesting that the behavioral changes are likely to increase predation non-specifically. These changes in behavior are striking because rodents infected with parasites other than *T. gondii* are typically more fearful of

predators than are controls (Thompson and Kavaliers 1994), probably because an immune challenge increases anxiety in rodents (Lacosta et al. 1999). Therefore the changes in host behavior are the opposite of what would be expected from a typical host-response to infection.

Some behavioral changes appear to bias the host towards a feline predator. Rodents infected with *T. gondii* lose their aversion to cat odor and become attracted to it (Webster et al. 2006), but not to the odor of non-predatory mammals (Berdoy et al. 2000; Vyas et al. 2007a), or those of non-feline predators (i.e., mink, Lamberton et al. 2008). Unfortunately actual predation studies are lacking (Lamberton et al. 2008), so it is unknown whether these changes in preference are strong enough to promote predation by a definitive host. Moreover, whether cat odor is attractive to infected rodents depends on the concentration. When cat odor concentrations are high, predicting the greatest opportunity for predation, infected rats show the same aversion to cat odor as uninfected controls (Vyas et al. 2007b).

Most behaviors are normal in infected rodents. For example, feeding (Vyas et al. 2007) and mating success (Berdoy et al. 1995) are unaffected by the parasite. There is also no evidence of sickness behaviors, once the acute phase of infection has passed (Vyas et al. 2007a).

How would a neuroscientist produce these effects? Rodents have innate defensive reactions to predator odors and these defensive behaviors are subserved by specific neural circuits (Blanchard et al. 2003, 2005; Staples et al. 2005). A neuroscientist who wanted to induce rats to approach cat odor would probably begin by finding the neural circuits responsible for predator odor processing and anti-predator behavior and interrupt these circuits. Is there evidence *T. gondii* does the same?

Predator odor processing
Several neural circuits are involved in rodent avoidance of cat odor (Staples et al. 2005). Circuitry activated by cat odor includes limbic, hypothalamic, and brainstem regions (Staples et al. 2005). The hypothalamus, especially the medial hypothalamic defensive system, centered around the dorsal premammillary nucleus, and the dorsomedial part of

the ventromedial hypothalamus, show strong activation to cat odor (Staples et al. 2005). Staples et al.
(2005) found no significant activation of the amygdala in response to cat odor, despite its importance in mediating fear behavior (de la Mora et al.
2010). However, other researchers, using different
neurobiological methods, have found that some
amygdalar nuclei are involved in the response to
cat odor (e.g., Takahashi et al. 2007). The exact role
of the different nuclei of the amygdala in defensive
behaviors remains under debate (Blanchard et al.
2005; Staples et al. 2005). However, the role of some
hypothalamic nuclei seems clear. For example,
lesions of the dorsal premammillary nucleus reduce
defensive behaviors to both a live cat and to cat
odor, but have a minimal effect on non-predator
threat stimuli (Blanchard et al. 2003).

The locus ceruleus is also activated by cat odor
(Staples et al. 2005). The locus ceruleus is a small
nucleus rich in norepinephrine containing cells
(Bear et al. 2007). This nucleus responds to novel,
unexpected stimuli and sends processes throughout the brain (Bear et al. 2007). It is thought to have
a general arousal effect, making animals more alert
(Bear et al. 2007). DoMonte et al. (2008) found that
blocking ß-adrenergic receptors in the dorsal premammillary nucleus prevented cat odor from
increasing neural activity in this brain region.
Behaviorally, blocking ß-adrenergic receptors in the
dorsal premammilary nucleus increased the time
rats spent exploring a cat odor source, and decreased

the time they spent hiding (DoMonte et al. 2008),
very similar to the effect of *T. gondii* on rodent
behavior.

Given this information, a neurobiologist might
start by targeting nuclei involved in the medial
hypothalamic defensive system, especially specific
dorsal premammilary nuclei and/or noradrenergic
transmitter systems innervating this brain area
(Table 3.1). Olfactory areas are also plausible targets, although smell itself is not affected by infection (e.g., Lamberton et al. 2008).

What effects does T. gondii *have on the rodent brain?*
T. gondi is found in both glia and neurons (Gonzalez
et al. 2007). Parasites are found both as isolated
cysts and in clusters within the infected rodent
brain (Gonzalez et al. 2007). Parasites are found
throughout brain, although there is some increased
density in amygdalar regions (Gonzalez et al. 2007;
Vyas et al. 2007a, although see Gulinello et al. 2010).
Parasites also tend to be in more medial portions of
the brain (e.g., prefrontal cortex, accumbens nuclei,
and hypothalamus; Gonzalez et al. 2007). These
brain areas subserve many functions. Strikingly,
T. gondii does NOT selectively infect those areas of
the brain thought to mediate the response to cat
odor, such as the medial hypothalamic defensive
pathway. Therefore, one of the key methods used
by neuroscientists, precise neuroanatomical specificity, does not appear to be used by *T. gondii*.
However, Gonzalez et al. (2007) point out that

Table 3.1 Differences between the mechanisms used by *T. gondii* and those of a Neuroscientist to induce the specific behavioral changes observed in Toxoplasmosis (References in text, unless identified by subscript).

Altered host behavior	Possible advantage to parasite	Neurobiologist's method	Possible *T. gondii* method
↓ Anxiety,	↑ transmission to definitive host, but also to incompatible hosts	Lesion amygdala	↑ dopamine levels widely throughout brain (esp. amygdala)
↓ Neophobia		Enhance GABAergic neurotransmission [1]	
↓ Aversion to cat odour	↑ transmission to definitive host	Lesion dorsal premammillary nucleus	↑ dopamine levels widely throughout brain (esp. amygdala)
		Block ß-adrenergic receptors in dorsal premammilary nucleus	

1. Bear et al. 2007

among the numerous brain areas attacked are two areas important for the control of emotional behavior such as anxiety: the amygdala and hypothalamus. Therefore, although the parasite is not selective, it does reliably target those structures involved in emotional behavior. However, such a broad attack would not produce the specific behavioral changes observed in infected rats.

Neuropharmacology is another key method neurobiologists use to modify behavior. Because chemicals can diffuse and influence large numbers of neurons, drugs do not need to be given at a precise anatomical location to influence behavior. Vyas and Sapolsky (2010) and Webster and McConkey (2010) suspect that the behavioral effects caused by *T. gondii* are due, at least in part, to substances secreted from the cysts.

There is evidence to support this hypothesis. A recent study has shown that during the formation of bradyzoite cysts, the parasite expresses genes for tyrosine hydroxylases (Gaskell et al. 2009). This enzyme can convert phenylalanine to tyrosine and tyrosine to L-Dopa (Gaskell et al. 2009). L-Dopa is the precursor to two potent neuromodulators: dopamine and norepinephrine (Iverson et al. 2009). This enzyme is uncommon in protozoans, and seems special to *T. gondii* (Gaskell et al. 2009). The genetic sequence for the enzyme also suggests that it may be excreted (Gaskell et al. 2009). The enzyme may play a role in the formation of the cyst wall (Gaskell et al. 2009), in the same way that tyrosine derivatives are important for the formation of structural components in insects (Nation 2008). But there is also the possibility that the enzyme, if secreted into the host's neurons, could raise dopamine levels in the host (Webster and McConkey 2010). Increases in dopamine could have a large impact on a variety of behaviors (Iverson et al. 2009). If this hypothesis is correct, it would be an interesting example of a compound evolved for one role (formation of the cyst wall), fortuitously being able to fulfill a second function (parasitic manipulation). In support of this hypothesis, Stibbs (1985) found that dopamine is increased in concentration in the brain of mice chronically infected with *T. gondii*. This effect was specific for dopamine; Stibbs (1985) found no increase in serotonin or norepinephrine in the infected brain. Even more compelling is the ability of haloperidol, a dopamine receptor blocker (Iverson et al. 2009), to prevent attraction to cat odor (Webster et al. 2006). However, haloperidol also has anti-parasitic effects and may exert its effect by reducing the number of parasites in the brain (Jones-Brando et al. 2003). Therefore, a more specific dopamine receptor blocker would provide a better test of the hypothesis.

As exciting as these results are, it is unclear whether raising dopamine levels haphazardly across the mammalian brain would produce the specific behavioral effects seen during infection. First, if cysts are secreting enzymes that result in an increase in dopamine throughout the brain, then infection should also induce motor effects due to excitation of the extrapyramidal motor circuits (see Bear et al. 2007). There is evidence that infected mice do suffer motor impairment during chronic infection. For example, mice with chronic *T. gondii* infections are more likely to fall off a revolving platform than are controls (Hutchinson et al. 1980). However, elevated dopamine signaling should induce excessive grooming and perseverative gnawing (see Taylor et al. 2010 for the behavioral effects of excessive dopamine). It is possible that *T. gondii* produces increases in dopamine that are below the threshold for motor effects, but this remains to be demonstrated. Second, if *T. gondii* increases dopamine in rodent brains, then treatments that increase dopamine levels in a normal rodent brain should make uninfected rodents behave like *T. gondii* infested rodents. In mice, infection with *T. gondii* increases exploratory behavior in a holeboard test (e.g., increased sniffing; Skallova et al. 2006). Increasing dopamine levels by inhibiting reuptake had no effect on this behavior in uninfected mice (Skallova et al. 2006). Webster et al. (2006) found that depressing dopaminergic signaling in normal rats, that is inducing the opposite effect on the dopaminergic system that the parasite is thought to, induced behavior similar to that of infected animals. In summary, neuropharmacological manipulations of dopaminergic pathways produce complex effects on the behavior of infected and control rodents (e.g., Skallova et al. 2006; Webster et al. 2006), most of which do not support the hypothesis that excess dopamine is causing the change in host behavior.

Moreover, the brain area that is most heavily infested with cysts is the amydgala (Vyas et al. 2007a), which is not the brain area critical for the behavioral response to cat odor according to Staples et al. (2005). The amygdala is a brain area long identified as central for anxiety and learned fear behaviors (see ref. in de la Mora et al. 2010). However, learned fear is intact in chronically infected rodents, suggesting that the amygdala is still functional, despite the preponderance of cysts (Vyas et al. 2007a). Dopamine receptors exist in the amygdala (de la Mora et al. 2010), and, therefore it could be directly affected if cysts are capable of inducing increases in dopamine synthesis inside neurons. However, elevated levels of dopamine within the amygdala tend to increase anxiety and accentuate fearful responses, which is the opposite of the effect of the parasite (de la Mora et al. 2010).

Neurobiologists can also alter behavior by activating immune-neural connections (Dantzer 2004). Ongoing immune activity is required to prevent the cysts from reactivating, and this activity could influence neural function (e.g., by the release of cytokines, Webster and McConkey 2010). However, cytokine release itself does not explain the change in host behavior, as cytokine release in the brain typically results in sickness behaviors (Dantzer 2004), which are not seen during the latent phase of the infection (Vyas and Sapolsky 2010). Moreover, sickness behavior should lead to more, not less, anxiety-like behaviors in rodents (Thompson and Kavaliers 1994). However, immune-related damage might contribute to the effect. Gonzalez et al. (2007) suggest that immunological damage to the limbic system might reduce anxiety and that this change could account for much of the change in infected hosts.

Parasites that do not reside in the brain can also suppress rodent aversion to cats. For example, mice infected with *Eimeria vermiformis*, a parasite that resides in the gut (Todd and Lepp 1971), no longer avoid cats (Kavaliers and Colwell 1995). Unlike the situation with *T. gondii*, this host behavioral change is likely to reduce parasite transmission in *E. vermiformis*, because it requires only a rodent host (Kavaliers et al. 1999). Increasing the chance of being eaten by a cat would reduce the parasite's ability to transmit itself to additional rodent hosts (Kavaliers et al. 1999). Therefore, in this system, the reduction in anxiety is likely to be a byproduct of the parasite's interactions with its host. Similarly, mice infested with the cestode *Taenia crassiceps* show a reduced stress response in the presence of a cat (Gourbal et al. 2001). *T. crassiceps* is a tapeworm (Cestoda) that typically resides solely in the abdomen, although there are rare cases of it infesting the brain (Wünschmann et al. 2003). Transmission from the rodent intermediate host to a final host (usually a canid) is via predation (Wünschmann et al. 2003). Gourbal et al. (2001) argues that the decline in a predator-produced stress response is likely a byproduct of the immune manipulation necessary for the parasite's survival inside the rodent. These observations suggest that reduced anxiety may be a common side effect of parasitic immune modulation, and may partly explain the changes in behavior in rodents infected with *T. gondii* (Table 3.1).

Many gaps remain in our understanding of the neural substrate mediating avoidance of cats, as well as how *T. gondii* alters neural function. Given what we know about the neural control of antipredator behavior, *T. gondii* seems to use an indirect method of redirecting its host towards its definitive host. However, this conclusion may change as both topics are understood in more detail.

Parasitic manipulation of humans?

It is estimated that 15–85% of the adult human population is chronically infected with *T. gondii*, depending on factors such as geography, diet, and proximity to cats (Costa da Silva and Langoni 2009). *T. gondii* can induce serious neurological abnormalities if the infection is acquired during fetal development (e.g., see Costa da Silva and Langoni 2009). In adults, the parasite typically causes only a minor, transient illness. However, as in rodents, *T. gondii* forms cysts in the human brain and these cysts last for many years (Montoya and Liesenfeld 2004). In immunocompromised persons the parasite can become reactivated and re-enter an active phase (Montoya and Liesenfeld 2004). If *T. gondii* cysts alter rodent behavior by secreting neuromodulatory substances and/or enzymes, then these substances could also influence human behavior (see also Chapter 10). This is an issue of some importance given the millions of people infected with *T. gondii.*

Although *T. gondii* causes specific behavioral changes in rats, in humans the effects appear to be a sex-specific shift in some personality traits (Flegr 2007). There is also evidence that *T. gondii* infection can increase the propensity for mental disorders such as schizophrenia (Yolken et al 2009). However, the data are sometimes hard to interpret. For example, despite the evidence that latent infections of *T. gondii* can predispose individuals towards schizophrenia, the prevalence of schizophrenia does not correlate with the prevalence of *T. gondii* in different countries (Yolken et al. 2009). For ethical reasons it is not possible to test the behavior of individuals before and after a controlled infection. Therefore, although the results found to date are intriguing, they rely entirely on correlative evidence (Flegr 2007).

3.2.2 Neuroviruses

Some neuroparasites cause behavioral changes beyond the standard headache, malaise, nausea, and lethargy that are the common effects of neuroparasites (Tomonaga 2004; Walker and Zunt

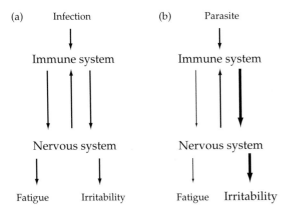

Figure 3.1 A. During an infection, the immune system becomes activated, prompting it to release signaling molecules (e.g. cytokines). These can alter neural function and induce sickness behaviors such as fatigue and irritability. B. A parasite may manipulate this system and induce the host's immune system to secrete substances leading to abnormal behaviors.

2005). A few neuroviruses are spread by a bite, such as rabies and Borna disease virus, and one of the common features of infection with these viruses is an increase in aggression and tendency to bite (see Klein 2003; Lefèvre et al. 2009). Both rabies (Laothamatas et al. 2003) and Borna disease virus (Carbone et al. 1987) are widely disseminated throughout the brain, and they do not specifically target the areas of the brain thought to regulate aggressive behavior (e.g., the limbic system). Axonal and dendritic transport can carry viruses quickly throughout the brain (Mohammed et al. 1993), and this probably accounts for the wide distribution. However, although these viruses typically produce diffuse attacks on the brain, they do reliably attack certain brain areas. For example, in rabies, the limbic system is always involved (Lefèvre et al. 2009).

There is also a massive immune response, and, at least in rabies, this immune response appears to play a causal role in increasing aggressive behavior (Hemachudha et al. 2002). High levels of immune modulators can increase aggressive behavior (e.g., cytokines, Kraus et al. 2003) even without a viral infection. Irritability is one form of sickness behavior in most vertebrates (Hart 1988), suggesting the possibility that the increased aggression observed during rabies infections may be an example of augmenting specific types of sickness behavior (Fig. 3.1).

Not all hosts infected with rabies show increased aggression. Some become progressively more inactive, slip into a coma and die (paralytic rabies) (Hemadchuha et al. 2003). The difference does not depend on the strain of rabies virus (Hemachudha et al. 2003). The behavioral change in the host may depend on the type immune response the virus elicits (Laothamatas et al. 2008). This possibility makes sense if the parasite is using some aspects of the host's immune response to alter host behavior.

3.3 Invertebrate examples

Invertebrates are invaluable model systems for neuroscientists (Chase 2002). Given the complexity of vertebrate nervous systems, understanding how parasites manipulate host behavior may be easier when the host is an invertebrate.

3.3.1 Gammarids—don't go into the light!

Gammarids are small crustacea that are attacked by parasites. Some parasites (e.g., acanthocephalans) reside in the abdomen, while others (e.g., trematodes) reside within the central nervous system. Members of both these parasitic groups are capable of altering their host's behavior (see Lefèvre et al. 2009). These parasites use the gammarid as an intermediate host, and, as with *T. gondii*, need the host to be eaten by a specific vertebrate in order to complete its life cycle (see Lefèvre et al. 2009). Once the parasite reaches the infective stage, the escape behavior of its host changes (e.g., from swimming away from the light to swimming towards it, Lefèvre et al. 2009; see also Chapter 2; Fig. 2.7). These changes in escape behavior can enhance parasitic transmission (Cézilly et al. 2010). Additional behavioral changes also occur, but whether they benefit host, parasite, or neither remains unknown (Cézilly and Perrot-Minnot 2005).

The details of how the different parasites produce changes in host escape behavior remain a topic of ongoing investigation. The evidence (see Lefèvre et al. 2009 for a review) suggests that manipulation of the host's serotonergic neuromodulatory system is involved (see Lefèvre et al. 2009 table 3.1). Only those parasitic infections that alter host serotonergic systems have an effect on host phototactic behavior (Lefèvre et al. 2009). The greater the increase in serotonergic immunohistochemical staining within the central nervous system, the greater is the change in host behavior (Tain et al. 2006). Moreover, injecting serotonin can mimic some of the effects of parasitism and this effect is specific for serotonin; injections of other biogenic amines do not produce these effects in *Gammarus lacustris* (Helluy and Holmes 1990). An injection of serotonin not only increases movement towards the light in uninfected animals, it also augments it in parasitized animals (Tain et al. 2006, 2007). Immersion in the serotonin reuptake inhibitor, fluoxetine, increases phototaxis in a related crustacean, the marine amphipod *Echinogammarus marinus* (Guler and Ford 2010). Serotonin has no effect on geotactic behavior in *Gammarus pulex* (Tain et al. 2006), demonstrating that serotonin does not simply activate a wide range of non-specific behaviors. Proteomic studies of infected gammarids (*G. insensibilis*, *M. papillorobustus*) also suggest that changes in serotoninergic neurotransmission play a role in altering the host's response to light (Ponton et al. 2006). Furthermore, serotonin has been shown to induce relatively specific behavioral changes in other crustacea (Huber 2005). Biogenic amines like serotonin typically have broad connections throughout the invertebrate central nervous system (Huber 2005). They are thought to modulate the likelihood of producing different behaviors given the right context (Huber 2005).

There are some puzzling features, however. Mimicking the effect of parasitism in gammarids requires injections of supraphysiological levels of serotonin, resulting in concentrations 1,000 to 10,000 times above the likely physiological concentration. Such levels would be difficult for either host or parasite to synthesize and maintain. Furthermore, when crayfish are infused with high doses of serotonin for an extended period of time, the central nervous system adapts, and the elevated serotonin loses its behavioral effectiveness (Panskepp and Huber 2002). Another problem, as Tain et al. (2007) discuss, is that increased serotonin immunoreactivity typically means that neurons are releasing less serotonin. The reduction in release leads to an increase in the amount of serotonin inside the neuron available for immunohistochemical staining. However, the injection studies suggest that serotonin release is increased in parasitized hosts. Tain et al. (2007) suggest that infection induces a massive increase in serotonin synthesis, leading to both increased staining and increased release. Researchers have not yet tested this hypothesis (e.g., by using pharmacological blockers of serotonin synthesis).

There are other explanations that can account for both the immunohistochemical and serotonin injection results. For example, the parasite could secrete a compound that induces long-term activation of host serotonergic receptors. Such a compound would probably decrease the release of the host's own serotonin, leading to increased immunohistochemical staining, while at the same time increasing activity in the host's serotonergic system. There is evidence for a weak dose effect that suggests that a

THE STRINGS OF THE PUPPET MASTER **43**

secreted substance is involved. Behavioral manipulation is stronger if two parasites (acanthocephalans) infest *G. pulex* rather than one (Franceschi et al. 2008).

It is also possible that infected gammarids suffer from numerous changes within their central nervous system, serotonin being but one of them. When the brains of parasitized *Manduca sexta* (a caterpillar) were stained for a variety of different neuromodulatory substances, an increase in staining was seen for ALL of them (Zitnan et al. 1995). This increase in staining likely reflects a decrease in release, probably caused by the general decline in metabolism that occurs in parasitized *M. sexta* at the time its behavior changes (Alleyne et al. 1997). Like parasitized *M. sexta*, *G. pulex* also shows a decline in respiration after infection (Rumpus and Kennedy 1974). Therefore, some of the increased staining in parasitized gammarids may be due to a general decline in neural activity, but this possibility does not explain the injection results.

Surprisingly, the serotonergic pathways within the central nervous system of a behaviorally-altered, infected gammarid show increased immunohistochemical staining regardless of whether the parasite is a trematode developing in the brain or an acanthocephalan residing in the abdomen (Tain et al. 2006). Possibly, the serotonergic system is uniquely susceptible to modification by parasites. Other parasites are known to influence serotonergic neural pathways in their hosts (Shaw et al. 2009). It may be fortuitous that in these crustacea, serotonin is also important for anti-predator responses, thereby leaving these organisms pre-adapted for this type of parasitic manipulation.

The serotonergic system could be susceptible to manipulation by parasites if serotonin is involved in both activating the immune response and in neuromodulation. In that case, the effect on host behavior may be a fortuitous byproduct of parasitic manipulation of the host's immune system (Adamo 2002). Tain et al. (2007) suggest that one reason why the acanthocephalan *Pomphorhynchus laevis* alters the behavior of its host *G. pulex*, but not that of *G. roeseli*—a related host—is because *P. laevis* has a greater ability to manipulate the immune system of *G. pulex* compared with that of *G. roeseli*. However,

Cornet et al. (2009) found no relationship between the ability of parasites (acanthocephalans) to suppress phenoloxidase activity (an important immune-related enzyme) and the severity of the host's behavioral change. Moreover, in crustacea, there is no evidence that serotonin is involved in immune function, although this may be because no one has looked. Serotonin appears to play a role in immune function in other arthropods (e.g., insects, Baines et al. 1992).

Along with serotonin, immune responses themselves may play a role in altering host behavior. Glial cells in the central nervous system of gammarids (*G. insensibilis*) parasitized with the brain-dwelling trematode *Microphallus papillorobustus* show changes due to immune activation (Helluy and Thomas 2010). Proteomic results show that arginine kinase is differentially expressed in the brain of infected *G. insensibilis* and *G. pulex* compared to uninfected individuals (Ponton et al. 2006). This phosphotransferase is known to be one of the regulating factors in nitric oxide (NO) synthesis (Mori and Gotoh 2000). NO is liberated during immunological reactions, but it is also a neuromodulator coordinating many neuronal activities in insects (Bicker 2001). Thus, these proteomic results support the hypothesis that chronic neuroinflammation due to parasitism causes disruption of neural function leading to changes in host behavior (Helluy and Thomas 2010). However, this mechanism does not account for the relatively specific nature of the host's behavioral change.

3.3.2 Suicidal crickets

Adult nematomorphs are aquatic and free-living, but the juvenile stages are parasitic, often on terrestrial insects such as orthopterans (crickets and grasshoppers) (see references in Thomas et al. 2002). Therefore the parasites are stuck with a difficult dilemma at the end of their juvenile development—how to "persuade" a terrestrial host to enter the water to allow an aquatic adult to emerge? In many nematomorph–orthopteran systems, once the nematomorph is ready to exit its host, the host makes a suicidal leap into a body of water (see Chapter 2, Fig. 2.5; Thomas et al. 2002, 2003; Sanchez et al. 2008). The host does not seek out water *per se*, but it increases locomotion and becomes phototactic, bringing it towards bodies of water (Ponton et al. 2011). Then, infected individuals are much more likely to leap into the water, compared with uninfected controls (Thomas et al. 2002; Sanchez et al. 2008). However, not all parasitized crickets with mature worms jump into the water (Sanchez et al. 2008). About 80% of crickets (*Nemobius sylvestris*) infected with a mature nematomorph (*Paragordius tricuspidatus*) jump into the water, compared with 10% of uninfected crickets (Sanchez et al. 2008). Therefore, the parasite induces both a novel behavior (phototaxis) and dramatically increases the prevalence of jumping into water. Another interesting feature of this system is that hosts with mature nematomorphs jump into water only at night before 3 am (Thomas et al. 2002). This strange behavior reoccurs in infected hosts on a circadian basis (Sanchez et al. 2008).

How does the nematomorph, that exists in the abdominal cavity and does not directly contact the host's brain (Thomas et al. 2003; Biron et al. 2006), alter host behavior? The change in behavior does not occur early in infection when jumping into the water would not benefit the parasite (Thomas et al. 2002). Moreover, after the nematomorph exits the host, the increased locomotion and phototaxis decline to almost baseline levels (Ponton et al. 2011). Thomas and colleagues (e.g., Thomas et al. 2003; Biron et al. 2006) compared infected animals during the night with both uninfected animals and with infected animals that were still displaying normal behavior (i.e., infected animals during the day). Serotonin could not be measured, but the dopamine content of the brain did not change significantly with host behavioral change (Thomas et al. 2003).

Parasitized crickets did show an increase in neurogenesis in the mushroom bodies of the brain, although the significance of this remains unknown (Thomas et al. 2003). Thomas et al. (2003) suggest that increased neurogenesis could interfere with neural circuitry, and, as the mushroom bodies are the main sensory integrative centers of the insect brain (Nation 2008), this could lead to aberrant behavior. If this speculation is correct, then it would be a very different mechanism of behavioral control

than that used by a typical neuroscientist. However, it would not explain the circadian timing of the change, as neurogenesis should produce relatively permanent changes.

Using proteomics, Biron and colleagues (Biron et al. 2005; Biron et al. 2006) studied two different species of nematomorphs, *Paragordius tricuspidatus* and *Spinochordodes tellinii* that infest two different orthopterans (the cricket *N. sylvestris* and the grasshopper *Meconema thalassinum*). Both nematomorphs induce their hosts to jump into water. Despite the lack of a sequenced orthopteran genome, some protein families could be identified. The brains of grasshoppers and crickets are similar, and it might be expected that both nematomorphs would use the same mechanisms to induce the same changes in behavior. However, although both cricket and grasshopper brains showed changes in expression in Wnt-like proteins and actin concomitant with an increased tendency to jump into water, there were few other similiarities (Biron et al. 2005, 2006). One study used whole heads (Biron et al. 2006) and one used dissected brains (Biron et al. 2005) and this difference may have confounded the comparison. Also, crickets show a change in neurogenesis, while grasshoppers do not (Biron et al. 2005), demonstrating striking neurobiological differences in the two systems despite the same changes in host behavior.

In both systems there was overproduction of Wnt-like molecules at the time of host behavioral change. Wnt proteins have been implicated in neural development, and there is evidence that Wnt can influence neuronal stability in *Drosophila melanogaster* (Chiang et al. 2009). It appears to protect neurons from degeneration (Chiang et al. 2009). Whether it has direct effects on neuronal activity has not been demonstrated electrophysiologically. In the parasite itself, there was also an increase in Wnt-like protein production. Wnt is found in some non-insect invertebrates, including *Caenorhabditis elegans* (Hilliard and Bargmann 2006), although the sequence of the Wnt-like protein found in the nematomorphs seems more similar to those found in arthropods than in nematodes (Biron et al. 2005). Biron et al. (2005, 2006) suggest that nematomorphs have evolved to secrete mimics of the host's Wnt proteins. Further studies are needed to test this intriguing hypothesis. For example, it is unclear whether Wnt-like proteins synthesized in the hemocoel can reach the host's central nervous system.

In crickets two of the proteins that show altered expression concomitant with the change in host behavior had the protein domains GH16463p or CH4 protein (Biron et al. 2006). Proteins with these protein domains are found in a variety of systems, but are especially important in proteins in the visual system (e.g., GH164163p, Wang and Montell 2005). Ponton et al. (2011) speculate that changes in these proteins may play a role in inducing phototaxis, but the mechanism remains to be determined.

Almost as interesting was what was NOT found to change in the proteomic studies. Very few of the changes found in the brains of parasitized hosts directly relate to neural function. In other systems (e.g., *D. melangaster*), altering neural activity produces an increase in the expression of a variety of ion channels, neurotransmitter receptors, and other molecules that have an obvious and direct connection with neural function (Guan et al. 2005). However, these types of proteins did not appear to change in expression in parasitized hosts. Although Guan et al. (2005) examined changes in gene expression, not protein expression, proteomic studies of the central nervous system in *D. melanogaster* also find changes in the expression of proteins with both direct and indirect effects on central nervous system function in mutated flies with abnormal behavior (e.g., Zhang et al. 2005). There are a number of technical reasons why more neurally relevant molecules may have been missed in the nematamorph studies (Lefèvre et al. 2009), but their lack is striking.

Although we do not know how nematomorphs alter host behavior, two conclusions can be drawn. First, behavioral changes in the host correlate with changes in protein expression (Biron et al. 2005, 2006). Second, the parasites appear to be using obscure and indirect mechanisms of altering nervous system function, unlike the targeted approach of a neuroscientist. A neuroscientist would focus on changing proteins directly related to neural function, such as those related to ion channels and synaptic transmission. However, more mechanisms may be uncovered with further research.

3.4 How might parasites manipulate host behavior?

Parasites can manipulate the minds of their host. However, parasites do not appear to usurp control in the way a neuroscientist would (Table 3.2). For example, none of the parasites reviewed here target a specific neuroanatomical site, one of the most common methods used by neuroscientists to manipulate behavior. Although some parasites reliably strike certain brain areas, they invariably infect other parts of the brain as well. Parasites appear to

Table 3.2 How parasites manipulate behavior

System parasite/host	Neuroanatomical specificity	Neurochemical modulation	Neuroimmun-ological involvement	Manipulation-specific substances
T. gondii/Rodent	No	Yes—dopamine	Yes	?
Neuroviruses/Mammal	No	Yes—e.g. dopamine[1], serotonin[2]	Yes	?
Acanthocephalan +Trematode/Gammarids	No	Yes—serotonin	Yes	?
Nematomorph/Crickets +Grasshoppers	No	?	?	Wnt-like proteins?

?—unknown
1. Borna disease virus. Hornig et al. (2003)
2. Tomonaga (2004)

rely more on chemical manipulation of the host brain. Although the mechanisms remain unclear, biogenic amine (e.g., dopamine and serotonin) neuromodulatory pathways appear to be a common target of parasitic manipulation.

Parasites could neurochemically manipulate their host by secreting host neuromodulators. However, synthesizing physiologically relevant amounts of a host's neuromodulator is likely to be energetically expensive because these compounds are often rapidly metabolized (e.g., Goosey and Candy 1982), requiring constant synthesis to maintain high levels (see Adamo 1997, 2002 for a discussion). Some parasites appear to have overcome this difficulty. For example, the parasitic wasp *Cotesia congregata* reduces the breakdown of the neurohormone octopamine (also a biogenic amine) in its host (Adamo 2005). Another method a parasite could use to avoid having to synthesize large amounts of host neuromodulator is to secrete a substance that is not easily degraded and/or has long lasting effects at the receptor level (Weisel-Eichler and Libersat 2004). For example, some venoms can manipulate specific biogenic amine neuromodulatory systems using a variety of cellular mechanisms (Weisel-Eichler and Libersat 2004).

However, neuropharmacologists know that altering specific behaviors often requires more than manipulating the level of a single biogenic amine. The parasitic wasp *Ampulex compressa* uses a cocktail of compounds to manipulate the behavior of its cockroach prey (Libersat et al. 2009). The wasp delivers venom directly into the cockroach's brain using its stinger. Low weight components of the venom abolish neural activity by blocking acetylcholine and gamma-aminobutyric acid (GABA) mediated synaptic transmission (Libersat et al. 2009). The venom also activates dopaminergic neural circuits using either dopamine or a dopamine agonist, and directly or indirectly induces abnormal neural activity in octopamine-containing neurons (Libersat et al. 2009). Therefore, the wasp uses multiple compounds to produce coordinated changes in the host's neurophysiology and behavior (Gal and Libersat 2008; Libersat et al. 2009, see also Chapter 2, Fig. 2.6). Such a strategy may be common in parasitic manipulators.

Altered immune-neural interactions also appear to play a role in many parasite-induced changes in behavior (Thomas et al. 2005). This congruence could reflect a common evolutionary path, especially for those parasites not descended from animals that produce venom. Parasites must manipulate a host's immune system to survive. It may be a small evolutionary step to select for parasites that induce changes in host behavior concomitant with suppressing the host's immune response. Even under normal (i.e., uninfected) conditions, immune factors (e.g., cytokines) play a role in neural communication (e.g., Goshen et al. 2007). These interactions provide ample opportunity for parasites to manipulate behavior by interfering with the host's immune function.

Unfortunately, few neuroscientists are studying parasitic manipulation of behavior, despite its potential importance as a tool for understanding neural control (Libersat et al. 2009). Parasites can produce specific, long-lasting behavioral changes, something neuroscientists still struggle to do. Moreover, genomic and proteomic approaches are still being integrated into neuroscience. Our lack of a comprehensive understanding of neuronal regulation at a subcellular level is one of the reasons that it is difficult to interpret the proteomic results of Biron and colleagues on the hairworm-infested crickets. Studying the methods by which parasites manipulate host behavior may help resolve the long-standing problem of how molecular changes within the central nervous system produce behavioral changes. Solving this mystery is certain to identify novel molecular pathways capable of altering behavior; some of these may already be in use by parasites.

3.5 How can parasitic infections produce specific changes in host behavior without neuroanatomical specificity?

In humans, neuroactive drugs are usually given systemically (i.e., to the entire brain), yet they still produce relatively specific behavioral effects (e.g., L-Dopa, Iverson et al. 2009). These drugs also have effects on non-target behaviors, but these can be minimized by using carefully controlled dosages (Iverson et al. 2009). In addition, neurologists frequently prescribe additional drugs to help reduce side effects (Iverson et al. 2009). Neuroactive drugs can influence specific behaviors, even when delivered to the entire brain, because neurons have multiple types of receptors, each recognizing only one of the numerous neuromodulators in the brain (Bear et al. 2007). There are even multiple types of receptors for a single neuromodulator such as dopamine (Iverson et al. 2009). Each receptor has different pharmacological properties and produces different neuronal effects (Iverson et al. 2009). Therefore, not all dopaminergic pathways, for example, are sensitive to the same drug, and this feature reduces a drug's effects on non-target behaviors (Iverson et al. 2009).

Like neuroactive drugs, parasites that manipulate host behavior also affect "non-target" behaviors (i.e., host behaviors that appear to be unrelated to parasite fitness), although this issue tends to be understudied (Cézilly and Perrot-Minnot 2005, see also Chapter 2). The effects on non-target behaviors may be of little consequence to some parasites, as long as the net result is increased parasitic transmission. Other parasites may secrete additional substances that reduce their effects on non-target behaviors, in the same way that neurologists prescribe additional drugs to reduce behavioral side effects. Therefore, the neurobiological features that make clinical neuropharmacology possible may also allow parasites to induce relatively specific behavioral changes without neuroanatomical specificity. Searching for the parasites' neuropharmacological toolkit may lead to the discovery of novel substances and novel targets that will help in the treatment of human neurobiological disorders.

Acknowledgements

I thank M. Schmidt for critically reading the manuscript and the Natural Science and Engineering Research Council of Canada (NSERC) for financial support.

References

Adamo, S. A. (1997). How parasites alter the behaviour of their insect hosts. In: *Parasites and Pathogens. Effects on Host Hormones and Behavior*, (ed. Beckage, N.) pp. 231–245. Chapman and Hall, New York.

Adamo, S. A. (2002). Modulating the modulators: Parasites, neuromodulators and host behavioral change. *Brain, Behavior and Evolution* **60**, 370–377.

Adamo, S. A. (2005). Parasitic suppression of feeding in the tobacco hornworm, *Manduca sexta*: parallels with feeding depression after an immune challenge. *Archives of Insect Biochemistry and Physiology* **60**, 185–197.

Alleyne, M., Chappell, M. A., Gelman, D. B., and Beckage, N. E. (1997). Effects of parasitism by the braconid wasp *Cotesia congregata* on metabolic rate in host larvae of the tobacco hornworm, *Manduca sexta*. *Journal of Insect Physiology* **43**, 143–154.

Baines, D., DeSantis, T., and Downer, R. (1992). Octopamine and 5-hydroxytryptamine enhance the phagocytic and nodule formation activities of cockroach (*Periplaneta americana*) haemocytes. *Journal of Insect Physiology* **38**, 905–914.

Bear, M. F., Connors, B. W., and Paradiso, M. A. (2007). *Neuroscience: Exploring the Brain*. Lippincott, Williams and Wikins, Philadelphia, PA.

Berdoy, M., Webster, J. P., and MacDonald, D. W. (1995). Parasite altered behaviour: is the effect of *Toxoplasma gondii* on *Rattus norvegicus* specific? *Parasitology* **111**, 403–409.

Berdoy, M., Webster, J. P., and MacDonald, D. W. (2000). Fatal attraction in rats infected with *Toxoplasma gondii*. *Proceedings of the Royal Society B* **267**, 1591–1594.

Bicker, G. (2001). Nitric oxide: an unconventional messenger in the nervous system of an orthopteroid insect. *Archives of Insect Biochemistry and Physiology* **48**, 100–110.

Biron, D. G., Marché, L., Ponton, F., Loxdale, H. D., Galeotti, N., Renault, L., Joly, C., and Thomas, F. (2005). Behavioural manipulation in a grasshopper harbouring a hairworm: a proteomics approach. *Proceedings of the Royal Society B* **272**, 2117–2126.

Biron, D. G., Ponton, F., Marché, L., Galeotti, N., Renault, L., Demey-Thomas, E., Poncet, J., Brown, S. P., Jouin, P., and Thomas, F. (2006). "Suicide" of crickets harbouring hairworms: a proteomics investigation. *Insect Molecular Biology* **15**, 731–742.

Blanchard, C. D., Canteras, N. S., Markham, C. M., Pentkowski, N. S., and Blanchard, R. J. (2005). Lesions of structures showing FOS expression to cat presentation: effects of responsivity to a cat, cat odor and nonpredator threat. *Neuroscience and Biobehavioral Reviews* **29**, 1243–1253.

Blanchard, D. C., Li, C. I., Hubbard, D., Markham, C. M., Yang, M., Takahashi, L. K., and Blanchard, R. J. (2003).

Dorsal premammillary nucleus differentially modulates defensive behaviors induced by different threat stimuli in rats. *Neuroscience Letters* **345**, 145–148.

Carbone, K. M., Duchala, C. S., Griffin, J. W., Kincaid, A. L., and Narayan, O. (1987). Pathogenesis of borna disease in rats: evidence that intra-axonal spread is the major route for virus dissemination and the determinant for disease incubation. *Journal of Virology* **61**, 3431–3440.

Cézilly, F. and Perrot-Minnot, M. J. (2005). Studying adaptive changes in the behaviour of infected hosts: a long and winding road. *Behavioral Processes* **68**, 223–228.

Cézilly, F., Thomas, F., Médoc, V., and Perrot-Minnot, M. J. (2010). Host-manipulation by parasites with complex life cycles: adaptive or not? *Trends in Parasitology* **26**, 311–317.

Chase, R. (2002) *Behavior and its Neural Control in Gastropod Molluscs*. Oxford University Press, New York.

Chiang, A., Priya, R., Ramaswami, M., VijayRaghavan, K., and Rodrigues, V. (2009). Neuronal activity and Wnt signaling act through Gsk3-beta to regulate axonal integrity in mature *Drosophila* olfactory sensory neurons. *Development* **136**, 1273–1282.

Cornet, S., Franceschi, N., Bauer, A., Rigaud, T., and Moret, Y. (2009). Immune depression induced by acanthocephalan parasites in their intermediate crustacean host: consequences for the risk of super-infection and links with host behavioural manipulation. *International Journal for Parasitology* **39**, 221–229.

Costa da Silva, R. and Langoni, H. (2009). *Toxoplasma gondii*: host-parasite interaction and behavior manipulation. *Parasitology Research* **105**, 893–898.

Dantzer, R. (2004). Cytokine-induced sickness behaviour: a neuroimmune response to activation of the innate immunity. *European Journal of Pharmacology* **500**, 399–411.

De la Mora, M. P., Gallegos-Cari, A., Arizmendi-Garcia, Y., Marcellino, D., and Fuxe, K. (2010). Role of dopamine receptor mechanisms in the amygdaloid modulation of fear and anxiety: structural and functional analysis. *Progress in Neurobiology* **90**, 198–216.

DoMonte, F. H. M., Canteras, N. S., Fernandes, D., Assreuy, J., and Carobrez, A. P. (2008). New perspectives on ß-adrenergic mediation of innate and learned fear responses to predator odor. *Journal of Neuroscience* **28**, 13296–13302.

Dubey, J. P., Lindsay, D. S., and Speer, C. A. (1998). Structures of *Toxoplasma gondii* tachyzoites, bradyzoites and sporozoites and biology and development of tissue cysts. *Clinical Microbiology Reviews* **11**, 267–299.

Flegr, J. (2007). Effects of Toxoplama on human behavior. *Schizophrenia Bulletin* **33**, 757–760.

Franceschi, N., Bauer, A., Bollache, L., and Rigaud, T. (2008). The effects of parasite age and intensity on variability in acanthocephalan-induced behavioural manipulation. *International Journal for Parasitology* **38**, 1161–1170.

Gal, R. and Libersat, F. (2008). A parasitoid wasp manipulates the drive for walking of its cockroach prey. *Current Biology* **18**, 877–882.

Gaskell, E. A., Smith, J. E., Pinney, J. W., Westhead, D. R., and McConkey, G. A. (2009). A unique dual activity amino acid hyroxylase in *Toxoplasma gondii*. *PLoS One* **4**: e4801, 1–10.

Gonzalez, L. E., Rojnik, B., Urrea, F., Urdaneta, H., Petrosino, P., Colasante, C., Pino, S., and Hernandez, L. (2007). *Toxoplasma gondii* infection lower anxiety as measured in the plus-maze and social interaction tests in rats: a behavioral analysis. *Behavioural Brain Research* **177**, 70–79.

Goosey, M. W. and Candy, D. J. (1982). The release and removal of octopamine by tissues of the locust *Schistocerca americana*. *Insect Biochemistry* **12**, 681–685.

Goshen, I., Kreisel, T., Ounallah-Saad, H., Renbaum, P., Zalzstein, Y., Ben-Hur, T., Levy-Lahad, E., and Yirmiya, R. (2007). A dual role for interleukin-1 in hippocampal-dependent memory processes. *Psychoneuroendocrinology* **32**, 1106–1115.

Gourbal, B. E. F., Righi, M., Petit, G., and Gabrion, C. (2001). Parasite-altered host behavior in the face of a predator: manipulation or not? *Parasitology Research* **87**, 186–192.

Guan, Z., Saraswatic, S., Adolfsen, B., and Littleton, J. T. (2005). Genome-wide transcriptional changes associated with enhanced activity in the *Drosophila* nervous system. *Neuron* **48**, 91–107.

Guler, Y. and Ford, A. T. 2010. Anti-depressants make amphipods see the light. *Aquatic Toxicology* **99**, 397–404.

Gulinello, M., Acquarone, M., Kim, J. H., Spray, D. C., Barbosa, H. S., Sellers, R., Tanowitz, H. B., and Weiss, L. M. (2010). Acquired infection with *Toxoplasma gondii* in adult mice results in sensorimotor deficits by normal cognitive behavior despite widespread brain pathology. *Microbes and Infection* **12**, 528–537.

Hart, B. L. (1988). Biological basis of the behavior of sick animals. *Neuroscience and Biobehavioral Reviews* **12**, 123–137.

Helluy, S. and Thomas, F. (2003). Effects of *Microphallus papillorobustus* (Platyhelminthes: Trematoda) on serotonergic immunoreactivity and neuronal architecture in the brain of *Gammarus insensibilis*. *Proceedings of the Royal Society B* **270**, 563–568.

Helluy, S. and Thomas, F. (2010). Parasitic manipulation and neuroinflammation: evidence from the system

Microphallus papillorobustus (Trematoda)—Gammarus (Crustacea). *Parasites and Vectors* **3**, 38–49.

Hemachudha, T., Laothamatas, J., and Rupprecht, C. (2002). Human rabies: a disease of complex neuropathogenetic mechanisms and diagnostic challenges. *The Lancet Neurology* **1**, 101–109.

Hemachudha, T., Wacharapluesadee, S., Lumlertdaecha, B., Orciari, L., Rupprecht, C., La-ongpant, M., Juntrakul, S., and Denduangboripant, J. (2003). Sequence analysis of rabies virus in humans exhibiting encephalitic or paralytic rabies. *Journal of Infectious Diseases* **188**, 960–966.

Hilliard, M. A. and Bargmann, C. I. (2006). Wnt signals and frizzled activity orient anterior-posterior axon outgrowth in *C. elegans*. *Developmental Cell* **10**, 379–390.

Hornig, M., Briese, T., and Lipkin, W. I. (2003). Borna disease virus. *Journal of NeuroVirology* **9**, 259–273.

Hrdá, Š, Votypka, J., Kodym, P., and Flegr, J. (2000). Transient nature of *Toxoplasma gondii*-induced behavioral changes in mice. *Journal of Parasitology* **86**, 657–663.

Huber, R. (2005). Amines and motivated behaviors: a simpler systems approach to complex behavioral phenomena. *Journal of Comparative Physiology A* **191**, 231–239.

Hutchinson, W. H., Aitken, P. P., and Wells, B. W. P. (1980). Chronic *Toxoplasma* infections and motor performance in the mouse. *Annals of Tropical Medicine and Parasitology* **74**, 507–510.

Iverson, L. L., Iverson, S. D., Roth, R. H., and Bloom, F. E. (2009). *Introduction to Neuropsychopharmacology*. Oxford University Press, New York.

Jones-Brando, L., Torrey, F., and Yolken, R. (2003). Drugs used in the treatment of schizophrenia and bipolar disorder inhibit the replication of *Toxoplasma gondii*. *Schizophrenia Research* **62**, 237–244.

Kavaliers, M. and Colwell, D. D. (1995). Decreased predator avoidance in parasitized mice: neuromodulatory correlates. *Parasitology* **111**, 257–263.

Kavaliers, M., Colwell, D., and Choleris, E. (1999). Parasites and behavior: an ethopharmacological analysis and biomedical implications. *Neuroscience and Biobehavioral Reviews* **23**, 1037–1045.

Klein, S. L. (2003). Parasite manipulation of the proximate mechanisms that mediate social behavior in vertebrates. *Physiology and Behavior* **79**, 441–449.

Kraus, M. R., Schäfer, A., Faller, H., Csef, H., and Scheurlen, M. (2003). Psychiatric symptoms in patients with chronic hepatitis C receiving interferon Alfa-2b therapy. *Journal of Clinical Psychiatry* **64**, 708–714.

Lacosta, S., Merali, Z., and Anisman, H. (1999). Behavioral and neurochemical consequences of lipopolysaccharide in mice: anxiogenic-like effects. *Brain Research* **818**, 291–303.

Lamberton, P. H. L., Donnelly, C. A., and Webster, J. P. (2008). Specificity of the *Toxoplasma gondii*-altered behaviour to definitive versus non-definitive host predation risk. *Parasitology* **135**, 1143–1150.

Laothamatas, J., Wacharapluesadee, S., Lumlertdache, B., Ampawong, S., Tepsumethanon, V., Shuangshoti, S., Phumesin, P., Asavaphatiboon, S., Worapruekjaru, L., Avihingsanon, Y., Israsena, N., Lafon, M., Wilde, H., and Hemachudha, T. (2008). Furious and paralytic rabies of canine origin: Neuroimaging with virological and cytokine studies. *Journal of NeuroVirology* **14**, 119–129.

Laothamatas, J., Hemachudha, T., Mitrabhakdi, E., Wannakrairot, P., Tulayadaechanont, S. (2003) MR imaging in human rabies. *American Journal of Neuroradiology* **24**(6), 1102–1109.

Lefèvre, T., Adamo, S. A., Biron, D. G., Missé, D., Hughes, D., and Thomas, F. (2009). Invasion of the Body Snatchers: the diversity and evolution of manipulative strategies in host-parasite interactions. *Advances in Parasitology* **68**, 45–84.

Libersat, F., Delago, A., and Gal, R. (2009). Manipulation of host behavior by parasitic insects and insect parasites. *Annual Review of Entomology* **54**, 189–207.

Mohammed, A. H., Norrby, E., and Kristensson, K. (1993). Viruses and behavioural changes: A review of the clinical and experimental findings. *Reviews in the Neurosciences* **4**, 267–286.

Montoya, J. G. and Liesenfeld, O. (2004). Toxoplasmosis. *Lancet* **363**, 1965–1976.

Moore, J. (2002). *Parasites and the Behavior of Animals*. Oxford University Press, New York.

Mori, M. and Gotoh, T. (2000). Regulation of nitric oxide production by arginine metabolic enzymes. *Biochemical and Biophysical Research Communications* **275**, 715–719.

Nation, J. L. (2008). *Insect Physiology and Biochemistry*. CRC Press, Boca Raton.

Panskepp, J. B. and Huber, R. (2002). Chronic alterations in serotonin function: dynamic neurochemical properties in agonistic behavior of the crayfish, *Orconectes rusticus*. *Journal of Neurobiology* **50**, 276–290.

Ponton, F., Lefèvre, T., Lebarbenchon, C., Thomas, F., Loxdale, H. D., Marché, L., Renault, L., Perrot-Minnot, M. J., and Biron, D. G. 2006. Do distantly related parasites rely on the same proximate factors to alter the behaviour of their hosts? *Proceedings of the Royal Society B* **273**, 2869–2877.

Ponton, F., Otálora-Luna, F., Lefèvre, T., Guerin, P. M., Lebarbenchon, C., Duneau, D., Biron, D. G., and Thomas, F. (2011) Water-seeking behavior in worm-infested crickets and reversibility of parasitic manipulation. *Behavioral Ecology* **22**, 392–400

Roberts, L. S. and Janovy, J. (1996). *Foundations of Parasitology*. WCB, Boston.

Rumpus, A. E. and Kennedy, C. R. (1974). The effect of the acanthocephalan *Pomphorhynchus laevis* upon the respiration of its intermediate host, *Gammarus pulex*. *Parasitology* **68**, 271–284.

Sanchez, M. I., Ponton, F., Schmidt-Rhaesa, A., Hughes, D. P., Missé, D., and Thomas, F. (2008). Two steps to suicide in crickets harbouring hairworms. *Animal Behaviour* **76**, 1621–1624.

Shaw, J. C., Korzan, W. J., Carpenter, R. E., Kuris, A. M., Lafferty, K. D., Summers, C. H., and Øverli, Ø. (2009). Parasite manipulation of brain monoamines in California killifish (*Fundulus parvipinnis*) by the trematode *Euhaplorchis californiensis*. *Proceedings of the Royal Society B* **276**, 1137–1146.

Skallova, A., Kodym, P., Frynta, D., and Flegr, J. (2006). The role of dopamine in *Toxoplasma*-induced behavioural alterations in mice: an ethological and ethopharmacological study. *Parasitology* **133**, 525–535.

Staples, L. G., Hunt, G. E., Cornish, J. L., and McGregor, I. S. (2005). Neural activation during cat odor-induced conditioned fear and "trial 2" fear in rats. *Neuroscience and Biobehavioral Reviews* **29**, 1265–1277.

Stibbs, H. H. (1985). Changes in brain concentrations of catcholamines and indoleamines in *Toxoplasma gondii* infected mice. *Annals of Tropical Medicine and Parasitology* **79**, 153–157.

Tain, L., Perrot-Minnot, M. J., and Cézilly, F. (2006). Altered host behaviour and brain serotonergic activity caused by acanthocephalans: evidence for specificity. *Proceedings of the Royal Society B* **273**, 3039–3045.

Tain, L., Perrot-Minnot, M. J., and Cézilly, F. (2007). Differential influence of *Pomphorhynchus laevis* (Acanthocephala) on brain serotonerigic activity in two congeneric hosts. *Biology Letters* **3**, 68–71.

Takahashi, L. K., Hubbard, D. T., Lee, I., Dar, Y., and Sipes, S. M. (2007). Predator odor-induced conditional fear involves the basloateral and medial amygdala. *Behavioral Neuroscience* **121**, 100–110.

Taylor, J. L., Rajbhandari, A. K., Berridge, K. C., and Aldridge, J. W. (2010). Dopamine receptor modulation of repetitive grooming actions in the rat: potential relevance for Tourette syndrome. *Brain Research* **1322**, 92–101.

Thomas, F., Adamo, S. A., and Moore, J. (2005). Parasitic manipulation: where are we and where should we go? *Behavioral Processes* **68**, 185–199.

Thomas, F., Schmidt-Rhaesa, A., Martin, G., Manu, C., Durand, P., and Renaud, F. (2002). Do hairworms (Nematomorpha) manipulate the water seeking behaviour of their terrestrial hosts? *Journal of Evolutionary Biology* **15**, 356–361.

Thomas, F., Ulitsky, P., Augier, R., Dusticier, N., Samuel, D., Strambi, C., Biron, D. G., a ndCayre, M. (2003). Biochemical and histological changes in the brain of the cricket *Nemobius sylvestris* infected by the manipulative parasite *Paragordius tricuspidatus* (Nematomorpha). *International Journal for Parasitology* **33**, 435–443.

Thompson, S. N. and Kavaliers, M. (1994). Physiological bases for parasite-induced alterations in host behaviour. *Parasitology* **109**, S119–S138.

Todd, K. S. and Lepp, D. L. (1971). The life cycle of Eimeria vermiformis in the mouse Mus musculus. *Journal of Protozoology* **18**, 332–337.

Tomonaga, K. (2004). Virus-induced neurobehavioral disorders: mechanisms and implications. *Trends in Molecular Medicine* **10**, 71–77.

Vyas, A., Kim, S. K., Giacomini, N., Boothroyd, J. C., and Sapolsky, R. M. (2007a). Behavioral changes induced by *Toxoplasma* infection of rodents are highly specific to aversion of cat odors. *Proceedings of the National Academy of Sciences* **104**, 6442–6447.

Vyas, A., Kim, S. K., and Sapolsky, R. M. (2007b). The effects of *Toxoplasma* infection on rodent behavior and depedent on dose response of the stimulus. *Neuroscience* **148**, 342–348.

Vyas, A. and Sapolsky, R. M. (2010). Manipulation of host behaviour by *Toxoplasma gondii*: what is the minimum a proposed proximate mechanism should explain? *Folia parasitologica* **57**, 88–94.

Walker, M. D. and Zunt, J. R. (2005). Neuroparasitic infections: cestodes, trematodes and protozoans. *Seminars in Neurology* **25**, 262–277.

Wang, T. and Montell, C. 2005. Rhodopsin formation in *Drosophila* is dependent on the PINTA retinoid-binding protein. *Journal of Neuroscience*, **25**, 5187–5194.

Webster, J. P. (1994). The effect of *Toxoplasma gondii* and other parasites on activity levels in wild and hybrid *Rattus norvegicus*. *Parasitology* **109**, 583–589.

Webster, J. P., Brunton, C. F. A., and MacDonald, D. W. (1994). Effect of *Toxoplasma gondii* on neophobic behaviour in wild brown rats, *Rattus norvegicus*. *Parasitology* **109**, 37–43.

Webster, J. P., Lamberton, P. H. L., Donnelly, C. A.nd, a Torrey, E. F. (2006). Parasites as causative agents of human affective disorders? The impact of anti-psychotic, mood-stabilizer and anti-parasite medication on *Toxoplasma gondii*'s ability to alter host behaviour. *Proceedings of the Royal Society B* **273**, 1023–1030.

Webster, J. P. and McConkey, G. A. (2010). *Toxoplama gondii* -altered host behaviour: clues as to mechanism of action. *Folia parasitologica* **57**, 95–104.

Weisel-Eichler, A. and LIbersat, F. (2004). Venom effects on monoaminergic systems. *Journal of Comparative Physiology A* **190**, 683–690.

Wünschmann, A., Garlie, V., Averbeck, G., Kurtz, H., and Hoberg, E. P. (2003). Cerebral cysticercosis by *Taenia crassiceps* in a domestic cat. *Journal of Veterinary Diagnositic Investigation* **15**, 484–488.

Yolken, R., Dickerson, F. B., and Torrey, E. F. (2009). Toxoplasma and schizophrenia. *Parasite Immunology* **31**, 706–715.

Zhang, Y. Q., Friedman, D. B., Wang, Z., Woodruff, E., Pan, L., O'Donnell, J., and Broadie, K. (2005). Protein expression profiling of the *Drosophila* fragile X mutant brain reveals up-regulation of monoamine synthesis. *Molecular and Cellular Proteomics* **4**(3), 278–290.

Zitnan, D., Kingan, T. G., Kramer, S. J., and Beckage, N. E. (1995). Accumulation of neuropeptides in the cerebral neurosecretory system of *Manduca sexta* larvae parasitized by the braconid wasp *Cotesia congregata*. *Journal of Comparative Neurology* **355**, 83–100.

Afterword

Gene Robinson

After reading this broad and comprehensive chapter, there can be no doubt that parasites affect the behavior of their hosts in diabolically clever ways. But how do they do it? Apparently, not like a neurobiologist would, according to Adamo, who effectively employed this approach to explore what is known about the mechanisms of parasite manipulation of host behavior.

Take the case of the protozoan *Toxoplasma gondii*, which famously makes rats less scared of cats; this results in the inevitable, thus allowing the protozoa to pass from rat to cat to reproduce. Adamo reviewed our current understanding of the neurobiology of fear, involving specific neural circuits in limbic, hypothalamic, and brainstem regions. She pointed out that *T. gondii* is distributed far more broadly in the brain than would be expected if it targeted just these circuits; it thus lacks the "precise neuroanatomical specificity" that current neuroscience would predict for the observed change in behavior. Has *T. gondii* come up with a novel mechanism, distinct from endogenous control? I came away wondering whether the dissonances perceptively pointed out by Adamo actually reflect the fact that we still have a lot to learn about how the brain itself produces behavior. Parasites may have much to teach us about the regulation of behavior, particularly instincts, behavioral choices, and social behavior.

As covered in this chapter, the rabies virus makes animals aggressive and *T. gondii* eases their fears. Many parasites affect instincts such as these. Behaviors are either instinctive, reflect the effects of experience, or both. Experience-dependent changes in behavior, especially through learning, are thought to be evolutionarily derived traits that help modify and adapt instincts. It is therefore ironic that more is known about the molecular and neural basis of learned behaviors than of the instincts themselves. Instincts are said to be "hard-wired," but this heuristic has not yet helped elucidate the roots of some of the more spectacular instincts, such as the dance language of the honey bee or the migrations of Monarch butterflies and many species of birds. For learning and memory, there is an overarching theory that involves changes in synapses that lead to increases in the connectivity of neurons, complete with identification of some of the key molecules involved. For instincts there are several well-studied cases, for example courtship in fruitflies or mating in rats, but no overarching theory. Perhaps studying parasites will help us to one day develop a comprehensive theory of the neural and molecular basis of instincts.

Parasites may also teach us about behavioral choice. Animals must perform many different types of behavior to survive in nature, so the issue is not whether or not to perform a specific behavior, but which behavior to perform at a particular time. In some cases it appears that an animal performs whichever behavior it is most provoked to do at the time, the provocation coming from the perception of stimuli in the environment that elicit the performance of that behavior. In other cases, it appears that an animal performs whichever behavior will bring it the most reward. So how do parasites induce their host to choose a suicidal behavior, like the nematomorph-infected crickets that jump into a lake? An impressive body of research documents many neural and molecular changes associated with nematomorph infection in crickets. Adamo pointed out that this suicidal behavior is not unheard of in healthy crickets, it's just that parasite-infected individuals are more likely to do it. There is research on animal choice in a variety of species involving behavior, neuroscience, molecular biology, and theoretical modeling. Adding parasites to the mix might help further elucidate animal choice.

Parasites may also teach us about social behavior. Animal social behavior—behavior that involves intense interactions with conspecifics—is considered evolutionarily more derived than solitary behavior. Many social behaviors require cooperation, and cooperation requires coordination among individuals. Researchers are beginning to find genes involved in social behavior that also influence solitary behavior, and are using this information to develop insights into evolutionarily labile pathways that elaborate social behavior from solitary precur-

sors. Adamo did not include examples of parasites that make animals more social, but the converse of course is well known; sick individuals, whether humans or animals, tend to keep to themselves. Perhaps this effect is generally attributable to malaise, in which case it would not be particularly helpful. But if parasites really could make animals more or less social, they could be used as probes to help elucidate the mechanisms and evolution of sociality.

Do some parasites really manipulate behavior differently to what neurobiology would predict or does it just seem that way because of important gaps in our understanding of endogenous mechanisms regulating complex behavior? Adamo is bullish on parasites as inventive neurochemists, and this is a very attractive hypothesis. Either way, we are learning a lot from parasites about brain and behavior, and as this chapter indicates, the best is yet to come.

CHAPTER 4

Parasites discover behavioral ecology: how to manage one's host in a complex world

Bernard D. Roitberg

4.1 Introduction

Variation is a common theme in biology (Carmona et al. 2011). For any given species, one can document a variety of responses to environmental conditions (Franks and Weiss 2008). Even within the same individual, variation is easily observable within similar environments. It is the job of biologists to explain such variation particularly since variation is a key ingredient of evolution by natural selection.

When individuals express typical behaviors, the evolutionary biologist can readily explain them using standard theory, however when so-called abnormal behaviors are expressed, things get really interesting. Are such behaviors mistakes, are they maladaptive, or are they neutral? It often turns out that such responses are more readily explained by taking into account parameters or axes not normally considered. For example, superparasitism, the apparent wastage of eggs via oviposition of eggs in or on already parasitized hosts, can be shown as a relatively rare but adaptive behavior when the context of state dependency, in this case eggload, is considered (e.g., Mangel 1989; van Alphen and Visser 1990).

My point is that when apparently healthy individuals express rare, unusual, or apparently maladaptive traits, this presents a challenge for evolutionary biologists in general and adaptationists in particular (Nesse 2005). The key is to elucidate how suites of behavioral traits are likely to impact Darwinian fitness in the broad sense (i.e.,

within a full life history framework) rather than to study isolated phenomena and then judge them as adaptive or not.

In this chapter I consider behaviors expressed by unhealthy (i.e., parasitized individuals) in the context of the range of behaviors within the typical portfolio of healthy individuals but under the control of the parasite and not the host. In other words, I ask *why* a parasite would "want" its host to express a particular part of its repertoire. As such, I will employ evolutionary theory to help answer this question and I will use this chapter to point out the kinds of theories that are necessary for answering particular kinds of parasite–host questions (e.g., strict optimization, game theory, etc.).

I will not consider novel behaviors that are expressed by parasitized hosts. While interesting in their own right, such behaviors are often the byproduct of host–parasite interactions (see Ewald 1994) and beyond the scope of this chapter.

Let's take an example of abnormal behavior analysis. Suppose that a bloodfeeding organism has evolved to procure bloodmeals at some rate that maximizes its evolutionary fitness. Suppose that the metric that we use to measure this procurement rate is the organism's attack persistence when bloodhosts are defensive, for example when hosts swish their tails to deter landings by the bloodfeeder. When we run a series of carefully controlled experiments, we find that parasitized bloodfeeders have a higher persistence rate compared to healthy conspecifics (e.g., Anderson and Koella 2000). Why? How? A simple way to answer these questions is to ask why a healthy individual would express a higher than normal persistence rate. In fact, we have done such an analysis and complementary experiments and found that high persistence can be generated by individuals with both very high and very low energy reserves, but for different reasons; in the former case because they can afford to do so and in the latter because they cannot, that is they would starve otherwise (Roitberg et al. 2010a). Thus a fitness surface for persistence as a function of energy state can be derived showing that high persistence can be a viable tactic for a limited portion of the fitness landscape. The how question remains to be answered, but presumably there is some feedback between energy reserves and that area of the central nervous system where persistence decisions are made. (Note: I am not suggesting any cognitive function when I use the term "decision" though the term cognition is becoming increasingly accepted for describing information processing in invertebrates (e.g., Chittka and Skorupski 2011). Here I simply mean that animals possess a range of behaviors that they express in response to different situations.)

Now, recall that parasitized bloodfeeders also tend to express higher persistence on average than healthy feeders do. How do we explain this? The same approach can be applied, that is derive a fitness surface for the parasite (e.g., Plasmodium) that manipulates its host's behavior. The fact that parasitized individuals express greater persistence on average suggests that the fitness surface for the parasite, the one in control, contains more area in the high persistence zone than healthy individual does.

Again, the thought experiment above helps us answer the why question. As to how, it is easy to imagine that a parasite manipulates its host by modifying that host's sensitivity to energy state. A good example of manipulated sensitivity is the change in wavelength sensitivity to light as a function of change in infection state in the amphipod *Hyalella azteca* (Benesh et al. 2005).

In the exercise above, I argued that it is critical to somehow determine the fitness surface, and in this case it seems that the straight optimization approach has good utility. However, note that this optimiza-

tion has a state dependent component similar to that discussed earlier for the superparasitism problem. By state dependent, I mean that the solution to the problem depends upon one or more state variables that describe the focal organism, and by state variable I mean some variable(s) that describes the state of a dynamic system. State variable models seem particularly apt for studying parasite–host interactions where parasites draw on the energy reserves of their hosts and the performance of the host is (state) energy dependent. And, since I also wish to discuss state dependence with regard to environmental variation, it is appropriate to employ stochastic, state variable models also referred to as stochastic dynamic programming (SDP) models (Houston et al. 1988) and dynamic state variable (DSV) models (Clark and Mangel 2000). These kinds of models were first developed by Bellman to solve problems in economics (Bellman 1957) but have since demonstrated great utility in the biological sciences where the currency employed is Darwinian fitness (see Lalonde and Roitberg 2011).

Adding environmental variation, state dependence, and behavioral response to parasite–host interactions adds more complications to an already complex interaction. Why would I want to do that especially given the suite of excellent models that already exist for parasite–host coevolution? My answer is that many, if not most, biological entities have evolved phenotypically plastic traits that allow them to either mitigate or exploit the aforementioned variance, it is not clear how this plasticity will play out under parasite or host-induced stress, and this has received very little attention to date. Individuals have the ability to "work harder" (e.g., alter foraging tactics—Rosenberg and McKelvey 1999) or "work smarter" (e.g., reduce exposure to predators—Lima and Dill 1990) under different conditions and this ability can change the balance in an antagonistic relationship that would be unobvious, compared to what we might expect under constant conditions with hard-wired responses.

4.2 The problem

The question I will address is how feeding behaviors by a host might change as a function of its being parasitized, either via host response to parasitism or via host manipulation. With regard to the discussion above, an important aspect of host manipulation that has not received much attention is how stochastic variation of the host's nutrition resources can impact parasite–host interactions. If, by contrast, all food plants were nutritionally identical, then parasite manipulation traits may evolve to exploit host growth rates in some hardwired, deterministic fashion. Of course, this is not the case and that has not escaped the attention of ecologists who study herbivore feeding-strategies in response to nutritional variation within and among host plants (e.g., Karban et al. 1997; Roitberg et al. 1999).

My goal here is to initiate discussion on how theory might incorporate state dependence and environmental stochasticity regarding host manipulation problems, and as such I will not make any quantitative predictions about any one system. On the other hand, the models that follow could easily be parameterized and rigorously tested in a variety of systems.

The history of parasite manipulation models is a rather brief one and yet, as pointed out by Gandon (2005), such models have already been shown to be critical in helping us understand the unobvious, complex world of parasite–host interactions. These models are largely evolutionary in scope and they consider the fitness consequences of various tactics and strategies for manipulation and response to manipulation. As an example of unobvious insights, Lafferty (1992) showed that predators should not discriminate against parasite-manipulated (intermediate-host) prey even if it leads to higher rates of parasitism so long as the energetic benefits outweigh the costs. Similarly, Brown (1999), using a game theoretical model with explicit pay-off functions for the focal individual parasite versus from within-parasite groups, showed that the degree of host manipulation often depends upon intricacies of the parasite–host interaction. Of course, employing a game theoretic approach requires the researcher to consider the impact of alternative responses of conspecifics and heterospecifics, including cheating (e.g., hitch hiking Thomas et al. 1998). Parker et al. (2009) studied conditions that favored the spread of host mortality enhancement

versus host mortality suppression genes in a population of parasites. Interestingly, they refer to the potentially subtle impacts of such suppression when it affects growth rates of hosts and subsequently parasite growth, one key aspect of my models developed below. In that regard, in the text that follows I refer to new, dynamic models that are emerging that specifically consider states and growth rates. Finally, host manipulation models need not focus solely on the parasite–host interaction. Fenton and Rands (2006) showed that host manipulations can impact community dynamics and not only that details of the interactions among predator and prey may be important, but, more specifically, which details are critical to elucidate, such as predator functional responses. Given the large number of parameters inherent in parasite–host interactions, a key feature of the models cited here is their ability to identify key parameters; this may be the greatest strength of the theoretical approach. A list of the model parameters and their values can be found in Table 4.1.

4.2.1 A healthy caterpillar

Since we wish to explain variation in behavior as a function of parasitism, we must first define the optimal behavior of a healthy, unparasitized individual for comparison's sake. Suppose we study a generalist "caterpillar" that feeds on leaves from a variety of plants. The caterpillar has a window in time (T) within which feeding must be completed and pupation initiated (note: to keep things simple, I will work within the final larval stage and ignore moulting dynamics). The size of the caterpillar at time T translates directly into expected fitness as an adult and can be described by the typical curve of diminishing returns:

$$F(T) = \omega(1 - e^{\eta(T)\phi})$$ 4.1

where ω is the maximum expected fitness, $\eta(T)$ is the size at T and Φ is a shape parameter. Expected fitness of a caterpillar at some time t is also determined by survival to pupation and here there are two impediments to survival, death from starvation and death from predators. Below, I consider these impediments as future fitness discounting terms.

Table 4.1 Parameter definitions and values from the dynamic state variable models for healthy and parasitized herbivores.

Parameter	Definition	Value
t	Time steps (each step is 0.25 days)	(1,40)
η	Size state	(1, 20)
ε	Energy reserve state	(1,20)
ρ	Host plant nutrient value state	(0.75,1.25)
$\tau,$	Plant site nutrient value state	(1,5)
δ	Parasite size state	(1,20)
Φ	Shape parameter for fitness function for size at T	0.15
ε_{met}	Minimum energy reserve for pupation	4
μ_ψ	Mortality rate while moving between plants	0.008
μ_θ	Mortality rate while moving within plants	0.006
μ_α	Mortality rate while feeding	0.005
C	Host head capsule	4
H	Half saturation	1.5
v_h	Conversion rate of energy to growth for hosts	0.04
v_p	Conversion rate of energy to growth for parasites	0.1
α	Foraging effort	(0,1)
χ	Proportional allocation of energy reserve to growth	(0,1)
E	Extraction rate relative to baseline	(0, 1.1)

For our focal caterpillar, feeding on high quality leaves provides considerable energy and nutrients that can be allocated to growth and/or energy reserves. Recent papers by Hall and associates (e.g., Hall et al. 2009) show how one can calculate the dynamic optimal allocation to growth and reserves in a globally heterogeneous but locally deterministic world. These decisions maximize fitness by optimizing size and survival with the constraint that a growing individual can assimilate nutrients at a higher rate as it grows but also pays greater maintenance costs. Of course, the real world is not deterministic and leaf quality can vary both within and among plants (Karban et al. 1997). Here I consider how caterpillars might use their foraging behaviors to mitigate and/or exploit natural variation in host quality; rather than optimizing growth based upon some fixed nutrient intake at some site, they can change sites or foraging effort. Let's assume that caterpillars can assess plant quality and, from there, they can make several foraging decisions in addition to allocation, (1) whether to remain on the current plant, (2) whether to continue to feed at the current site, and (3) how much effort to expend to feed.

As noted above, an appropriate approach to evaluating foraging and allocation decisions in this nutrient-stochastic environment is via the use of dynamic state variable models (Mangel and Clark 1988). DSV models make explicit use of a focal organism's states (e.g., size, energy, etc.), the probability of relevant events occurring (e.g., encounters with particular resources) and the repercussions from making particular decisions (e.g., accept vs. reject specific foodstuffs). One feature to note is that state variable models are not simulation models even though the DSV equations are usually solved numerically at a computer. In fact, DSV models provide a deterministic, single-best strategy for living in a stochastic world.

To develop a state variable model of caterpillar decisions in a stochastic world, I characterize the caterpillar by body size at time t, $\eta(t)$, energy reserves $\varepsilon(t)$, mean nutritional quality of the current host plant $\rho(t)$ and nutritional value at the current plant-site $\tau(t)$. Thus, the expected fitness for a caterpillar at time t is: $F(\eta, \varepsilon, \rho, \tau, t)$ (note: fitness cannot be accrued by the caterpillar but rather only through reproduction once it has achieved its butterfly form).

The caterpillar grows at rate (γ) as a function of nutrient intake and nutritional value at site τ (t) and allocation to growth (χ) with remaining nutrients allocated to energy reserves after paying a metabolic cost $\varsigma(\eta,\alpha)$ which is size and activity dependent, respectively. Here are the activity decisions, that the caterpillar must make: (1) Leave from or stay on the current plant—if the decision is to leave then the caterpillar abandons the current plant and arrives at a new plant with mean nutrient quality (ρ) according to the distribution of such qualities in the environment. I assume that seeking out new plants is inherently dangerous, thus I also include a feature that any benefit from arriving at a new plant must be discounted by the probability of arriving intact, μ_ψ^κ where μ_ψ is the probability of surviving per unit time while moving off plant and κ is the number of time units required to find a new plant. (2) Stay on the current plant ρ but move to a new site with future discounting, μ_θ^σ where μ_θ refers to survival while moving within plants and σ the time to find a new site, as above. (3) If the caterpillar chooses to remain at the current site, for a single time step, then it must choose a feeding effort α. I assume that foraging (working) harder leads to increased rate of nutrient intake but also leads to greater risk of predation because of greater exposure thus, we have μ_α for that single time step.

Before describing the dynamics of the state variables, first I state my assumptions:

1. The caterpillar is a strict herbivore.
2. The caterpillar can carry a maximum energy reserve ε_{max}.
3. If energy reserves fall below some critical level ε_{min}, the caterpillar can no longer maintain somatic function and dies of starvation.
4. Moving between plants costs more energy than moving within plants, which is energetically more expensive than remaining at a given site.
5. Moving between plants is inherently more dangerous than moving within plants, which is more dangerous than remaining at a given site. Thus, $\mu_{\alpha>}\mu_\theta^\sigma > \mu_\psi^\kappa$.
6. Survival from predation is an inverse function of foraging effort. Greater effort leads to greater exposure to predators.

7. Caterpillar encounters with food plant quality are stochastic with known plant nutrition mean and variance.
8. There is much lower variance in nutritional quality within than among plants, that is CV within plants is less than CV among plants.
9. Caterpillars consume individual leaves, which vary in nutrient content but not in size. Thus, I assume that during any one feeding period that the same amount of leaf material is consumed for a given level of feeding effort, regardless of leaf nutrient content.
10. Feeding at a site does not significantly deplete resources at that site, that is future probabilities of encounter with leaves do not change as a function of current feeding decisions.
11. Caterpillars do not compete for nutrition on any one plant, that is they essentially forage alone on plants.
12. Adult fitness is a positive, decelerating function of larval size at time of final molt (t = T). There is, however, a minimal energy reserve value ε_{met} below which the caterpillar will not possess sufficient energy to metamorphose into a butterfly (Fig. 4.1). Thus, fitness at (t = T) is:

$$F(\eta,\varepsilon,\rho,\tau,T) = \begin{cases} \omega(1-e^{-\eta(T)\phi}) >= \varepsilon_{met} \\ 0 \end{cases} \qquad 4.2$$

Below are the dynamics that ensue from the various caterpillar decisions:

First, if the decision is to move to another plant, then the caterpillar arrives at new plant ρ and position τ on that plant at time t + κ with probability dependent upon the distribution of plants and feed-

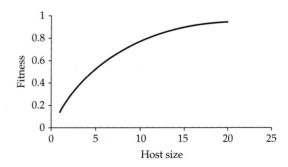

Figure 4.1 Host fitness as a function of size when sufficient energy reserves are present to moult to an adult.

ing sites, respectively. Here, following Hall et al. (2009), growth during this search is:

$$\eta(t+\kappa) = \eta(t)e^{(\chi - \Omega_h)v_h\kappa} \qquad 4.3$$

where Ω_h is the size-dependent growth cost and v_h is the conversion rate of energy into body mass. The h subscript indicates the caterpillar (host) and the p subscript, when employed, refers to the parasite as discussed below.

Energy reserves are reduced according to, (i) allocation to growth per unit time and (ii) size-dependent costs of plant search $\varsigma_\psi(\eta)$ per unit time. Thus:

$$\varepsilon(t+\kappa) = \varepsilon(t)e^{-(\chi\varsigma_\psi)\kappa} \qquad 4.4$$

with concomitant lowest survival, μ_ψ, because seeking new host plants renders foragers at their most conspicuous from all activities.

Second, if the decision is to move to another position on the current plant, to seek better quality food, then the caterpillar arrives at a new position ρ at time $t + \sigma$ with probabilities of arriving at sites of particular quality dependent upon the distribution of feeding sites of quality ρ. Here, growth is:

$$\eta(t+\sigma) = \eta(t)e^{(\chi-\Omega_h)\vartheta_h\sigma} \qquad 4.5$$

As above, no feeding occurs during this activity, thus energy reserves decline as:

$$\varepsilon(t+\sigma) = \varepsilon(t)e^{-(\chi\varsigma_\psi)\sigma} \qquad 4.6$$

Finally, if the decision is to remain at the current feeding site then the caterpillar grows as follows:

$$\eta(t+1) = \eta(t)e^{(\chi-\Omega_h)\vartheta_h} \qquad 4.7$$

Along with the decision to stay, the caterpillar must also decide how much effort to invest in foraging. As such, energy reserves change according to forage quality, foraging effort, and leaf assimilation as a function of caterpillar size, allocation of energy to growth, and size-dependent metabolic costs.

I follow Hall et al.'s (2009) lead and apply a Type II functional response (FR) for acquisition and assimilation as a function of size.

$$\varepsilon(t+1) = \varepsilon(t)e^{-(\chi\varsigma_\psi)+FR} \qquad 4.8$$

and the functional response, FR, is:

$$FR = \eta\left(\frac{\dfrac{\alpha}{C}}{\dfrac{\alpha}{C}+H}\right) \qquad 4.9$$

where C is the head capsule or gape size and H is the half saturation constant.

Survival from predators, μ_α, is set at a lower value than above because the caterpillar is less conspicuous while feeding than while moving within or between plants.

This leads to the state variable model:

$$F(\eta,\varepsilon,\rho,\tau,t) = max\ \chi\ max \begin{bmatrix} \displaystyle\sum_{\rho=1}^{P}\sum_{\tau=1}^{T}\mu_\psi^\kappa F(\eta',\varepsilon',\rho,\tau,t+\kappa); \\[1.5em] \displaystyle\sum_{\rho=1}^{P}\mu_\theta^\sigma F(\eta'',\varepsilon'',\rho,\tau,t+\sigma); \\[1em] max\ \alpha\ \mu_\alpha F(\eta''',\varepsilon''',\rho,\tau,t+1); \end{bmatrix} \qquad 4.10$$

The model can be read as follows: The left-hand side is the expected future fitness for a caterpillar of size η, energy ε, on plant ρ at site τ at time t. The left-hand side has two maximization requirements, the first, χ for the optimal allocation of energy to growth and the second is to choose the optimal activity. These three activities and their expected fitness payoffs and new state variable values are shown on the right-hand side, line-by-line, with future discounting indicated by the μ with subscript indicating the particular activity: seek out a new food plant, seek out a new feeding site, and stay and feed wherein the optimal feeding level is denoted by max α. The primes above the symbols refer back to the dynamics discussed above with a single, double, and triple primes for the among plants search, the within plant search, and feeding activities, respectively. Note: the optimal feeding rate could be zero, which could be interpreted as a resting decision. The model is solved by backwards induction starting with F(η, ε, ρ, τ, T) and iterating backwards until the appropriate parameter values are reached (Mangel and Clark 1988). In other words, one solves for the optimal behavior and allocation for all combinations of size, energy reserve, feeding site, and plant across all times. I solved for optimal decisions for discrete integer values of the state variables (at time t) and

I used single or two-dimensional linear interpolation when calculating fitness when future state values were non-integer for energy and size state, respectively. The optimal decisions for allocation, activity, and foraging effort were then entered into optimal decisions matrices, one for allocation and the other for activity.

The figures below show optimal activity and allocation decisions for a caterpillar that finds itself at a low-quality site, early on in its life (t = 0.25T) at various combinations of size and energy state. At this point, the optimal responses are no longer changing for a given set of state values. Though these two optimal decision matrices are shown separately it should be remembered that, in fact, they are integrated to maximize lifetime fitness. For example, when a decision is made to leave a site, then less energy should be allocated to growth because of the loss of energy during site search. In Fig. 4.2, the three aforementioned activities of leave the plant, leave the site and feed are shown as light gray, dark gray, and black, respectively under two within-plant resource distributions, skewed to the left and right, respectively. Under such conditions, there is an 87% congruence in activity decisions with very few decisions to exit the plant. When I decreased the value of the current plant state or skewed plant quality distributions to the right (not shown), cater-

pillars were more willing emigrants from those plants.

In Fig. 4.3, the darker the cell, the greater allocation of energy reserves to growth, that is allocation is in proportional amounts of energy reserves (e.g., 35%), not absolute. We see two major tactics employed within the overall growth strategy that maximizes fitness in a stochastic world: (1) remain feeding at the current site and allocate as much energy to growth as possible without starving, based upon certain expected returns from feeding or (2) move to a new site but refrain from allocating very much (often no) energy to growth because of the cost of moving and the stochasticity regarding nutrient value at the new arrival site. To show that this interpretation is correct, as above I set the caterpillar up in two worlds, a poor one where nutrient quality within plant is skewed toward the left and a rich one where nutrient quality within plant is skewed toward the right. The results, as expected, are that caterpillars are more conservative when resources are left skewed than right with only a 67% congruence on allocation values. Again, we can see that the variance in resources plays a role in decisions.

These optimal decision matrices were derived with an assumption of a moderately steep fitness function (0.25 maximum size returns 0.5 maximum

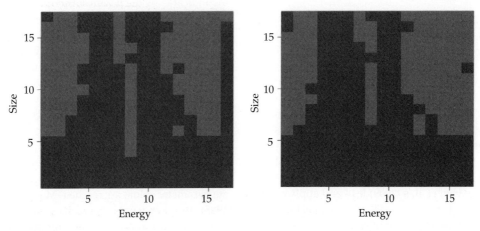

Figure 4.2 Optimal activities as a function of energy (x axis) and body size (y axis) for a healthy "caterpillar" when feeding on plants with two different within-plant distributions of nutrient quality: in 4.2a, the distribution of resources is left skewed toward low values and in 4.2b, the distribution of resources is left skewed toward high values. Three activities are possible: lightest shade is for leave the plant, medium shade is to move within the plant, and darkest shade is for a remain-at-the-site decision. For the set of values shown, plant emigration (lightest shade) is never an optimal decision. See also Plate 4.

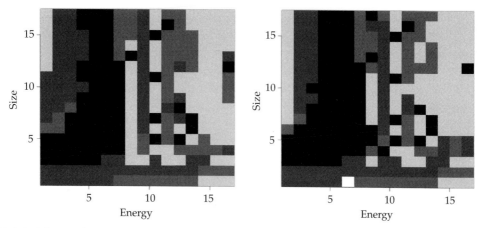

Figure 4.3 Optimal allocation of energy to growth as a function of energy (x axis) and body size (y axis) for a healthy "caterpillar" when feeding on plants with two different within-plant distributions of nutrient quality: in 4.3a, the distribution of resources is left skewed toward low values and in 4.3b, the distribution of resources is left skewed toward high values. The darker the cell the greater the proportion of energy that is allocated to growth. See also Plate 5.

fitness, Fig. 4.1) for those individuals that harbor more than the minimal energy reserves. How would the decisions change if the fitness functions differed in their slopes? This question is not easy to answer because caterpillars have three means of addressing their nutritional needs—movement decisions, feeding rate decisions, and allocation decisions. Thus a change in any one of those parameters can impact the effect from the other two. Investigations that I have made do confirm that caterpillars act according to expectations, that is they become less conservative as the pay-off for doing so increases. Keep in mind, however, that such responses depend upon the distribution of resources, thus abandoning poor sites or plants only makes sense when resources are distributed to favor an increase in fitness from moving, on average, even when moving costs are factored in.

Another simplifying assumption I have made is that plant nutrition can be described by a single parameter whereas plant nutrition may comprise disparate components that require optimal balance. Recent papers on this topic shed some light on this problem (e.g., Rosenheim et al. 2010; Peterson and Roitberg 2010) but will not be discussed further here.

Finally, optimal decision matrices can be a little misleading in that they display all possible combi-

nations of state variables while in reality, some of these combinations will never be experienced. To better visualize what a population of caterpillars is likely to express, decision-wise, one can incorporate the decisions matrices into a forward moving, stochastic simulation. In these simulations, encounters with sites of different nutrition values are randomly drawn from known distributions (see Mangel and Clark (1988) for an excellent description on how to incorporate this approach). As an example, we recently employed a mosquito optimal decision matrix within a simulation to test the effects of various types of bednets to elucidate their effectiveness at reducing disease transmission in east Africa (Roitberg and Mangel 2010).

4.2.2 A parasitized caterpillar

Now, I develop theory for the behavior of a parasitized caterpillar. The kind of parasite that I have in mind is not essential to the argument below but it might be a fungus, a nematode, or a parasitic wasp that functions as a koinobiont (i.e., it lives and grows within a growing, developing host) that can manipulate its host (Fenton et al. 2011). It is not so important whether the parasite acts as a parasitoid (i.e., eventually kills its host) so long as there is some possibility that immune responses can be mounted and are sometimes successful. Further, the host need not be a caterpillar but rather any insect herbivore (e.g., aphid, beetle, etc.) whose nutritional needs are met by feeding on plants and whose behavior can be manipulated (Libersat 2009). However, I will continue to refer to the host as a "caterpillar." There are at least three possible parasitism scenarios:

1. The caterpillar is in control of its body and responds to the energetic demands of its parasite.
2. The parasite is in control of its host's behavior.
3. The caterpillar and parasite are engaged in a dynamic game.

Scenario I: caterpillar is in full control
The theory for this scenario is similar to that for the healthy caterpillar with one key modification: the parasite extracts energy from its host at some fixed rate per unit parasite mass, that is the parasite does not dynamically adjust its energy extraction. All other assumptions from above hold, including the assumption that the caterpillar can estimate the rate of growth of both itself and its parasite.

The new dynamics are shown below for the three activities, moving between plants, moving within plants, and feeding at the current site, respectively.

First, the growth equations for the host do not change, that is they are based upon allocation of energy reserves to growth.

By contrast, energy reserves are modified according to, (1) allocation to growth per unit time, (2) parasite size-dependent energy extraction $E(\delta)$, and (3) size-dependent costs of between plant search $\varsigma_{\psi}(\eta)$ per unit time and energy intake during feeding activities. Thus:

$$\varepsilon(t+\kappa) = \varepsilon(t)e^{-(\chi\varsigma_{\psi}+E(\delta))\kappa} \qquad 4.11$$

$$\varepsilon(t+\sigma) = \varepsilon(t)e^{-(\chi\varsigma_{\psi}+E(\delta))\sigma} \qquad 4.12$$

$$\varepsilon(t+1) = \varepsilon(t)e^{(-(\chi\varsigma_{\psi}+E(\delta)))+FR} \qquad 4.13$$

We must include a new state variable, parasite size, δ.

$$\delta(t+\Delta t) = \delta(t)e^{(E(\delta)-\Omega_p)\vartheta_p\Delta t} \qquad 4.14$$

where Ω_p and v_p refer to parasite growth costs and conversion rates, respectively.

This leads to the new state variable model:

$$F(\eta,\varepsilon,\rho,\tau,\delta,t) = max \, \chi \, max \begin{bmatrix} \sum_{\rho=1}^{P}\sum_{\tau=1}^{T} \mu_{\psi}^{\kappa} F(\eta',\varepsilon',\rho,\tau,\delta',t+\kappa); \\ \sum_{\tau=1}^{T} \mu_{\theta}^{\sigma} F(\eta'',\varepsilon'',\rho,\tau,\delta'',t+\sigma); \\ max \, \alpha \, \mu_{\alpha} F(\eta''',\varepsilon''',\rho,\tau,\delta''',t+1); \end{bmatrix} \qquad 4.15$$

Here, the caterpillar must make activity and allocation decisions based upon expected loss of energy to a growing parasite. This creates an added pressure to retain greater energy in reserves to avoid starvation. Thus, I predict more conservative growth, that is the caterpillar should choose to allocate less energy to growth and more to reserves due to the parasite extraction constraint. To test this assumption, one can again use backwards induction to generate optimal decisions matrices. This was carried out as above except that, under some conditions, three-dimensional linear interpolation was required when caterpillar size and energy and parasite size reached non-integer values.

The results from backward induction can be seen below (Fig. 4.4, 4.5). Again, note that there are both size state and energy state effects. First, compare activity decisions between a healthy caterpillar and one parasitized with a moderate-sized parasite (35% of maximum parasite size) on a plant with left-skewed resources. Indeed, parasitized caterpillars are more conservative in their foraging behavior both in terms of their choice of activities (fewer movements within plants) with a congruence of 63% between the two caterpillar types. However, more sick caterpillars leave plants when at low energy states simply because they cannot improve their lot by remaining, given the parasite load they are carrying and the nutrition values that are left skewed, that is only slight improvement in nutrition can be expected within plants (cf. energy deprived mosquitoes in (Roitberg et al. 2010a)). Similarly (Fig. 4.5), sick caterpillars allocate less energy to growth compared to healthy individuals, thus our interpretation is that caterpillars *choose* to grow more slowly and protect their current size and

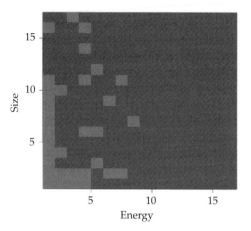

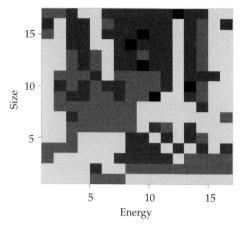

Figure 4.4 Optimal activities as a function of energy (x axis) and body size (y axis) for a parasitized "caterpillar", in control of its body, when feeding on a plant where the distribution of nutrients is left skewed toward low values. Lightest shade is for leave the plant, medium shade is to move within the plant, and darkest shade is for a remain-at-the-site decision. See also Plate 6.

Figure 4.5 Optimal allocation of energy to growth as a function of energy (x axis) and body size (y axis) for a parasitized "caterpillar" in control of its own body, when feeding on a plant where the distribution of nutrients is left skewed toward low values. The darker the cell the greater the proportion of energy that is allocated to growth. See also Plate 7.

energy assets (see Clark's asset protection principle (Clark 1994)), given the uncertainty of their future and their parasite load as opposed to growing slowly because they are sick. Here, there is a congruence of 38% in allocation between the two classes. Finally, (not shown) if one compares decisions by sick caterpillars on right versus left-skewed plants, we see 65% and 78% congruence in activities and allocation, respectively.

To demonstrate the importance of state once more, I now compare activity and allocation decisions between caterpillars harboring a large parasite (60% of maximum size versus 35%) (Fig. 4.6). Caterpillars harboring large parasites at poor sites on left-skewed low-quality plants are again more likely to leave those plants when energetically deprived because they are in dire need of nutrients, with about 80% congruence between the two classes. By contrast, when nutrition is right-skewed, caterpillars with small parasites can take advantage and seek out better sites within plants much better than caterpillars with large parasite loads who risk starvation too much to leave feeding sites when at low energy states (congruency of 76%). These examples demonstrate an intricate balance of state dependence and the stochastic nature of resources.

Another version of this model, which space prevents me from discussing in further detail, is one in which the caterpillar mounts an immune response, which, if successful with probability ξ, can rid the host of its parasite. I assume that there is a fixed cost to initiate and maintain immune function. When

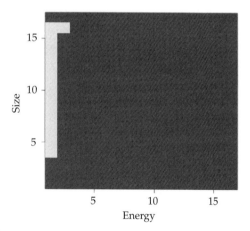

Figure 4.6 Optimal activities for a "caterpillar" as a function of its energy (x axis) and body size (y axis), where the parasite controls its host body, when the host is feeding on a plant where the distribution of nutrients is left skewed toward low values. Lightest shade is for leave the plant, medium shade is to move within the plant and darkest shade is for a remain-at-the-site decision. See also Plate 8.

the assumptions discussed above hold then this leads to the following DSV equation:

$$F(\eta,\varepsilon,\rho,\tau,\delta,t) = \max \chi \max$$

$$\begin{bmatrix} \xi \left(\sum_{\rho=1}^{P} \sum_{\tau=1}^{T} \mu_{\psi}^{\kappa} F(\eta',\varepsilon',\rho,\tau,0,t+\kappa) \right) + \\ (1-\xi)\left(\sum_{\rho=1}^{P} \sum_{\tau=1}^{T} \mu_{\psi}^{\kappa} F(\eta',\varepsilon',\rho,\tau,\delta',t+\kappa) \right); \\ \xi \left(\sum_{\tau=1}^{T} \mu_{\theta}^{\sigma} F(\eta'',\varepsilon'',\rho,\tau,0,t+\sigma) \right) + \\ (1-\xi)\left(\sum_{\tau=1}^{T} \mu_{\theta}^{\sigma} F(\eta'',\varepsilon'',\rho,\tau,\delta'',t+\sigma) \right); \\ \max \alpha \, \mu_{\alpha} \begin{pmatrix} \xi F(\eta''',\varepsilon''',\rho,\tau,0,t+1) + \\ (1-\xi)(F(\eta''',\varepsilon''',\rho,\tau,\delta''',t+1)) \end{pmatrix} \end{bmatrix} \quad 4.16$$

In equation 4.16, we now see two lines for each activity. In the first line, we have the outcome from a successful immune response (thus $\delta = 0$) and the second line shows the outcome from an unsuccessful immune response. Also, note that the energy dynamics would also include a loss of energy that is diverted to drive the immune response a factor that may turn out to have considerable impact (e.g., Diamond and Kingsolver 2010; Pursall and Rolff 2011).

Scenario II: the parasite is in charge
Here, the parasite must choose the energy extraction rate from its host to maximize its own fitness. This scenario is the same as that developed by Hall and associates (Hall et al. 2009) except this earlier theory dealt with a deterministic world whereas my caterpillar's world is stochastic, as discussed above, and includes parasite control over both host feeding and extraction rate. When examining the DSV equation below, it is important to keep in mind that the optimization now is from the parasite's perspective, thus the critical fitness parameter is parasite size $\delta(T)$ rather than host size as was explicated above (see equation 4.2). As such, another fitness-based decision is required: optimal extraction rate. And, as above for the caterpillar, parasite fitness is accrued as an adult and is a positive function of size at T. In order to be able to directly compare results from the two models, I assume that both caterpillar

and parasite accrue fitness from identical size-based functions. Now, assuming that no immune response is mounted, the dynamics from the three caterpillar behaviors and subsequent DSV models are discussed below.

First, as was the case for the scenario where the caterpillar was in control, host growth occurs simply through allocation of some proportion of the energy reserves to growth, thus equations 4.3, 4.5, and 4.7 still hold. For parasite growth, equation 4.14 holds except that the extraction rate is no longer a constant but rather a decision variable. Similarly, host energy state for the three activities can be described by equations 4.11, 4.12, and 4.13 with extraction rate as a decision variable.

If the decision is to feed then the host and parasite grow as above and the energy reserves changes according to assimilation as a function of host size.

If the decision is to move to another position on the plant, then the host arrives at a new position τ with probability dependent upon the distribution of feeding sites of quality τ. Allocation-based growth occurs, energy reserves change, and the energetic cost of moving is assumed to be higher than for resting or feeding.

If the decision is to move to another plant then the host arrives at a new plant ρ at position τ on the plant with probability dependent upon the distribution of feeding sites of quality τ. No growth occurs, energy reserves change, and energy cost of moving among plants is assumed to be higher than seeking a new site within a plant. Survival is assumed to be at its lowest when such a search is conducted.

The new DSV model with optimal extraction is:

$$F(\eta,\varepsilon,\rho,\tau,\delta,t) = \max \chi \max E \max \begin{bmatrix} \sum_{\rho=1}^{P} \sum_{\tau=1}^{T} \mu_{\psi}^{\kappa} F(\eta',\varepsilon',\rho,\tau,\delta',t+\kappa); \\ \sum_{\tau=1}^{T} \mu_{\theta}^{\sigma} F(\eta'',\varepsilon'',\rho,\tau,\delta'',t+\sigma); \\ \max \alpha \, \mu_{\alpha} F(\eta''',\varepsilon''',\rho,\tau,\delta''',t+1) \end{bmatrix} \quad 4.17$$

As above, I used backwards induction to solve for optimal behaviors for different distributions of resources across all possible state variable combinations when nutrition was skewed to the left. The

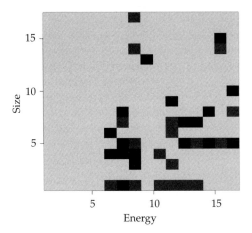

Figure 4.7 Optimal extraction rates as a function of host energy (x axis) and body size (y axis) for a parasite in control of its host's body when found on plants where the distribution of nutrients is left skewed toward low values. The darker the cell the greater the extraction rate. See also Plate 9.

first effect that I noticed was that the caterpillar rarely if ever was favored to move to new sites or plants (Fig. 4.6), far less so than healthy individuals with activity congruence of 54%. In addition, caterpillars controlled by parasites tended to eat less, though it is difficult to compare the two groups because the healthy individuals whose optimal behavior was to move within plants would not generate meaningful optimal feeding rates since they cannot feed while moving. This apparently lowered feeding rate by parasitized individuals has been documented in a wide variety of insect larvae (e.g., Scriber and Slansky 1981; Adamo 1998) without a good explanation as to why this is so. My model suggests that the parasite was acting in a conservative fashion to protect its assets. Recall that I assumed that it is more dangerous to feed more vigorously. In drosophila, parasitized larvae move to lower quality but safer feeding sites compared to healthy individuals and this has been interpreted as support for Clark's asset protection principle (Clark 1994; Robinson 1998). The same may apply here. Further support for this conclusion comes from the optimal extraction matrix (Fig. 4.7), here one can see that the parasite is extracted at the two lowest extraction rates 75% of the time. This is akin to work on optimal virulence wherein submaximal virulence will often be favored depending upon partic-

ular parameter values and assumptions including the mode of transmission (e.g., Day 2001).

When I ran the model a second time but now with within-plant nutrition values skewed to the right, I found a small shift in behavior and allocation and extraction. First, whereas it was never optimal to move in the former situation, now there is a small region, low energy and host-size state, where within-plant movement was advantageous from the parasite's perspective. This is analogous to the earlier situation for healthy energy bereft individuals. Still, there was a very high congruence (98%) in behaviors between parasitized caterpillars under these two resource distributions. Similarly, as expected given the above, optimal extraction rates were nearly identical under the two nutrition distribution scenarios as were optimal allocation values.

When I compared parasitized caterpillar behaviors and allocation when they were in charge of their own bodies versus when under control of the parasite, I found much more congruence compared to sick versus healthy caterpillars (93% vs. 63% congruence). In both cases, caterpillars acted in a conservative fashion, but recall that the benefits were to the host in the one case and the parasite in the other. In the latter, the parasite has control of the extraction rate and to some extent could drive host behavior simply through control of energy reserves. Overall, it appears that all else being equal (i.e., no immune response) that the damping down of behavioral sensitivity to conditions removes much of the impact of environmental stochasticity. Further, it appears that given the caveat above, that a parasite should "want" its host to act more like a sick host than a healthy one, in other words act as risk averse (Real and Caraco 1986) regarding the mean-variance pay-off.

Scenario III: a dynamic war
Results from the scenarios above suggest that there are many situations where there is conflict between the parasite and its host with regard to how the caterpillar should behave. In the scenarios above, I assumed the focal organism had control and that the other player had a fixed, hardwired growth-allocation strategy. Of course, this was a simplification and one would like to predict outcomes when

both players are able to respond to the tactics of its opponent. Below, I briefly describe two methodologies by which this can be accomplished while still retaining state dependence and environmental stochasticity, (1) dynamic games and (2) genetic algorithms.

A dynamic game includes state dependent fitness functions that depend both upon the decision(s) of the focal organism as well as the actions of other individuals, which may include both conspecifics and heterospecifics. By contrast, in the models outlined above, pay-offs for all decisions were independent of the actions of others. An example of how the non-game and game approaches contrast comes from some recent work we completed on sibling cannibalism on ladybird beetle eggs (Perry and Roitberg 2006). Imagine that an individual larva must choose to consume or not consume an about-to-hatch sibling. Eating a brother or sister provides energy and can be particularly important when prey are in short supply and energy reserves are low (the state dependent aspect) but those benefits must be considered in lieu of inclusive fitness. One can calculate the optimal sibling-attack decision of a focal individual based upon Hamilton's inclusive fitness theory (Hamilton 1964) wherein the benefit from not consuming that sibling comes from common genes that are added to the population gene pool. Imagine, however, that other siblings can also choose to eat the very sibling that our focal individual chooses to grant reprieve. Now, the expected pay-off for this reprieve decision may decline because of the possible actions of others and this might favor higher sibling cannibalism than expected otherwise; this is exactly what we found when we allowed for a sibling attack game among brother and sister ladybird larvae. Further, the game context often includes the frequency of behaviors expressed by a population; thus, if sibling cannibalism is rare then the pay-off for the reprieve is higher than it would be otherwise. Thus, rather than calculating the optimum, one seeks the Evolutionary Stable Strategy (ESS), an equilibrium condition, once adopted by the population, is not invasive to individuals with alternate behaviors. A dynamic game also retains the state dependence as explicated above when solving for the ESS.

How would we develop a dynamic game for parasite–host interactions with behavioral manipulation? The key here is to seek a state dependent Nash equilibrium wherein neither the host nor the parasite can perform any better by altering its strategy. A good example of this approach that is relevant to the problem here can be found in (Wolf and Mangel 2007) for a game between predatory orca whales and sea lions. Orcas are more successful the further they forage from haul out (terrestrial resting) areas so sea lions should forage close to haul out sites. On the other hand, access to food for sea lions increases with distance to haul out. Wolf and Mangel developed separate DSV equations for the predator and its prey taking the actions of the heterospecifics into account. From there, they plotted fitness curves from the optimal response of each player to different values expressed from the other player. Where the two lines crossed there is a Nash equilibrium.

A similar approach can be employed in our parasite–host system. One particular trade-off for the host would be the maintenance of an energy-reserved based immune function along with its concomitant cost. Here the strategy is to optimize the level of immune function based upon the level of host energy stores (see Lazzaro and Little 2009). By contrast, the parasite strategy would be size or energy dependent extraction rates. High virulence (extraction rates) by the parasite will favor higher investment in immunity and vice versa. A full version of this dynamic game will be published elsewhere.

Another approach that is gaining in popularity is the use of genetic algorithms or GAs (Goldberg 1989). GAs are search algorithms that are used to efficiently search complex fitness surfaces. Genetic algorithms are inspired by, and borrow terminology from, evolutionary biology but they do not contain genes as we know them and the terminology that is employed with GAs (e.g., mutation, crossover) is only analogous to DNA-based evolution; the underlying genetics is not reflected in these models. They are particularly useful for solving complex problems, especially those with frequency dependent pay-offs as might be the case in dynamic parasite–host games.

In GAs, strategies are encoded in binary strings or "genotypes," which are then subjected to some fitness determining simulation. The better performing genotypes contribute more to the next generation via a mating routine that allows for crossing over and mutation. As such, heritable variation is generated and the introduction of this variability in strategies via these pathways ensures that regions of solution space around fitness optima are explored. GAs can rapidly evolve toward stable optima or stable strategies, often in less than 1,000 generations.

Determining the outcome of host manipulation in an evolutionary game would require that we construct strategies for both the host and the parasite as discussed above for the dynamic game. For example, for the host, there would be a state-dependent immunity-investment strategy with each "allele" being a different investment value. The problem that one faces when including state dependence is the vast, almost impossible, number of combinations of alleles that could play against one another in a game within and between the two species. For example, suppose there were ten different possible energy-reserve states and four different proportional investment alleles for each state. For this simple state-dependent problem, we already have 4^{10} or more than 1,000,000 possible unique strategies. And, even though the GA only considers a subset of all possible strategies during any one generation, one can see that the problem will get out of hand very quickly. The way that we deal with this problem is to consider the immune-response curves themselves and not the individual points along the curve; now it is the sign and shape parameters that are the strategies under selection. For example, for the problem above, we might have four different curves that describe the immune response to energy state. This approach reduces the size of the problem but retains its essence. In the assessment section of the simulation, hosts and parasites interact and realize fitness based upon their relative performance. The models are run until evolutionary stability occurs, stable cycles are generated, or until some maximum number of generations is run without reaching a stable solution.

The final step in the process is to examine the process itself to gain insight. For example, as the simulation proceeds and poor strategies are weeded out, one can track performance of any given strategy for particular frequencies and against different genetic backgrounds. For example, we might find that high virulence is favored when that strategy is rare but not when common depending upon the distribution of immune strategies in the host population. Again, a full-scale analysis of this approach is beyond the scope of this chapter. A warning to novices: try to keep the problem simple. While it is possible to include all sorts of details in GAs, like any experiment, the more details that are permitted the lower will be the resolution.

4.3 Discussion

Host manipulation by parasites can generate a wide array of spectacular extended phenotypes (Dawkins 1982) from hairworm-induced hydrophilia in crickets (Thomas et al. 2002a, see Chapter 2) to nematode-induced feminization of mayflies (Vance 1996), to bodyguard induction in caterpillars by parasitic wasps (Harvey 2008). These examples seem obvious and greatly help us understand parasite–host interactions both from proximate and ultimate perspectives (Lefèvre et al. 2009) but there are many more examples where the effects are much more subtle and may be more difficult to untangle.

In this chapter, I explored an area that has received limited attention to date, the interplay between environmental heterogeneity and state dependence from the parasite, host, and behavioral ecology perspective (also, see Lazzaro and Little (2009) for discussion on environmental variation and immunity). Nearly ten years ago, Thomas and colleagues (Thomas et al. 2002b) issued a call for state dependent approaches to parasite–host interactions, and since then progress has been made, most notably the dynamic energy budget work of Hall, Nisbet, and associates as discussed earlier. What I have added to this discussion is an allowance for the caterpillar either under its own control or under control of its parasite to either alter its use of incoming nutrients or alter its rate of nutrient acquisition by altering foraging rate or moving to new resource sites. As noted earlier, the behavioral ecology literature provides many examples of such tactics for healthy hosts so it seems reasonable that parasites could usurp (Brodeur and Vet 1994) such typical behavior.

The approach I have taken in this chapter is to show how one might think (and model) about these sorts or problems and as such I have avoided details associated with any one system. There is value in this first step but I also acknowledge that any predictions that emerged from my analysis may be difficult to test. On the other hand, it should not be so difficult to parameterize these models and test them as we have recently done with a dynamic danger–reward trade-off problem for parasitic wasps (Roitberg et al. 2010b).

As our models of animal behavior become more realistic they also become more complicated and more difficult to analyze. Recall that analytical solutions to DSV models are often not possible and thus, we rely on numerical solutions via backwards induction. At first glance, it is hard to imagine how one might incorporate such models into general, analytical theories regarding higher levels of organization, for example at the population or community level. On the other hand, if generalizations can be found in these dynamic models, these relationships can be easily incorporated into our theories of higher level processes. Similar approaches have been taken with regard to the response of animals to danger and their non-consumptive effects (e.g., Preisser and Bolnick 2008).

Finally, though I presented my results as single point optima this can be problematic when it comes to making predictions regarding patterns that we might predict in nature. Recall, the discussion in Section 4.1 when I argued that it is important to understand the fitness surface regarding trait expression. In some cases, I expect that the optimum lies on a relatively flat fitness surface and there would be little cost in making sub-optimal energy extraction or allocation decisions (e.g., Roitberg and Mangel 1997; Lewis et al. 2010). Also, note that when I provided the infected caterpillars with different options for managing energy budgets, the environmental heterogeneity imposed much lower penalties for making errors and in essence levels the playing field.

The world is complicated and we still have some way to go in understanding complex interactions among organisms. An acknowledgement of these complexities is a good first step in deriving theory that captures the essence of the problem in a crisp, cogent manner.

Acknowledgements

This work was supported by a Discovery Grant from NSERC, Canada. Thanks to Alex Chubaty and Mick Wu for help with computer code.

References

Adamo, S. A. (1998). Feeding suppression in the tobacco hornworm, Manduca sexta: costs and benefits to the parasitic wasp *Cotesia congregata*. *Canadian Journal of Zoology* **76**, 1634–1640.

Alphen, van J. J. and Visser, M. (1990). Superparasitism as an adaptive strategy for insect parasitoids. *Annual Review of Entomology* **35**, 59–79.

Anderson R. A. and Koella J. C. (2000). *Plasmodium falciparum* sporozoites increase feeding-associated mortality of their mosquito hosts *Anopheles gambiae* s.l. *Parasitology* **120**, 329–333.

Bellman, R. (1957). *Dynamic Programming*. Princeton University. Press, Princeton.

Benesh, D., Duclos, B., and Nickol, B. (2005). The behavioral response of amphipods haboring *Corynosoma constrictum* to various components of light. *Journal of Parasitology* **91**, 731–736.

Brodeur, J. and Vet, L. E. M. (1994). Usurpation of host behaviour by a parasitic wasp. *Animal Behaviour* **48**, 187–192.

Brown, S. P. (1999). Cooperation and conflict in host-manipulating parasites. *Proceedings of the Royal Socieity of London B* **266**, 1899–1904.

Carmona, D., Lajeunesse, M., and Johnson, M. (2011). Plant traits that predict resistance to herbivores. *Functional Ecology* **25**, 358–367.

Chittka, L. and Skorupski, P. (2011). Information processing in miniature brains. *Proceedings of the Royal Society of London B* **278**, 885–888.

Clark, C. W. (1994). Antipredator Behavior and the Asset-Protection Principle. *Behavioral Ecology* **2**, 159–170.

Clark C. W. and Mangel, M. (2000). *Dynamic State Variable Models in Ecology—Methods and Applications*. Oxford University Press, New York.

Dawkins, R. K. (1982). *The Extended Phenotype*. Oxford University Press, Oxford.

Day, T. (2001). Parasite transmission modes and the evolution of virulence. *Evolution* **55**, 2389–2400.

Diamond, S. and Kingsolver, J. (2010). Host plant quality, selection history and trade-offs shape the immune responses of *Manduca sexta*. *Proceedings of the Royal Society of London B* **278**, 289–297.

Ewald, P. (1994). *Evolution of Infectious Diseases*. Oxford University Press, Oxford.

Fenton, A., Magoolagan, L., Kennedy, Z., and Spencer, K. (2011). Parasite-induced warning coloration: a novel form of host manipulation. *Animal Behaviour* **81**, 417–422.

Fenton, A. and Rands, S. A. (2006). The impact of parasite manipulation and predator forging behavior on predator-prey communities. *Ecology* **87**, 2832–2841.

Franks, S. and Weiss, A. (2008). A change in climate causes rapid evolution of multiple life-history traits and their interactions in an annual plant. *Journal of Evolutionary Biology* **21**, 1321–1334.

Gandon, S. (2005). Parasite manipulation: A theoretical framework might help. *Behavioral Processes* **68**, 247–248.

Goldberg, D. E. (1989). *Genetic Algorithms in Search, Optimization and Machine Learning*. Reading, Mass, Addison-Wesley.

Hall, S., Simonis, J., Nisbet, R., Tessier, A., and Caceres, C. (2009). Resource ecology of virulence in a planktonic host-parasite system: An explanation using dynamic energy budgets. *American Naturalist* **174**, 149 162.

Hamilton, W. D. (1964). The genetical evolution of social behavior, I, II. *Journal of Theoretical Biology* **7**, 1–16, 17–52.

Harvey, J., Kos, M., Nakamatsu, Y., Tanaka, T., Dicke, M., Vet, L. E. M., Brodeur, J., and Bezemer, M. (2008). Do parasitized caterpillars protect their parasitoids from hyperparasitoids? A test of the "usurpation hypothesis." *Animal Behaviour* **76**, 701–708.

Houston, A., Clark, C., McNamara, J., and Mangel, M. (1988). Dynamic models in behavioural and evolutionary ecology. *Nature* **332**, 29–34.

Karban, R., Agrawal, A. A., and Mangel, M. (1997). The benefits of induced defenses against herbivores. *Ecology* **78**, 1351–1355.

Lafferty, K. D. (1992). Foraging on prey that are modified by parasites. *American Naturalist* **140**, 854–867.

Lalonde, R. and Roitberg, BD (2011). Behavioral ecology. In: *Encyclopedia of Theoretical Ecology* (ed. Hastings, A.) University of California Press, Berkeley.

Lazzaro, B. and Little, T. (2009). Immunity in a variable world. *Philosophical Transactions of the Royal Society of London B* **364**, 15–26.

Lefèvre, T., Adamo, S. A., Biron, D., Missé, D., Hughes, D., and Thomas, F. (2009). Invasion of the Body Snatchers: The Diversity and Evolution of Manipulative Strategies in Host–Parasite Interactions. *Advances in Parasitology* **68**, 46–83.

Lewis, H., Tosh, C., O'Keefe, S. Shuker, D. West, S., and Mayhew, P. (2010). Constraints on adaptation: explaining deviation from optimal sex ratio using artificial neural networks. *Journal of Evolutionary Biology* **23**, 1708–1719.

Libersat, F., Delago, A., and Gal, R. (2009). Manipulation of host behavior by parasitic insects and insect parasites. *Annual Review of Entomology* **54**, 189–207.

Lima, S. and Dill, L. (1990). Behavioral decisions made under the risk of predation: a review and prospectus. *Canadian Journal of Zoology* **68**, 619–640.

Mangel, M. (1989). Evolution of host selection in parasitoids: Does the state of the parasitoid matter? *Amercian Naturalist* **133**, 688–705.

Mangel M. and Clark, C. (1988). *Dynamic Modeling in Behavioral Ecology*. Princeton University Press, Princeton NJ.

Nesse, R. (2005). Maladaptation and natural selection. *Quaterly Review of Biology* **80**, 62–70.

Parker, G. A., Ball, M., Chubb, J., Hammerschmidt, K., and Milinski, M. (2009) When should a trophically transmitted parasite manipulate its host? *Evolution* **63**, 448–458.

Perry, J. and Roitberg, B. (2006). Sibling rivalry and the evolution of trophic eggs. *Journal of Evolutionary Biology* **18**, 1523–1533.

Peterson, J. and Roitberg, B. (2010). Egg maturation, nest state and sex ratios: A DSV model. *Evolutionary Ecology Research* **12**, 347–361.

Preisser, E. and Bolnick, D. (2008). The many faces of fear: Comparing the pathways and impacts of nonconsumptive predator effects on prey populations. *PLoS One* **3**, e2465.

Pursall, E. and Rolff, J. (2011). Immune responses accelerate ageing: Proof-of-principle in an insect model. *PLoS One* **6**, e19972.

Real, L. and Caraco, T. (1986). Risk and foraging in stochastic environments. *Annual Review of Ecology and Systematics* **17**, 371–390.

Robinson, M. (1998). Investigation of behaviour changes in parasitized *Drosophila melanogaster* larvae. *BioSciences*. Burnaby, Simon Fraser University. MSc: 92.

Roitberg, B., Keiser, Z., and Hoffmeister, T. (2010a). State dependent attack persistence in a mosquito. *Physiological Entomology* **35**, 46–51.

Roitberg, B. and Mangel, M. (1997). Individuals on the landscape: behavior can mitigate differences among habitats. *Oikos* **80**, 234–240.

Roitberg, B. and Mangel, M. (2010). Mosquito biting and movement rates as an emergent community property and the implications for malarial interventions. *Israel Journal Ecology and Evolution* **56**, 297–312.

Roitberg, B., Robertson, I., and Tyreman, J. (1999). Vive la variance: A new theory for the evolution of host selection. *Entomologia Experimentalis et Applicata* **91**, 187–194.

Roitberg, B., Zimmerman, K., and Hoffmeister, T. (2010b). Dynamic response to danger in a parasitoid wasp. *Behavioral Ecology Sociobiololy* **64**, 627–637.

Rosenberg, D. and McKelvey, K. (1999). Estimation of habitat selection for central-place foraging animals. *Journal of Wildlife Manage* **63**, 1028–1038.

Rosenheim, J. A., Alon, U., and Shinar, G. (2010). Evolutionary balancing of fitness-limiting factors. *American Naturalist* **175**, 662–674.

Scriber, M. and Slansky, F. Jr. (1981). The nutritional ecology of immature insects. *Annual Review of Entomology* **26**, 183–211.

Thomas, F., Brown, S., Sukhdeo, M., and Renaud, F. (2002b). Understanding parasite strategies: a state-dependent approach? *Trends in Parasitology* **18**, 387–390.

Thomas, F., Renaud, F., and Poulin, R. (1998). Exploitation of manipulators: hitch-hiking as a parasite transmission strategy. *Animal Behaviour* **56**, 199–206.

Thomas, F., Schmidt-Rhaesa, A., Martin, G., Manu, C., Durand, P., and Renaud, F. (2002a). Do hairworms (Nematomorpha) manipulate the water seeking behaviour of their terrestrial hosts? *Journal of Evolutionary Biology* **15**, 356–361.

Vance, S. (1996). Morphological and behavioural sex reversal in mermithid-infected mayflies. *Proceedings of Royal Society of London B* **263**, 907–912.

Wolf, N. and Mangel, M. (2007). Strategy, compromise, and cheating in predator–prey games. *Evolutionary Ecology Research* **9**, 1293–1304.

Afterword

Frédérique Dubois

Host–parasite relationships are usually quite complex, becoming difficult to examine and quantify. Yet, understanding and predicting the dynamics of parasites and their hosts requires knowledge about which mechanisms cause behavioral changes in parasitized hosts, as well as how the interplay between host defences and parasite manipulation strategies contributes to maintaining variation among host and parasite individuals. Model predictions in the work developed by Bernard D. Roitberg strongly suggest that dynamic models that take into account physiological differences among individuals could be highly relevant for addressing both issues. As a consequence, this chapter represents an important contribution that opens exciting new avenues of investigation towards a better understanding of the complex interactions between hosts and parasites.

When hosts are parasitized, they almost inevitably modify their behavior compared to healthy individuals in various ways. In particular, infection may affect the hosts' feeding behavior, their susceptibility to predators, as well as their reproductive capacity. It follows that in many situations, the changes induced by the presence of a parasite appear maladaptive, as they negatively impact the fitness of the hosts, and researchers have favored the parasite manipulation hypothesis to account for these alterations. However, even when the host behavioral changes benefit the parasite, they are not necessarily induced only by the parasite that manipulates the behavior of its host, but can also reflect, partly or exclusively, an adaptive response of the host aimed at reducing or preventing the deleterious effect of the parasite (Poulin 2007). Accordingly, there is increasing evidence that several physiological traits, such as feeding rate, might evolve as a consequence of parasite pressure, even if the parasites do not manipulate the behavior of their host. Among other mechanisms, immunological defences, for example, can result in reduced feeding when there is a tradeoff between digestion and immune function (Adamo et al. 2010).

Because the mechanisms potentially involved in host manipulation have different evolutionary consequences, a key element is to evaluate their relative importance in order to make reliable predictions about host–parasite dynamics over time. Determining which of these mechanisms are responsible for the host behavioural alteration, however, requires identifying and measuring both the costs and the benefits for the host, which may be a daunting task. In many situations, as a first approach, state-dependent models can provide a simpler way to discriminate among different hypotheses. Indeed, as they consider interindividual differences (notably in physiological state) they can in some cases generate different expectations regarding how each category of individuals should respond to changes in their environment. Results from the models developed by Roitberg in this chapter support this finding. Roitberg's models predict that infested caterpillars should be more conservative than healthy individuals, notably by allocating less energy to growth and more to reserves due to the parasite extraction constraint. However, he found that the optimal decision of host categories differed depending on whether they were in control of their own bodies or under control of the parasite.

The same approach could also be used to determine if hosts that tend to reduce their feeding when infected by a parasite do so because they are manipulated by the parasite or because they are following the activation of immune defences. Intuitively, we would expect that hosts with low energy reserves should modify their behavior less when compared to those with high energy reserves, if the change reflects a trade-off between digestion and immune function. This difference would arise because only individuals with high energy reserves can reduce their feeding rate without dying of starvation or impairing their reproductive success. Alternatively, if the induced-parasite alteration in host behavior results from parasite manipulation, hosts with low energy reserves, that can consequently invest less energy to resist manipulation, should be more affected by the parasite manipulation and hence should decrease their feeding rate to a greater extent compared to hosts with high energy reserves.

Considering inter-individual differences in host physiological state can, then, be relevant to evaluate the relative importance of each mechanism potentially involved, but can also provide a simpler explanation for why and how variance in host defences and parasite strategies are maintained within populations. Indeed, a commonly proposed hypothesis for variance in host defences considers that it may occur when hosts face many different parasite species, if protection against one parasite does not provide protection to the others (Parker et al. 2011). Variance in parasite strategies, however, could simply arise because their hosts are capable of dynamically adjusting their behavior to local conditions (e.g., depending on their physiological state or the level of environmental heterogeneity), thereby forcing the parasites to modify their behavior as well in order to maximize their fitness. In particular, given that parasite manipulation is costly, it has

been suggested that the amount of effort a parasite should expend on host manipulation should vary according to its expected success (in terms of transmission rate). For instance, when hosts are infected by several parasites, intra-host competition leads to a decrease in the parasite investment because parasites that do not manipulate host behavior can exploit the manipulative effort of others without paying the cost of manipulation (Vickery and Poulin 2010). Similarly, inter-individual differences in host behavior within a population should influence the mean parasite investment in host manipulation and could even favor, under certain conditions, the persistence of different parasite phenotypes. This could be so notably when environmental conditions change frequently, thereby leading to frequent changes in the relative proportion of hosts with high and low energy reserves. Dynamic models will be useful to address this issue and make predictions regarding the conditions under which variation among individuals should be maintained, not only in host–parasite systems but more generally in any system in which individuals may have conflicting interests.

References

Adamo, S. A., Bartlett, A., Le, J., Spencer, N., and Sullivan, K. (2010). Illness-induced anorexia may reduce trade-offs between digestion and immune function. *Animal Behaviour* **79**, 3–10.

Parker B. J., Barribeau, S. M., Laughton, A. M., de Roode, J. C., and Gerardo, N. M. (2011). Non-immunological defense in a evolutionary framework. *Trends in Ecology and Evolution* **26**, 242–248.

Poulin, R. (2007). *Evolutionary Ecology of Parasites*, 2nd edition. Princeton University Press, Princeton.

Vickery, W. L. and Poulin, R. (2010). The evolution of host manipulation by parasites: a game theory analysis. *Evolutionary Ecology* **24**, 773–788.

CHAPTER 5

Manipulation of plant phenotypes by insects and insect-borne pathogens

Mark C. Mescher

5.1 Introduction

Over the past few decades, the manipulation of host phenotypes by parasites has come to be recognized as a widespread phenomenon with potentially far-reaching implications for ecology (Lefèvre et al. 2009), human health (Hurd 2003; Lefèvre et al. 2006), and agriculture (Eigenbrode et al. 2002; Mauck et al. 2010). There are now numerous well-documented examples of the manipulation of plant phenotypes by various antagonists (Roy 1994; Lill and Marquis 2007; Grant and Jones 2009; Verhage et al. 2010). But the broader literature on parasite manipulation to date has tended to focus on effects on animal hosts and especially on the manipulation of animal behavior (e.g., Poulin 2010), so that the occurrence of manipulation in plant systems has not always received attention as a general phenomenon commensurate with its potential significance for both basic and applied science. For example, Lefèvre et al. (2009) recently published an—otherwise excellent and insightful—paper addressing the potentially far-reaching ecological significance of manipulative parasites without discussing a single, real or hypothetical, example of parasitic manipulation of plant phenotypes. Yet, plants undeniably play a central role in each of the processes through which these authors propose that parasites may influence key ecological and evolutionary processes—including impacts on the outcome of interspecific competition, changes in food-web structure and energy and nutrient flows, and habitat creation.

Plants dominate the biomass of terrestrial ecosystems and reside at the center of most community interaction webs; and their immense taxonomic and ecological diversity is dwarfed only by the diversity of the interactions in which they engage with one another and with other organisms. Beyond their direct interactions with other species, plants play key roles in structuring the environments in which those species reside. As primary producers, plants obviously play a major role in regulating the flow of energy and nutrients within ecosystems. And vegetation structure largely delineates the three-dimensional environment in which inter and intraspecific interactions among other organisms take place. Furthermore, because plants engage in continuous gas exchange with the surrounding atmosphere and release complex blends of volatile compounds that vary in systematic ways with a variety of biotic and abiotic factors, they play key roles in shaping the olfactory landscape for other organisms (e.g., Pickett et al. 1992; De Moraes et al. 1998). And their impact on visual environments is similarly profound (Döring and Chitka 2007). Consequently, manipulative parasites of plants are particularly well placed to provide examples of ecosystem engineering—a term coined by Jones et al. (1994, 1997) to describe the activities that have strong impacts on the availability of resources to other species and play key roles in the creation, modification, and maintenance of habitats (see also Chapter 9).

Plants typically engage simultaneously in multiple interactions with a broad range of other organisms, and a significant proportion of these interactions are essentially parasitic. For example, Shivas and Hyde (1997) estimated that there may be 270,000 species of plant pathogenic fungi in the tropics alone. Moreover, there are likely at least several million species of arthropod herbivores on earth, the vast majority of which are fundamentally parasitic in their interactions with host plants, along with untold numbers of microbial plant pathogens, including nematodes, bacteria, viruses, and others. There are even more than 4,000 species of flowering plants (around 1% of all angiosperms) known to be parasitic on other plants (Parker and Riches 1993). It is likely that a large proportion of the interactions of these organisms with their host plants involves some degree of manipulation, broadly defined as changes in the phenotype of the host that confer fitness advantages upon the parasite (Brown 2005; Thomas et al. 2005). Yet our attempts to catalogue these manipulative interactions remain in their infancy, let alone our understanding of their potentially far-reaching effects on biological communities and ecosystems.

As the manipulation of plant phenotypes by other organisms has rarely been addressed as a general phenomenon, this chapter provides an overview of some of the clearest and best-studied examples of parasite manipulation in plant systems. Despite how much remains to be discovered about such interactions, it is beyond the scope of a single book chapter to document the fascinating detail of manipulative interactions that have been described for plants and their parasites. Thus, to narrow the focus somewhat, I will specifically address examples of manipulation that involve plant–insect interactions, including both those in which the insects themselves are manipulators of their host plants (as in the case of gall-inducing herbivores) and those in which the interaction between plant and insect is the target of manipulation (as in the case of insect-vectored plant pathogens).

5.2 Plant manipulation by insect herbivores

Insects are among the most important plant antagonists, and plants and insect herbivores are engaged in an ancient and seemingly endless struggle of adaptation and counter adaptation. Plants defend themselves against insects through physical defenses such as trichomes and spines—which in

addition to directly deterring feeding may also play a role in sensing the presence of herbivores and inducing appropriate defenses—and also deploy an array of constitutive and induced chemical defenses to poison or otherwise deter their arthropod enemies. Plants even respond to herbivory through the emission of volatile signals that serve as foraging cues for the natural enemies of feeding herbivores. In response, insects have developed their own mechanisms for overcoming or evading plants physical and chemical defenses, for example by detoxifying the chemical defenses of plants or in some cases even sequestering them for their own defense against predators. But beyond simply countering the defenses of host plants, many insects actively alter host plant phenotypes in ways that improve the quality of the host as a resource.

In the insect literature, the general phenomenon in which herbivore feeding causes plants to become better hosts for the herbivores has been called induced plant susceptibility (Karban and Baldwin 1997), which includes effects such as the regulation of plant ontogeny by herbivores (e.g., Fisher 1987). In this section, I review some of the most prominent and best-studied examples of herbivore modification of host plant phenotypes, including the induction of galls, direct modifications of plant structure by arthropods, and the manipulation of phytohormone signaling pathways.

Where these modifications of host plants enhance the fitness of the herbivores they constitute examples of adaptive manipulations and, to the extent that they have been shaped by selection acting on herbivore genes, as extended phenotypes (Dawkins 1982). They can also profitably be viewed within the broader context of adaptive effects of insects on their biotic and abiotic environments (e.g., Danks 2002). And to the extent that such modifications have broader, community and ecosystem-level implications—for example, where induced changes in plant chemistry affect their suitability for other organisms (Agrawal 1999) or rates of litter decomposition and nutrient cycling (Choudhury 1988)—they may represent potentially important examples of ecosystem engineering (Jones et al. 1994, 1997).

5.2.1 Gall-inducing insects

Plant galls are abnormal tissue growths (Fig. 5.1) that are induced by disparate groups of insects and mites, as well as by some other plant parasites, including fungal and bacterial plant pathogens and even some parasitic vascular plants (mistletoe). In some gall systems multiple plant antagonists act in concert, as in the case of ambrosia galls in which the internal walls of galls induced by a gall midge are lined by the mycelia of a fungal symbiont that is inoculated into plant tissue by the wasp at the time of egg-laying and serves as a food source for the developing wasp larvae (Bissett and Borkent 1988; Rohfritsch 1997). In insects, the galling habit has evolved numerous times in different groups, occurring in at least six insect orders and exhibiting multiple independent origins within most of these (Meyer 1987; Dreger-Jauffret and Shorthouse 1992). In total as many as 21,000 insect species may induce plant galls (Espirito-Santo and Fernandes 2007). Gall-inducing insects are invariably specialists on particular plant species and generally also specialize on particular tissues and structural locations within host plants—some gall-inducing aphids even exhibit territoriality and intense competition for prime galling sites (e.g., Whitham 1979).

The structures of insect-derived galls range from relatively simple tissue invaginations to tumor-like growths that completely enclose the gall-inducer and often exhibit highly differentiated layers of plant tissue development and complex morphologies (reviewed in Stone and Schönrogge 2003). Despite their diverse origins, gall-inducing insects share a number of common characteristics and face similar ecological challenges. The sedentary lifestyle of gall-inducers during the developmental period—and their physical localization within the structure of the gall—requires that gall-inducing species satisfy all of their nutritional and other requirements in a highly-constrained environment, in turn requiring galls to be highly efficient and sophisticated at exploiting host resources. Given

(a)   (b)

Figure 5.1 Plant galls induced by two different herbivores, the tephritid fly *Eurosta solidaginis* and the gelechiid moth *Gnorimoschema gallaesolidaginis*, on the same host plant, *Solidago altissima* (goldenrod). Used with permission from Dr. John Tooker.

these challenges and the intimacy of their physiological relationships to host plants it is perhaps not surprising that the apparent far-reaching effects of gallers on host plant phenotypes are unparalleled among other classes of in sect herbivores (Tooker et al. 2008).

Galling as an adaptive manipulation

As early as 1889, Romanes suggested that plant galls should be interpreted as manipulations benefiting the parasitic organisms that elicit them. As discussed by Weis and Abrahamson (1986; Abrahamson and Weis 1997), this interpretation was proposed in response to the contention that plant galls provided an example of a structure produced by one species exclusively for the good of another (Mivart 1889)—something Darwin (1859) had explicitly stated could not be explained by his theory of natural selection. A competing adaptationist explanation posited a defensive role for galls as plant adaptations for isolating and encapsulating plant antagonists (Cockerell 1890). The hypothesis that galls might primarily be defensive adaptations of plants retained currency well into the twentieth century (Mani 1964; Price et al. 1987), but there is now considerable evidence that gall development is largely controlled by the genes of the galling insects and that galling indeed constitutes an adaptive manipulation of the host phenotype (reviewed in Price et al. 1987; Stone and Schönrogge 2003).

For example, phylogenetic analyses of a number of galling taxa suggest that patterns in gall morphology correspond to patterns of genetic relatedness among gallers rather than similarities in the host plant (or plant tissues) galled—that is, closely related gallers often induce similar galls even on unrelated host plants (Stern 1995; Nyman et al. 2000; Cook et al. 2002). Conversely, different species of gallers often produce very different but characteristic gall morphologies on the same host (Nyman et al. 2000). The induction of galls also typically depends on chemical stimuli that are actively conveyed to plants by the galling insects (e.g., in secretions injected during oviposition by sawflies or by salivary secretions in aphids; Shorthouse and Rohfritsch 1992)—although this alone doesn't rule out primary agency by the host

plant as other plant defense responses are similarly elicited by herbivore-derived compounds (e.g., Alborn et al. 1997), which presumably serve some other essential function for the herbivores. Nevertheless, many features of galls seem unambiguously adaptive from the gallers' point of view. For example, galls often exhibit highly differentiated tissue types that are not present in ungalled plants, including nutritive tissues that provide high quality food sources for the developing insect (Shorthouse and Rohfritsch 1992; Stone and Schönrogge 2003; Raman et al. 2005). Moreover, these interior gall tissues often contain few defensive compounds, while the exterior of the galls often have a high concentration of secondary metabolites (Abrahamson et al. 1991; Hartley 1998; Nyman and Julkunen-Tiitto 2000; Allison and Schultz 2005). Some species even induce the production of extrafloral nectaries, absent from ungalled tissues, which function to recruit ants that defend the gall from natural enemies (Seibert 1993). Additional evidence that galls constitute valuable resources for their residents comes from the observation that they are often occupied for more than one generation and may be defended, or even repaired, by their residents (Crespi 1992; Inbar 1998; Kurosu et al. 2003).

In addition, while gall-inducing species appear in general to be somewhat less damaging to their hosts than other endophagous insects such as leaf-miners and wood borers (Raman et al. 2005), there is ample evidence that galls harm their host plants (Stone and Schönrogge 2003; Raman et al. 2005). Gallers can alter plant biomass allocations, elevate photosynthesis rates in affected tissues, and mobilize resources from neighboring tissues (Hartnett and Abrahamson 1979; Fay et al. 1993; McCrea et al. 1985; Whitham 1992). Consequently galls are typically energy and nutrient sinks (Raman and Abrahamson 1995), which deprive the host plant of resources that it could otherwise employ for growth and reproduction (McCrea et al. 1985; Abrahamson and Weis 1997; Bronner 1992). Furthermore, indirect evidence of the costs imposed by galls is provided by the documentation of specific host plant responses to galler attack (e.g., Whitham 1992; Fernandes 1998; Rausher 2001).

Adaptive features of plant galls

The benefits of the galling habit for insect parasites have generally been thought to fall in three categories: nutritional enhancement, creation of controlled microenvironments, and protection from natural enemies (Price et al. 1987; Stone and Schönrogge 2003). These posited benefits are obviously not mutually exclusive, and while there is some debate about the relative importance of these functions in driving the evolution of the galling habit, it seems clear that various features of the design of plant galls have been modified for each of these purposes.

Enhanced nutritional quality is widely cited as an adaptive feature of plant galls (Price et al. 1987; Hartley and Lawton 1992). As noted above, many galls exhibit differentiated nutritive tissues that appear to be more nourishing and less well defended than ungalled tissues on the same plant. These tissues can exhibit characteristic cytological features including cell wall modifications, cytoplasm enrichment, hypertrophied nuclei, and abundant organelles (Bronner 1992; Raman et al. 2005), and their expression appears to be tightly linked to the ongoing feeding activity of the galler: in the absence of feeding the nutritive tissue ceases activity and is replaced by hypertrophied parenchyma tissue (Raman 1987). Galls inhabited by social thrips and aphids (Foster and Northcott 1994; Crespi et al. 1998) also frequently exhibit morphological characteristics, such as internal tissue folds, that apparently function to increase the surface area available to feeding insects (Crespi and Worobey 1998). Enhanced nutritional quality of gall tissues is not universally reported (e.g., Anderson and Mizell 1987; Brewer et al. 1987; Hartley and Lawton 1992), and relatively few studies have explored the adaptive significance of gall nutritional characteristics through controlled experiments, but those that have frequently found results consistent with the hypothesis that gallers successfully manipulate the nutritional quality of galled tissues (e.g., Larson and Whitham 1991; Hartley and Lawton 1992; Hartley 1998; Diamond et al. 2008). For example, Larson and Whitham (1991) used ^{14}C labeling to show that aphid-induced galls on cottonwood leaves are strong physiological sinks that draw resources from surrounding plant tissues. More recently, Diamond et al. (2008) transplanted larvae of the non-galling but endophagous lepidopteran herbivore *Hellinsia glenni* into the nutrient-rich galls induced on that plant by the tephritid fly *Eurosta solidaginis* and found that the larvae attained greater final mass than controls transplanted into ungalled stems (their normal developmental environment).

Shelter from adverse abiotic conditions is another factor that has often been cited as an adaptive function of galls. Protection from the direct effects of temperature is not likely to be a major factor, however, as galls, like other interior plant tissues, appear to closely track ambient temperatures and do so on relatively rapid time scales (Levitt 1980; Price et al. 1987). Moreover, gall-inducing species themselves seem to be as well adapted in their overwintering strategies as free-living species (Baust et al. 1979). In contrast, protection from desiccation and from the adverse effects of low humidity on herbivore feeding efficiency—especially in hot dry environments—are suspected to be important advantages of the galling habit (Price et al. 1987). Insects feeding in enclosed galls are bathed in host-plant fluids, and partially closed galls lie within the boundary layers of relative moist air surrounding plant structures (Shorthouse and Rohfritsch 1992; Stone and Schönrogge 2003). Other potential benefits include protection from UV radiation, flooding, or physical detachment from the host plant (e.g., by wind, precipitation, or other types of physical disturbance) (Price et al. 1987; Kurosu and Aoki 1998; Miller et al. 2009). To an even greater extent than for nutritional features, relatively few studies have explicitly tested the hypothesis that galls provide adaptive benefits to their residents by providing controlled microenvironments. However, Miller et al. (2009) found that larvae of the cynipid wasp galler *Andricus quercuscali fornicus* (Bassett) on oak trees in California, *Quercus lobata Née* (Fagaceae), were able to consistently maintain the interior humidity of their galls near saturation over the course of development and demonstrated experimentally that reduced humidity levels adversely impacted survival.

Protection from natural enemies is a third frequently suggested adaptive benefit of the galling habit. Galls clearly provide concealment and some degree of physical protection against generalist predators since, at least in the case of enclosed galls, attackers must first penetrate the tissues of the gall itself. Moreover, gall tissues are frequently enriched in secondary metabolites (e.g., phenolics and tannins) relative to non-galled tissues, and the accumulation of these defensive compounds has been interpreted as an adaptive defense of the gall-inducer (e.g., Abrahamson et al. 1991; Hartley 1998; Allison and Schultz 2005). Various structural characteristics of galls also appear to enhance defense, including increased tissue thickness and hardness and the presence of defensive hairs (reviewed in Stone and Schönrogge 2003). As noted above, some galls also exhibit extrafloral nectaries that recruit ants as defensive mutualists (e.g., Seibert 1993). It has even been proposed that the bright, conspicuous coloration of many galls may serve an aposematic function, as a warning to visually foraging predators such as birds (Inbar et al. 2010).

There is also evidence that gall-inducing herbivores can influence indirect plant defenses mediated by the emission (in response to herbivory) of plant volatiles that serve as key foraging cues for natural enemies of the feeding herbivores. In at least one system, volatiles induced by galling were found to be attractive to the galler's natural enemies (Tooker and Hanks 2006)—and, given that galls suffer high rates of successful parasitism and that hymenopteran and dipteran parasitoids are typically sophisticated in their exploitation of olfactory cues, it seems likely that such interactions are widespread. Nevertheless, in two other systems, gall-inducing dipteran larvae were found not to significantly induce volatile emissions (Tooker and De Moraes 2007; Tooker et al. 2008); moreover, in each of these systems galled plants also failed to emit volatiles in response to herbivory by generalist herbivores that induced strong responses on ungalled plants, suggesting that gall inducers might actively suppress induced volatile responses.

Despite these apparently adaptive defensive traits, galls clearly do not represent an "enemy free space," as gallers are targeted by a variety of natural enemies including fungi, hymenopteran and dipteran parasitoids, and beetles, which often inflict high levels of mortality (Price et al. 1987; Waring and Price 1989; Zwölfer and Arnold-Rinehart 1994; Abrahamson and Weis 1997; Stone and Schönrogge 2003). There is evidence that gall-inducing sawflies are attacked by fewer parasitoid species and experience lower mortality than non-galling species (Price and Pschorn-Walcher 1988), but this pattern does not appear to hold generally as many gall-inducing species are attacked by more parasitoids than their free-living relatives (Tscharntke 1994). Consequently, Stone and Schönrogge (2003) concluded that there was insufficient evidence to support the hypothesis that defense against natural enemies is a major factor driving the evolution of the galling habit, though they suggest that secondary defensive adaptations are a likely explanation for the evolution of structural diversity among galls.

5.2.2 Structural modification of host plants

Shelter-building herbivores
Many non-galling insect herbivores actively manipulate the phenotypes of their host plants in slightly less dramatic but analogous ways. Some modify the architecture of plant foliage to construct shelters—often by rolling or folding leaves, binding leaves together with silk, or enclosing them within webs. In addition to creating protected feeding sites, such activities can induce changes in plant tissues that enhance their quality as a food source for the herbivores. Diverse groups of arthropods use plant foliage for shelter construction, including sawflies, beetles, ants, spiders, thrips, and tree crickets, but the shelter-building habit appears to be most widespread among Lepidopterans and is particularly prominent among the Microlepidoptera (Lill and Marquis 2007). The adaptive benefits proposed to derive from shelter building fall into the same broad categories as those attributed to gall induction, including the enhancement of plant tissue quality for feeding, amelioration of harsh environmental conditions, and protection from natural enemies (reviewed in Lill and Marquis 2007)—and as in the

case of galls these potential benefits are not mutually exclusive.

One interesting feature of leaf-rolling and some other forms of shelter construction is the exclusion of light, which can result in advantageous (to the insect) changes in plant tissues located in the interior of the shelter. For example, caterpillars from the families Psyralidae and Ctenuchidae roll leaves around an expanding bud of their host plants. In the neotropical shrub, *Psychotria horizontalis*, the resultant shading—light levels in the interior of the leaf roll were reduced by as much as 95%—was found to significantly reduce the toughness and tannin concentrations of expanding leaves, while nutritional quality (measured as nitrogen concentration and water content) was unaffected (Sagers 1992). Light-mediated impacts on structural defenses (e.g., lignins and leaf toughness) and some other constitutive chemical defenses (e.g., polyphenolics and tannins) may be maximized when shelters are constructed with young, expanding foliage, because the light environment can have significant impacts on the development of these defense traits (Lill and Marquis 2007). Light exclusion can also allow herbivores to evade the effects of plant defenses that are directly mediated by light. For example, St. John's wort, *Hypericum perforatum* (Guttiferae), produces a compound (hypericin) that acts as a photosensitizing agent and thus as a "phototoxin" that is effective against both vertebrate and invertebrate herbivores. But larvae of the tortricid moth *Platynota flavedana* are able to develop on this plant by feeding within shelters constructed by tying leaves (Sandberg and Berenbaum 1989).

In addition to effects mediated through interaction with the host plant, shelter building arthropods may benefit directly from reduced exposure to intense solar radiation or other harsh abiotic conditions. There is considerable evidence that both architectural features of shelters and their siting and orientation on the host plant may act to regulate microenvironmental conditions (Alonso 1997; Lill and Marquis 2007). The threat of desiccation poses a significant challenge for many insects, and shelters can both ameliorate the drying effects of wind and create humid microenvironments through the transpiration of the foliage from which they are constructed, suggesting that regulation of relative humidity may be a major benefit of shelter construction (Lill and Marquis 2007); however, the prevalence of shelter building in humid tropical environments (e.g., Janzen 1988; Diniz and Morais 1997) argues that this is not the primary factor driving the evolution of the shelter building habit. Temperature is another obvious microclimatic factor influencing insect performance and there is evidence that features of shelter construction can influence thermal dynamics (Hensen 1958). Some species, such as tent-building social caterpillars (Costa 1997) construct large and complex shelters that exhibit significant internal variation in temperature and other conditions that may be behaviorally exploited by occupants.

Protection from natural enemies is another frequently cited potential benefit of shelter construction, and a large number of studies have documented the protective functions of shelters against diverse natural enemies (reviewed in Lill and Marquis 2007). Avoidance of detection by predators is the most frequently discussed factor, and a key benefit of feeding within shelters may be the concealment of feeding behaviors from visually oriented predators that are highly attuned to movement (Bernays 1997). Consequently, shelter building species may often be able to engage in more continuous feeding than free-living herbivores (Wagner 2005), allowing rapid development which has the further benefit of reducing the temporal window of exposure to natural enemies (Lill and Marquis 2001, 2007). But while shelters may frequently provide protection against generalist predators, shelter residents can also suffer high rates of predation and parasitism by natural enemies that exploit shelter-associated cues in order to locate prey. For example, many bird species actively forage on leaf rolls and other insect-constructed structures recognized on the basis of visual cues (e.g., Greenberg 1987; Murakami 1999). Shelter dwelling insects can also suffer high rates of attack by specialized insect parasitoids (e.g., Lill 1999), and as in the case of gall-induction (discussed above) there is currently little evidence to suggest that rates of parasitism are generally lower on shelter-dwelling herbivores relative to free-living forms (Lill and Marquis 2007).

As with plant galls, evidence that arthropod-constructed shelters provide a valuable resource can be inferred from the observation that they are frequently colonized by members of subsequent generations or other species. Shelters constructed by one insect herbivore are often secondarily occupied by another (or by predators), and there is evidence that shelter builders have positive effects on arthropod diversity. Cappuccino and Martin (1994) found that removing early season shelter builders on paper birch enhanced the abundance of later season species—which tended to colonize existing shelters when available in preference to constructing their own. This positive effect on abundance carried over into the next year. Martinsen et al. (2000) also reported significantly higher levels of herbivore richness and abundance on cottonwood shoots bearing shelters constructed by the leaf-roller *Anacampsis niveopulvella* than on adjacent shoots without shelters.

Other structural changes imposed by insect herbivores
There are a number of other ways in which insect herbivores actively modify structural features of the host plant in order to enhance its quality as a resource. Canal cutting insects strategically sever leaf veins containing secretory canals that transport defensive exudates including toxins, antifeedants, and resins or other sticky secretions. These insects, which include multiple lineages of lepidopteran larvae, beetles, and katydids then feed on leaf tissues distal to the site of the cut—thus, canal cutting allows insects to access plant resources, and perhaps to exploit host species, that would otherwise be inaccessible (Dussourd 1999). Dussourd and Denno (1991) examined the canal cutting behavior of 33 species on diverse canal-bearing plants and found a very tight correlation between the insects' behavior and the architecture of the canal system: vein cutting occurs on plants that exhibit an arborescent (i.e., branching) pattern of secretory canals, while trenching—in which the feeding insect makes a linear cut through a portion or all of the leaf blade—is typical on plants exhibiting net-like canal systems. Many insects facultatively exhibit vein

cutting or trenching behaviors only on plants possessing secretory canals; where these behaviors occur on non-canal bearing plants they are most often associated with the folding or rolling of leaves to construct shelters, as discussed above (Dussourd 1993; Fitzgerald 1995). Some herbivores exhibit similarly complex behavioral adaptations to overcome physical plant defenses such as trichomes and spines. For example, some Lepidopteran larvae mow (i.e., remove) plant trichomes prior to feeding (e.g., Hulley 1988) and others create a silk scaffolding to cover them (e.g., Rathcke and Poole 1975).

5.2.3 Green islands

The creation of so-called green islands by some endophagous, leaf-mining lepidopteran larvae (Wild 1976; Walters et al. 2008) represents another class of manipulations that is somewhat analogous to the induction of galls—the similarity in this case residing in effects on the distribution and mobilization of plant resources (and in the apparent biochemical mechanisms of induction) rather than on the modification of plant structure. Green islands are characterized by photosynthetically active patches of green tissue surrounding feeding sites that can be observed in otherwise senescing leaves and which appear to provide an enriched nutritional environment for the feeding herbivores which induce them, prolonging the availability of the food source and potentially enabling the completion of a supplementary generation (Giron et al. 2007). It has been suggested that green islands surrounding leafminer feeding sites might reflect inputs of nitrogen from frass and respiration (Wild 1976), but this does not seem to explain why only some leaf-mining Lepidoptera induce green island formation (Plant 1984). Similar phenomena are induced by the presence of some other, non-insect plant parasites, including fungal and bacterial pathogens (Engelbrecht et al. 1969; Walters and McRoberts 2006; Giron et al. 2007; Walters et al. 2008; Kaiser et al. 2010). Recent work also suggests that the formation of green islands in conjunction with the feeding of leaf-mining insects may be mediated by bacterial symbionts: Kaiser et al. (2010) used antibiotics to eliminate endosymbiotic bacteria from leaf-mining larvae of the moth *Phyllonorycter blancardella* and observed a subsequent failure to induce green islands. These authors pointed to *Wolbachia* as the bacterial partner most likely responsible for mediating the induction of green islands and suggested that the symbiotic partnership facilitates exploitation of the limited resources presented by the spatially constrained feeding style of leaf-miners, which must complete development within the epidermal layers of a single leaf.

While the detailed molecular and biochemical mechanisms of green island formation are not well known, there is evidence that the accumulation of cytokinins plays a key role in the formation of green islands and furthermore that these cytokinins are produced by the herbivore (or associated symbionts) rather than being plant derived (Englebrecht et al. 1969; Giron et al. 2007). Cytokinins have also frequently been implicated in gall formation (Elzen 1983). Their potential involvement in the manipulation of plant phenotypes by herbivores is intriguing given that cytokinins are plant hormones implicated in the inhibition of senescence, maintenance of chlorophyll and the regulation of nutrient mobilization (Gan and Amasino 1995; Balibrea Lara et al. 2004; Walters and McRoberts 2006).

5.2.4 Manipulation of phytohormones

A detailed discussion of the biochemical and molecular mechanisms underlying each of the manipulations above is beyond the scope of the current chapter—and, as noted, our current knowledge of such mechanisms is limited—but it is likely that these mechanisms often involve the disruption (or co-option) of internal plant signaling mechanisms, as suggested by the involvement of cytokinins in the induction of green islands and galls. Moreover, given the complexity of the biochemical adaptation and counteradaptation apparent in interactions between herbivores and host plants, it is likely that biochemical manipulation of host plants is widespread in such interactions and not restricted only to the more spectacular and obvious examples discussed above. To date, the clearest examples of herbivore manipulation of plant signaling responses have been documented for phloem-feeding insects.

Plants responses to herbivore attack (or pathogen infection) are regulated by signal transduction pathways that regulate changes in the expression of suites of genes that provide appropriate responses to the particular type of antagonist encountered. These signaling pathways are governed by phytohormones, including salicylic acid (SA), jasmonates (JA), and ethylene, each of which has been shown to play a key role in the regulation of plant defense responses (Bari and Jones 2009)—a variety of other signaling molecules, including abscisic acid, auxins, gibberellins, and brassinosteroids interact with these primary defense signals in order to regulate and direct defense responses (see Verhage et al. 2010 for further discussion and additional references).

There is considerable antagonistic and synergistic interaction between these signaling pathways (Koornneef and Pieterse 2008), which presumably allows plants to more effectively fine tune defense responses to specific antagonists and to avoid the costly expression of ineffective defenses that divert resources away from growth and reproduction. But interaction and cross-talk among these signaling pathways also creates opportunities for manipulation by insect herbivores and pathogens (Walling 2008; Grant and Jones 2009; Verhage et al. 2010). One of the clearest examples of such manipulation involves the exploitation of mutual inhibition between SA- and JA-mediated signaling pathways by silverleaf whitefly, *Bemisia tabaci* biotype B (Kempema et al. 2007; Zarate et al. 2007). By inducing SA, this insect, which feeds on plant phloem, appears to suppress the induction (by *Arabidopsis*) of JA-mediated defenses, which mediate effective defenses against this herbivore (Zarate et al. 2007) and have been broadly implicated in the induction of direct and indirect defenses against herbivory (Walling 2008). Similar forms of manipulation may be widespread among phloem-feeding insects (Thompson and Goggin 2006; Walling 2008). For example, the salivary secretions of the aphid *Myzus persicae* contain glucose oxidase (Harmel et al. 2008), which can directly suppress JA production by inhibiting lipoxygenase activity but also converts glucose to gluconic acid and hydrogen peroxide, which may stimulate SA accumulation (Vandenabeele et al.

2003). Glucose oxidase is also present in the saliva of some leaf-chewing insects, and similar suppression of JA-dependent defenses through the activation of the SA pathway has been reported for the beet army worm (*Spodoptera exigua*)—glucose oxidase in the saliva of the caterpillar Helicoverpa zea was also previously shown to suppress induced resistance in tobacco (*Nicotiana tabacum*). Recent work even suggests that elicitors from insect eggs may induce SA responses with consequent benefits for the feeding larvae that emerge from them (Bruessow et al. 2010). Thus, there is considerable evidence that insect herbivores actively manipulate plant defense responses through "decoy" strategies. Many analogous examples of apparent manipulation of phytohormone defense responses have been reported for microbial parasites (see Grant and Jones 2009; Verhage et al. 2010).

5.3 Plant manipulation by insect-borne pathogens

The sessile lifestyles of plants pose challenges for sexual reproduction and for the dispersal of offspring. Some plants overcome these challenges by exploiting features of the abiotic environment, producing pollen that is readily dispersed by the wind and seeds that can drift on the wind or be carried by water away from the parent—a few species even engage in explosive seed dispersal in which specialized organs are employed to launch seeds as reproductive ballistic missiles. But many other plant species have developed adaptations that allow them to harness the mobility and dispersing capacity of animals (often insects), either through mutualistic interactions—as in the case of plants which provide rewards to insect pollinators that subsequently carry plant DNA from one individual to another, or to fruit-feeding animals that disperse seeds—or though more exploitative interactions such as deceptive pollination or the production of seeds that attach themselves to mammalian fur.

The immobility of plants poses similar problems for parasitic organisms that live obligately in or on plants and need to be transmitted (or to convey gametes) from one host individual to another. Many

of these parasites have evolved similar strategies for overcoming these challenges, including both dispersal via wind and water and dispersal via biotic interactions with insects that are analogous to—and in some cases represent outright exploitation of—the strategies employed by plants.

5.3.1 Manipulation of plant–pollinator interactions by fungal parasites

Some of the most striking and well-documented examples of parasite manipulation of plant phenotypes are produced by fungal plant pathogens that exploit pre-existing plant–pollinator interactions in order to achieve either out-crossed fertilization through the transfer of gametes between different fungal genotypes or reproduction through the dissemination of infectious propagules by the insects—some non-plant-associated fungi such as stinkhorns also use pollinators to disperse their spores. The diverse mechanisms by which fungi exploit insect pollinators were reviewed by Roy (1994). Ngugi and Scherm (2006) later discussed many of these same systems in the context of mimicry (i.e., by fungal pathogens of cues normally presented to insect pollinators by plants).

Figure 5.2 Pseudoflowers (center) formed by the rust fungus *Puccinia monoica* on *Boechera drummondii* (= *Arabis drummondii*). The (true) flowers to the left and right are *Thlaspi spp*. Used with permission from Dr. Bitty Roy.

In some cases fungi themselves produce the visual and olfactory cues that mimic—or at least function analogously to—the pollinator attractions systems of the host (or other) plants (Roy 1994). For example, rust fungi often exhibit bright red, orange, or yellow coloration (Roy 1993, 1994), while stinkhorns can be bright orange-red or exhibit high-contrast patterns of dark and light coloration (Roy 1994). Fungi may also produce sweet-smelling, or otherwise attractive, scents and may even produce nectar-like rewards for insect pollinators (Roy 1993).

But fungal parasites also frequently engage in active manipulation of host plant phenotypes in order to achieve pollinator attraction, as exemplified by rust fungi in the genera *Puccinia* and *Uromyces*, which induce modifications in floral and vegetative structures of their host plants to produce pseudoflowers (Fig. 5.2) that function to transmit fungal gametes rather than host pollen (Roy 1993; Pfunder and Roy 2000; Naef et al. 2002). Heterothalic rust species have multiple mating types and require the transfer of haploid gametes (spermatia) for outcrossing. In heterothalic species that cause non-systemic plant infections, different mating types may be present in close proximity to one another on the same leaf so that the transfer of spermatia between plants is not required for sexual reproduction—instead fertilization may occur when insects walk across the leaf or even by direct contact of fungal hyphae (Craigie 1927). But in species that induce systemic infections all the spermatial lesions on the host plant are typically generated by the same fungal genotype and thus have the same mating type. Thus, heterothalic species causing systemic infections require the transmission of spermatia among plants, and these fungi have evolved complex adaptions to achieve transmission via insects (Pfunder and Roy 2000), including mimicry of plant floral structures in which the fungi present their spermatia in a sugary nectar-like secretions that are attractive to insects (Roy 1993).

For example, the rust *Puccinia monoica* induces its host, the herbaceous mustard *Arabis holboellii*, to cease flowering—parasitic castration is a common feature of plant fungal infections—and instead produce elongated stems crowned by rosettes of bright yellow infected leaves that bear little resemblance to the flowers of the host plant but do resemble those of unrelated but co-occurring yellow-flowered species such as buttercups (Roy 1993). Another rust species, *Uromyces pisi* induces the production of similar pseudoflowers by its host *Euphorbia cyparissias* (Euphorbiaceae); Pfunder and Roy (2000) confirmed the function of these pseudoflowers as fungal adaptations for insect attraction through insect exclusion experiments that demonstrated the dependence of the fungus on insect visitation for cross-fertilization of mating types. Using a similar approach, Naef et al. (2002) found that the exclusion of flying visitors significantly reduced the reproductive success of *P. arrhenatheri* on common Barberry (*Berberis vulgaris*) compared to natural insect visitation, while the exclusion of crawling insects had no significant effect, suggesting that flying insects (mainly dipterans) are the most important gamete vectors, and that crawling insects, such as ants, are of minor importance.

In addition to being similar in form to co-occurring flowers, the pseudoflowers induced by these rust fungi reflect light at wavelengths that are indistinguishable in both the ultra-violet (UV) and visible spectra from flowers of co-occurring species (Ngugi and Scherm 2006). They also attract insects through the production of "floral" scents (Roy 1993; Raguso and Roy 1998; Pfunder and Roy 2000; Naef et al. 2002). Raguso and Roy (1998) analyzed the floral emissions of pseudoflowers induced by rusts in the *Puccinia monoica* complex on several species of cruciferous plants in the mustard genus *Arabis* and found them to be composed primarily of aromatic alcohols, aldehydes, and esters. Interestingly, while the fragrances identified bore similarity to those of some noctuid-moth-pollinated flowers, they were not similar either to the fragrances of host plant flowers or to those of co-blooming flowers of non-host angiosperms. Thus, the scents produced by such pseudoflowers appear to mimic those of true flowers functionally, but not chemically (Ngugi and Scherm 2006). A similar pattern was reported for the pseudoflowers of *P. arrhenatheri*-infected *Berberis vulgaris*, which emitted a blend of compounds including indole, methyl nicotinate, α-phellandrene, carvacryl methyl ether, and jasmin lactone (Naef et al. 2002). While many of the fragrant compounds emitted by pseudoflowers may be produced by the fungal parasite, at least some compounds in the blends emitted are suspected to be of host origin (Naef et al. 2002). Characteristic olfactory cues have also been shown to mediate the transfer of gametes among heterothallic endophytes in the genus *Epichloe* by Anthomyiid flies that are attracted to fungal stromata presented on host stems (Schiestl et al. 2006; Steinebrunner et al. 2008a). Two specific compounds isolated have been identified as attractive to flies in this system, methyl (Z)-3-methyldodec-2-enoate, and chokol K. The latter was present in volatile collections from fungi grown in culture, confirming its fungal origin, while the former was not, suggesting the involvement of host tissues—or at least of stimuli or precursors provided by the host—in its production (Steinebrunner et al. 2008b).

Roy and Raguso (1997) explored the relative importance of olfactory and visual cues in attracting insects to pseudoflowers of *Puccinia monoica* Arth. on *Arabis drummondii* Gray. They demonstrated that pseudoflower fragrance alone—presented in the absence of visual cues—was attractive. In further experiments employing arrays of artificial flowers with and without fragrance they determined that the relative importance of olfactory and visual cues varies among taxa of insect visitors: Halictid bees (*Dialictus* sp.) had a somewhat greater visual than olfactory response, whereas flies (muscids and anthomyiids) were more dependent on olfactory cues. There is also evidence that visual cues may be relatively more important for long-distance attraction, while olfactory cues play a stronger role in short-distance homing responses (Roy and Raguso 1997; Raguso and Roy 1998).

The olfactory and visual cues induced by heterothalic rusts function in the transmission of fungal gametes between mating types. But, some fungi accomplish similar manipulations in order to transmit infectious propagules that will establish novel infections when transmitted to true flowers or other uninfected host tissues. For example, the fungal pathogen *Monilinia vaccinii-corymbosi* which causes mummy berry disease in blueberries (*Vaccinium* spp.) and huckleberries (*Gaylussacia* spp.) initially infects the vegetative tissues of its hosts via windborne sexual ascospores that are released in the early spring from fungal sclerotia that overwinter in the soil in mummified berries, but asexual spores (conidia) on infected leaves must thus then be transferred to flowers to complete the life cycle of the parasite. Batra and Batra (1985) demonstrated that the discolored leaves of host plants infected by *Monilinia ascospores* become UV reflective at wavelengths that are similar to blueberry floral calcyes and present a high contrast to surrounding (UV-absorbent) leafy vegetation. Moreover, infected leaves emit fragrant odors and provide sugary rewards to insect pollinators. Exclusion of insects resulted in a dramatic reduction in the incidence of subsequent fruit infection (Batra and Batra 1985). Similar flower mimicry facilitating transmission of fungal conidia from leaves to flowers appears to occur in a number of other host-specific species in the *Disjunctorae* species group of *Monilinia* (Batra 1991; Ngugi and Scherm 2006).

Fungal parasites that depend on insects to transmit spores (rather than gametes) from infected to healthy flowers may gain no advantage in rendering infected flowers more attractive than—or even distinguishable from—healthy flowers, since effective transmission occurs when insect pollinators frequent flowers of both types (Roy 1994). It is thus not surprising that some fungal pathogens exhibiting this lifestyle exploit the pre-existing cues that host flowers present to foraging insects. Anther smut fungi sporulate in the anthers of dicot flowers and are subsequently transmitted by insect pollinators (Roy 1994). The most studied anther smut, *Ustilago violacea* Pers (= *Microbotryum violaceum* Pers), is an obligate parasite of many species in the Caryophyllaceae. Flowers infected by this fungus retain their attractiveness to insect pollinators, but the anthers bear fungal teliospores in place of pollen. Thus, the fungus appears to mimic only the anthers of the host rather than the entire flower (Ngugi and Scherm 2006). But while insect attraction to flowers infected by *U. violacea* and other anther smuts is often elicited by the same cues that are presented by healthy flowers, these fungi have been found to have a variety of apparently manipulative effects on the phenotypes of their hosts (Roy 1994; Shykoff and Kaltz 1998; Ngugi and Scherm 2006). Anther smuts partially or completely sterilize their hosts, and can alter host flowering phenology and floral morphology (e.g., Lee 1981; Biere and Honders 1996; Shykoff and Kaltz 1998). For example, smut-infected plants often flower earlier and produce more blossoms than uninfected ones (Jennersten 1988; Alexander and Maltby 1990), and infected flowers often remain open longer than uninfected ones (Jennersten 1988). In the case of *U. violacea*, Shykoff and Kaltz (1998) found that infection accelerated flowering in the weed *Silene latifolia* (Caryophyllaceae) and reduced investment in root biomass. Moreover, the flowers of infected plants were smaller and exhibited altered nectar sugar production compared to those of healthy plants. In a different *Silene* species, Altizer et al. (1998) found that bumblebees preferentially visited healthy flowers—though the strength of this preference was reduced if bees were given prior experience with diseased flowers—while nocturnal moths showed no significant preference for healthy or infected plants.

Finally, it is worth noting that in systems where parasite-induced changes of plant traits influencing the attraction of insects is not so dramatic as to obviously provide evidence for manipulation, interpretation of observed differences in attractiveness can be challenging, since infection could plausibly be either the cause of such differences or an effect (e.g., if more attractive individuals become infected at higher rates)—an observation discussed for fungal pathogens (Shykoff and Bucheli 1995; Shykoff et al. 1997) that likely holds equally well for other classes of putative manipulation by vector-borne parasites.

5.3.2 Pathogen manipulation of plant–herbivore interactions

In addition to exploiting plant–pollinator interactions, many fungal and (other) microbial pathogens are transmitted from one plant to another by insect herbivores that feed on infected plants, and these plant–insect interactions are likewise likely targets for manipulation by pathogens. It is clear that pathogen effects on host phenotypes can have important implications for the frequency and nature of interactions between host and host-feeding insect vectors, and such interactions have recently received increasing attention within the literature addressing parasite manipulation in animal systems (e.g., Hurd 2003; Lefèvre et al. 2006), although to date sig-

nificantly more attention has been paid to manipulation leading to trophic transmission (e.g., Berdoy et al. 2000).

In both animal and plant disease systems, pathogens transmitted by insects that feed on the primary host may potentially enhance their transmission by influencing (i) host traits (such as nutritional quality) that influence direct interactions with the vector (Ebbert and Nault 2001; Taylor and Hurd 2001; Mauck et al. 2010); and (ii) traits that influence the host-location cues (attractants) presented to foraging vectors (Eigenbrode et al. 2002; McLeod et al. 2005; Lacroix et al. 2005; Mauck et al. 2010). For plant pathogens, the latter are most likely to involve visual and olfactory cues—heat being a third sensory mode that is often critical for animal feeding insects but presumably less important for plant-feeders, while sound, humidity, or other factors might also be involved. Olfactory cues presented by hosts would seem to be a particularly likely target for manipulation by insect-borne pathogens as both plant- and animal-feeding insects typically use volatile chemical cues to locate their hosts (Pickett et al. 1992; De Moraes et al. 1998; Takken and Knols 1999; De Moraes et al. 2001; O'Shea et al. 2002), and pathogen infection is known to alter host odor profiles in both plants and animals with implications for odor-mediated interactions between infected plants and other organisms (Penn and Potts 1998; Cardoza et al. 2002; Kavaliers et al. 2005; McLeod et al. 2005). Effects on host quality for vectors and on the signals or cues presented are likely to be tightly linked: in the fungal systems discussed above, where transmission involves insect pollinators, pathogens can exploit (or elaborate) the existing signaling and reward systems that mediate pre-existing plant–insect interactions. In the case of pathogens transmitted by insect herbivores, patterns of vector attraction and dispersal mediated by changes in visual or olfactory cues will likely depend critically on how infection impacts the value of the host as a resource for the vector (e.g., Mauck et al. 2010).

A relatively small number of studies to date have examined pathogen effects on host-derived foraging cues within the context of adaptive manipulation. In animal systems, odor cues have been implicated in vector attraction to individuals infected with *Leishmania* (in an animal model; O'Shea et al. 2002) and malaria (Lacroix et al. 2005). Among plant systems, one of the clearest examples involves the fungal pathogen *Ophiostoma novo-ulmi* the causative agent of Dutch elm disease, which has severely impacted elm populations in North America where it is vectored by the elm bark beetles *Hylurgopinus rufipes* and *Scolytus multistriatus*. McLeod et al. (2005) examined the volatiles emitted by infected American elms, *Ulmus americana,* and identified four characteristic compounds from infected trees—the monoterpene (-)-β-pinene and three sesquiterpenes—that were attractive to *H. rufipes*. Subsequent laboratory and field assays revealed that the presence of each compound, and at appropriate relative ratios, was required for vector attraction, suggesting that *O. novo-ulmi* induces the emission of a characteristic volatile signature that beetles recognize and use to locate infected trees.

Similar effects have been reported for plant viruses vectored by aphids. Virus-induced changes in leaf color (e.g., yellowing of leaves) have been implicated in aphid attraction to infected plants (Pickett et al. 1992; Eckel et al. 1996; Hooks and Fereres 2006; Döring and Chittka 2007), as have induced changes in plant volatile emissions (Eigenbrode et al. 2002; Jiménez-Martínez et al. 2004; Mauck et al. 2010). A potentially interesting pattern of variation in virus effects may exist between viruses exhibiting different transmission mechanisms. Persistent (circulative) viruses form intimate associations with vector insects—which remain infective over long time periods, but also require prolonged aphid feeding in the phloem of infected plants for effective transmission to the insect. In contrast, non-persistent viruses form only transitory associations with the vector—binding to specific sites on the aphid stylet during probes of infected leaf tissues—but are rapidly acquired and transmitted. Two persistently transmitted viruses *Potato leaf roll virus* (PLRV) and *Barley yellow dwarf virus* (BYDV) have been reported to induce characteristic volatile emissions from host plants (potato and wheat, respectively) that are attractive to aphid vectors, and these viruses were also found to increase plant quality for aphids (e.g., as measured by colony growth rates) (Eigenbrode et al. 2002; Jiménez-Martínez et al. 2004; Ngumbi et al. 2007). In contrast, Mauck et al. (2010) found that aphid population growth on squash plants infected with the

non-persistent *Cucumber mosaic virus* (CMV) were strongly reduced compared to healthy plants and that aphids rapidly dispersed from infected to healthy plants in choice assays—subsequent (unpublished) work showed that CMV infection reduces levels of key amino acids in the phloem and stimulates the induction of anti-herbivore defense responses. Surprisingly, aphids (*Myzus persicae* and *Aphis gossypii*) were nevertheless attracted to the odors of infected plants, a finding that Mauck et al. (2010) attributed to the induction by CMV of greatly elevated emissions of a volatile blend otherwise similar to that of healthy plants. Thus, the very few studies currently available have reported very different effects on plant-vector interactions for persistent and non-persistent viruses, but the pattern of effects observed in each case seems highly conducive to virus transmission—suggesting that viruses may manipulate plant traits that influence interactions with vectors and that variation in virus transmission mechanisms may be an important factor shaping the evolution of such effects. It is too early to tell whether this pattern will hold more generally, but an analysis of existing literature on virus effects on vector attraction, dispersal, and performance on infected plants (Mauck et al. forthcoming) indicates that persistent and non-persistent viruses tend to have relatively positive and negative effects (respectively) on host plant quality (e.g., Ajayi and Dewar 1983; Blua and Perring 1992a,b; Castle and Berger 1993; Belliure et al. 2005; Donaldson and Gratton 2007; Jiu et al. 2007; Mauck et al. 2010), while positive aphid attraction to viruses of both types has been reported (e.g., Ajayi and Dewar 1983; Eckel et al. 1996; Eigenbrode et al. 2002; Jiménez-Martínez et al. 2004; Mauck et al. 2010)—interestingly, reports of aphid discrimination against virus-infected plants (of either type) via long-range olfactory or visual cues appear to be quite rare.

5.4 Conclusion

Manipulation of plant phenotypes is less well known than that in animal systems but likely to be of broad ecological significance, given the key ecological role of plants in almost all terrestrial ecosystems. Many plant antagonists modify plant phenotypes in ways that enhance their own fitness, and in so doing either directly manipulate or indirectly influence interactions between their host plants and other organisms. Beyond relatively conspicuous examples of manipulation, such as fungal pseudoflowers and plant galls, unseen manipulation of plant biochemistry by herbivores and pathogens is likely widespread, as suggested by the examples of herbivore (and pathogen) effects on plant signaling pathways discussed above. In addition to influencing resource availability and habitat quality for other organisms, the physical and chemical changes in host plants induced by manipulative parasites likely have significant implications for flows of energy and nutrients within ecosystems (see also Chapter 9). But, despite diverse and well-documented examples of manipulation in plant systems, our current knowledge of the prevalence of such interactions—and the specific biochemical and molecular mechanisms involved—is limited. And we are only beginning to explore the broader implications for biological communities and ecosystem-level processes. Future efforts will likely benefit from increased attention to manipulation as a general phenomenon in plant ecology and from conceptual integration of the literature addressing manipulation in plant and animal systems.

Acknowledgements

I wish to thank Consuelo De Moraes, Bitty Roy, and John Tooker for feedback and contributions to the text.

References

Abrahamson, W. G. and Weis, A. E. (1997). *Evolutionary Ecology Across Three trophic Levels: Goldenrods, Gallmakers, and Natural Enemies.* Princeton University Press, Princeton, NJ.

Abrahamson, W. G., McCrea, K. D., Whitwell, A. J., and Vernieri, L. A. (1991). The role of phenolics in goldenrod ball gall resistance and formation. *Biochemical Systematics and Ecology* **19**, 615–622.

Agrawal, A. (1999). Induced responses to herbivory in wild radish: Effects on several herbivores and plant fitness. *Ecology* **80**, 1713–1723.

Ajayi, O. and Dewar, A. M. (1983). The effect of barley yellow dwarf virus on field populations of the cereal aphids, *Sitobion avenae* and *Metopolophium dirhodum*. *Annals of Applied Biology* **103**, 1–11.

Alborn, H. T., Turlings, T. C. J., Jones, T. H., Stenhagen, G., Loughrin, J. H., and Tumlinson, J. H. (1997). An elicitor of plant volatiles from Beet Armyworm oral secretion. *Science* **276**, 945–949.

Alexander, H. M., and Maltby, A. (1990). Anther-smut infection of Silerle alba caused by Ustilago violacea: factors determining fungal reproduction. *Oecologia* **84**, 249–253.

Allison, S. D. and Schultz, J. C. (2005). Biochemical responses of chestnut oak to a galling cynipid. *Journal of Chemical Ecology* **31**, 151–166.

Alonso, C. (1997). Choosing a place to grow. Importance of within-plant abiotic microenvironment for Yponomeuta mahalabella. *Entomologia Experimentalis et Applicata* **83**,171–180.

Altizer, S. M., Thrall, P. H., and Antonovics, J. (1998). Vector behavior and the transmission of anther-smut infection in Silene alba. *American Midland Naturalist* **139**, 147–163.

Anderson, P. C. and Mizell, R. F. (1987). Physiological effects of galls induced by *Phylloxera notabilis* (Homoptera: Phylloxeridae) on Pecan foliage. *Environmental Entomology* **16**, 264–268.

Balibrea Lara, M. E., Gonzalez Garcia, M. C., Fatima, T., Ehness, R., Lee, T. K., Proels, R., Tanner, W., and Roitsch, T. (2004). Extracellular invertase is an essential component of cytokinin-mediated delay of senescence. *The Plant Cell* **16**, 1276–1287.

Bari, R. and Jones, J. D. G. (2009). Role of plant hormones in plant defense responses. *Plant Molecular Biology* **69**, 473–488.

Batra, L. R. (1991). World Species of Monilinia (Fungi): Their Ecology, Biosystematics and Control. Mycologia Memoir, No. 16. J. Cramer, Berlin.

Batra, L. R. and Batra, S. W. T. (1985). Floral mimicry induced by mummy-berry fungus exploits host's pollinators as vectors. *Science* **228**, 1011–1013.

Baust, J. G., Grandee, R., Condon, G., and Morrissey, R. E. (1979). The diversity of overwintering strategies utilized by separate populations of gall insects. *Physiological Zoology* **52**, 572–580.

Belliure, B., Janssen, A., Maris, P. C., Peters, D., and Sabelis, M. W. (2005). Herbivore arthropods benefit from vectoring plant viruses. *Ecology Letters* **8**, 70–79.

Berdoy, M., Webster, J. P., and Macdonald, D. W. (2000). Fatal attraction in rats infected with *Toxoplasma gondii*. *Proceedings of the Royal Society B: Biological Sciences* **267**, 1591–1594.

Bernays, E. A. (1997). Feeding by lepidopteran larvae is dangerous. *Ecological Entomology* **22**, 121–123.

Biere, A. and Honders, S. (1996). Host adaptation in the anther smut fungus *Ustilago violacea* (*Microbotryum violaceum*): infection success, spore production and alteration of floral traits on two host species and their F1-hybrid. *Oecologia* **107**, 307–320.

Bissett, J., and Borkent, A. (1988). Ambrosia galls: the significance of fungal nutrition in the evolution of the Cecidomyiidae (Diptera). In: *Coevolution of Fungi with Plants and Animals* (eds Pirozynski, K. A., and Hawksworth, D. L.). Academic Press, London.

Blua, M. J. and Perring, T. M. (1992a). Alatae production and population increase of aphid vectors on virus-infected host plants. *Oecologia* **92**, 65–70.

Blua, M. J. and Perring, T. M. (1992b). Effects of zucchini yellow mosaic virus on colonization and feeding behavior of *Aphis gossypii* (Homoptera: Aphididae) alatae. *Environmental Entomology* **21**, 578–585.

Brewer, J. W., Bishop, J. W., and Skuhravy, V. (1987). Levels of foliar chemicals in insect-induced galls (Diptera: Cecidomyiidae). *Journal of Applied Entomology* **104**, 504–510.

Bronner, R. (1992). The role of nutritive cells in the nutrition of cynipids and cecidomyiids. In: *Biology of Insect-Induced Galls* (eds Shorthouse, J. D. and Rohfritsch, O.) Oxford University Press, New York.

Brown, S. P. (2005). Do all parasites manipulate their hosts? *Behavioural Processes* **68**, 237–240.

Bruessow, F., Gouhier-Darimont, C., Buchala, A., Métraux, J. P., and Reymond, P. (2010). Insect eggs suppress plant defence against chewing herbivores. *The Plant Journal* **62**, 876–885.

Cappuccino, N., and Martin, M.-A. (1994). Eliminating early-season leaf-tiers of paper birch reduced abundance of mid-summer species. *Ecological Entomology* **19**, 399–401.

Cardoza, Y. J., Alborn, H. T., and Tumlinson, J. H. (2002). In vivo volatile emissions of peanut plants induced by fungal infection and insect damage. *Journal of Chemical Ecology* **28**, 161–174.

Castle, S. J. and Berger, P. H. (1993). Rates of growth and increase of *Myzus persicae* on virus-infected potatoes according to type of virus-vector relationship. *Entomologia Experimentalis et Applicata* **69**, 51–60.

Choudhury, D. (1988). Herbivore induced changes in leaf-litter resource quality: a neglected aspect of herbivory in ecosystem nutrient dynamics. *Oikos* **51**, 389–393.

Cockerell, T. D. K. (1890). Galls. *Nature* **42**, 344.

Cook, J. M., Rokas, A., Pagel, M., and Stone, G. N. (2002). Evolutionary shifts between host oak sections and host plant organs in Andricus gallwasps. *Evolution* **56**, 1821–1830.

Costa, J. T. (1997). Caterpillars as social insects. *American Scientist* **85**, 150–159.

Craigie, J. H. (1927). Discovery of the function of pycnia of the rust fungi. *Nature* **120**, 765–767.

Crespi, B. J. (1992). Eusociality in Australian gall thrips. *Nature* **359**, 724–726.

Crespi, B. J. and Worobey, M. (1998). Comparative analysis of gall morphology in Australian gall thrips: the evolution of extended phenotypes. *Evolution International Journal of Organic Evolution* **52**, 1686–1696.

Crespi, B. J., Carmean, D. A., Mound, L. A., Worobey, M., and Morris, D. (1998). Phylogenetics of social behaviour in Australian gall-forming thrips: evidence from mitochondrial DNA sequence, adult morphology and behaviour, and gall morphology. *Molecular Phylogenetics and Evolution* **9**, 163–180.

Danks, H. V. (2002). Modification of adverse conditions by insects. *Oikos* **99**, 10–24.

Darwin, C. (1859). *On The Origin of Species by Means of Natural Selection*. John Murray, Albermarle Street, London.

Dawkins, R. (1982). *The Extended Phenotype: The Gene as the Unit of Selection*. W. H. Freeman and Company, Oxford.

De Moraes, C. M., Mescher, M. C., and Tumlinson, J. H. (2001). Caterpillar-induced nocturnal plant volatiles repel conspecific females. *Nature* **410**, 577–580.

De Moraes, C. M., Lewis, W. J., Paré, P. W., Alborn, H. T., and Tumlinson, J. H. (1998). Herbivore-infested plants selectively attract parasitoids. *Nature* **393**, 570–573.

Diamond, S. E., Blair, C. P., and Abrahamson, W. G. (2008). Testing the nutrition hypothesis for the adaptive nature of insect galls: does a non-adapted herbivore perform better in galls? *Ecological Entomology* **33**, 385–393.

Diniz, I. R. and Morais, H. C. (1997). Lepidopteran caterpillar fauna of cerrado host plants. *Biodiversity and Conservation* **6**, 817–836.

Donaldson, J. R. and Gratton, C. (2007). Antagonistic effects of soybean viruses on soybean aphid performance. *Environmental Entomology* **34**, 918–925.

Döring, T. F. and Chittka, L. (2007). Visual ecology of aphids—a critical review on the role of colours in host finding. *Arthropod-Plant Interactions* **1**, 3–16.

Dreger-Jauffret, F. and Shorthouse, J. D. (1992). Diversity of gall-inducing insects and their galls. In: *Biology of Insect-Induced Galls* (eds Shorthouse, J. D. and Rohfritsch, O.). Oxford University Press, Oxford.

Dussourd, D. E. (1999). Behavioral sabotage of plant defense: do vein cuts and trenches reduce insect exposure to exudate? *Journal of Insect Behavior* **12**, 501–515.

Dussourd, D. E. (1993). Foraging with finesse: caterpillar adaptations for circumventing plant defenses. In: *Caterpillars: Ecological and Evolutionary Constraints on Foraging* (eds Stamp, N. E. and Casey, T. M.) Chapman and Hall, New York.

Dussourd, D. E. and Denno, R. F. (1991). Deactivation of plant defense: correspondence between insect behavior and secretory canal architecture. *Ecology* **72**, 1383–1396.

Ebbert, M. A. and Nault, L. R. (2001). Survival in Dalbulus leafhopper vectors improves after exposure to maize stunting pathogens. *Entomologia Experimentalis et Applicata* **100**, 311–324.

Eckel, W., Randi, V., and Lampert, E. P. (1996). Relative attractiveness of tobacco etch virus-infected and healthy flue-cured tobacco plants to aphids (Homoptera: Aphididae). *Journal of Economic Entomology* **89**, 1017–1027.

Eigenbrode, S. D., Ding, H., Shiel, P., and Berger. P. H. (2002). Volatiles from potato plants infected with potato leafroll virus attract and arrest the virus vector, *Myzus persicae* (Homoptera; Aphididae). *Proceedings of the Royal Society London B: Biological Sciences* **269**, 455–460.

Elzen, G. W. (1983). Cytokinins and insect galls. *Comparative Biochemistry and Physiology Part A: Physiology* **76**, 17–19.

Engelbrecht, L., Orban, U., and Heese, W. (1969). Leaf-miner caterpillars and cytokinins in the green islands of autumn leaves. *Nature* **223**, 319–321.

Espirito-Santo, M. M. and Fernandes, G. W. (2007). How many species of gall-inducing insects are there on Earth, and where are they? *Annals of the Entomological Society of America* **100**, 95–99.

Fay, P. A., Hartnett, D. C., and Knapp, A. K. (1993). Increased photosynthesis and water potentials in Silphium integrifolium galled by cynipid wasps. *Oecologia* **93**, 114–120.

Fernandes, G. W. (1998). Hypersensitivity as a phenotypic basis of plant induced resistance against a galling insect (Diptera: Cecidomyiidae). *Environmental Entomology* **27**, 260–267.

Fisher, M. (1987). The effect of previously infested spruce needles on the growth of the green spruce aphid, *Elatobium abietinum*, and the effect of the aphid on the amino acid balance of the host plant. *Annals of Applied Biology* **111**, 33–41.

Fitzgerald, T. D. (1995). Caterpillars roll their own. *Natural History* **104**, 30–37.

Foster, W. A. and Northcott, P. A. (1994). Galls and the evolution of social behaviour in aphids. In: *Plant Galls: Organisms, Interactions, Populations* (ed. Williams, M. A.J.) Clarendon Press, New York.

Gan, S. and Amasino, R. M. (1995). Inhibition of leaf senescence by autoregulated production of cytokinin. *Science* **270**, 1986–1988.

Giron, D., Kaiser, W., Imbault, N., and Casas, J. (2007). Cytokinin-mediated leaf manipulation by a leafminer caterpillar. *Biology Letters* **3**, 340–343.

Grant, M. R. and Jones, J. D. G. (2009). Hormone (dis)harmony moulds plant health and disease. *Science* **324**, 750–752.

Greenberg, R. (1987). Development of dead leaf foraging in a tropical migrant warbler. *Ecology* **68**,130–141.

Harmel, N., Letocart, E., Cherqui, A., Giordanengo, P., Mazzucchelli, G., Guillonneau, F., De Pauw, E., Haubruge, E., and Francis, F. (2008). Identification of aphid salivary proteins: a proteomic investigation of *Myzus persicae*. *Insect Molecular Biology* **17**, 165–174.

Hartley, S. E. (1998). The chemical composition of plant galls: are levels of nutrients and secondary compounds controlled by the gall former? *Oecologia* **113**, 492–501.

Hartley, S. E. and Lawton, J. H. (1992). Host-plant manipulation by gall-insects: a test of the nutrition hypothesis. *Journal of Animal Ecology* **61**, 113–119.

Hartnett, D. C. and Abrahamson, W. G. (1979). The effect of stem gall insects on life history patterns in Solidago canadensis. *Ecology* **60**, 910–916.

Hensen, W. R. (1958). The effects of radiation on the habitat temperatures of some poplar-inhabiting insects. *Canadian Journal of Zoology* **36**, 463–478.

Hooks, C. R. R. and Fereres, A. (2006). Protecting crops from non-persistently aphid-transmitted viruses: A review on the use of barrier plants as a management tool. *Virus Research* **120**, 1–16.

Hulley, P. E. (1988). Caterpillar attacks plant me- chanical defence by mowing trichomes before feeding. *Ecological Entomology* **13**, 239–241.

Hurd, H. (2003). Manipulation of medically important insect vectors by their parasites. *Annual Review of Entomology* **48**, 141–161.

Inbar M. (1998). Competition, territoriality and maternal defense in a gall-forming aphid. *Ethology Ecology and Evolution* **10**, 159–170.

Inbar, M., Izhaki, I., and Koplovich, A., Lupo, I., Silanikove, N., Glasser, T., Gerchman, Y., Perevolotsky, A., and Lev-Yadun, S. (2010). Why do many galls have conspicuous colors? A new hypothesis. *Arthropod-Plant Interactions* **4**, 1–6.

Janzen, D. H. (1988). Ecological characterization of a Costa Rican dry forest caterpillar fauna. *Biotropica* **20**, 120–135.

Jennersten, O. (1988). Insect dispersal of fungal disease: effects of Ustilago infection on pollinator attraction in *Viscaria vulgaris*. *Oikos* **51**, 163–170.

Jiménez-Martínez, E. S., Bosque-Pérez, N. A., Berger, P. H., Zemetra, R., Ding, H. J., and Eigenbrode, S. D. (2004). Volatile cues influence the response of *Rhopalosiphum padi* (Homoptera: Aphididae) to barley yellow dwarf virus-infected transgenic and untransformed wheat. *Environmental Entomology* **33**, 1207–1216.

Jiu, M., Zhou, X. P., Tong, L., Xu, J., Yang, X., Wan, F. H., and Liu, S-S. (2007). Vector-virus mutualism accelerates population increase of an invasive whitefly. *PLoS One* **2**, e182.

Jones, C. G., Lawton, J. H., and Shachak, M. (1994). Organisms as ecosystem engineers. *Oikos* **69**, 373–386.

Jones, C. G., Lawton, J. H., and Shachak, M. (1997). Positive and negative effects of organisms as ecosystem engineers. *Ecology* **78**, 1946–1957.

Kaiser, W., Huguet, K., Casas, J., Commin, C., and Giron, D. (2010). Plant green-island phenotype induced by leaf-miners is mediated by bacterial symbionts. *Proceedings of the Royal Society B: Biological Sciences* **277**, 2311–2319.

Karban, R. and Baldwin, I. T. (1997). Induced Responses to Herbivory. University of Chicago Press, Chicago, IL.

Kavaliers, M., Choleris, E., and Pfaff, D. W. (2005). Genes, odours and the recognition of parasitized individuals by rodents. *Trends in Parasitology* **21**, 423–429.

Kempema, L. A., Cui, X., Holzer, F. M., and Walling, L. L. (2007). *Arabidopsis Transcriptome* changes in response to phloem-feeding Silverleaf Whitefly nymphs. Similarities and distinctions in responses to aphids. *Plant Physiology* **143**, 849–865.

Koornneef, A. and Pieterse, C. M. J. (2008). Cross talk in defense signaling. *Plant Physiology* **146**, 839–844.

Kurosu, U. and Aoki, S. (1998). Long-lasting galls of Ceratoglyphina styracicola, a host-alternating subtropical aphid species. Nieto Nafría, J. M. and Dixon, A. F. G., (eds). *Aphids in Natural and Managed Ecosystems*. Universidad de León (Secretariado de Publicaciones), Leon (Spain).

Kurosu, U., Aoki, S., and Fukatsu, T. (2003). Self-sacrificing gall repair by aphid nymphs. *Proceedings of the Royal Society London B: Biological Sciences* **270**, S12–S14.

Lacroix, R., Mukabana, W. R., Gouagna, L. C., and Koella, J. C. (2005). Malaria infection increases attractiveness of humans to mosquitoes. *Public Library of Science Biology* **3**, e298.

Larson, K. C. and Whitham, T. G. (1991). Manipulation of food resources by a gall-forming aphid: the physiology of sink-source interactions. *Oecologia* **88**, 15–21.

Lee, J. A. (1981). Variation in the infection of Silene dioica (L.) clairv. by Ustilago violacae (pers.) fuckel in north west England. *New Phytologist* **87**, 81–89.

Lefèvre T., Koella, J. C., Renaud, F., Hurd, H., Biron, D. G., Thomas, F. (2006) New Prospects for Research on Manipulation of Insect Vectors by Pathogens. *Public Library of Science Pathogens*, **2**, e72.

Lefèvre, T., Lebarbenchon, C., Gauthier-Clerc, M., Misse, D., Poulin, R., and Thomas, F. (2009). The ecological significance of manipulative parasites. *Trends in Ecology & Evolution* **24**, 41–48.

Levitt, J. (1980). *Responses of Plants to Environmental Stresses (Physiological Ecology): Chilling, Freezing, and High Temperature Stresses*. Academic Press, New York.

Lill, J. T. (1999). Structure and dynamics of a parasitoid community attacking larvae of Psilocorsis quercicella (Lepidoptera: Oecophoridae). *Environmental Entomology* **28**, 1114–1123.

Lill, J. T. and Marquis, R. J. (2001). The effects of leaf quality on herbivore performance and attack from natural enemies. *Oecologia* **126**, 418–428.

Lill, J. T. and Marquis, R. J. (2007). Microhabitat manipulation: Ecosystem engineering by shelter-building insects. In: *Ecosystem Engineers—Plants to Protists*. Theoretical Ecology Series. Volume 4, 107–138. Academic Press, Elsevier B. V.

Mani, M. S. (1964). Ecology of Plant Galls. Dr. W. Junk Publishers, The Hague.

Martinsen, G. D., Floate, K. D., Waltz, A. M., Wimp, G. M., and Whitham, T. G. (2000). Positive interactions between leafrollers and other arthropods enhance biodiversity on hybrid cottonwoods. *Oecologia* **123**, 82–89.

Mauck, K. E., De Moraes Consuelo, M., and Mescher, M. C. (2010). Deceptive chemical signals induced by a plant virus attrack insect vectors to inferior hotst. *Proceedings of the National Academy of sciences of the United States of America* **107**, 3600–3605.

McCrea, K. D., Abrahamson, W. G., and Weis, A. E. (1985). Goldenrod ball gall effects on Solidago altissima: 14C translocation and growth. *Ecology* **66**, 1902–1907.

McLeod G, Gries, R., Von Reuß, S. H., Rahe, J. E., McIntosh, R., König, W. A., and Gries, G. (2005). The pathogen causing Dutch elm disease makes host trees attract insect vectors. *Proceedings of the Royal Society London B: Biological Sciences* **272**, 2499–2503.

Meyer, J. (1987). *Plant Galls and Gall Inducers*. Stuggart, Gebrüder Borntraeger, Berlin.

Miller, D. G., Ivey, C. T., and Shedd, J. D. (2009). Support for the microenvironment hypothesis for adaptive value of gall induction in the California gall wasp, *Andricus quercuscalifornicus*. *Entomologia Experimentalis et Applicata* **132**, 126–133.

Mivart, S. G. (1889). Professor Weisman's essay. *Nature* (Lond.) **41**, 41.

Murakami, M. (1999). Effect of avian predation on survival of leaf-rolling lepidopterous larvae. *Researches on Population Ecology* **41**, 135–138.

Naef, A., Roy, B. A., Kaiser, R., and Honegger, R. (2002). Insect-mediated reproduction of systemic infections by Puccinia arrhenatheri on Berberis vulgaris. *New Phytologist* **154**, 717–730.

Ngugi, H. K. and Scherm, H. (2006). Mimicry in plant-parasitic fungi. *FEMS Microbiology Letters* **257**, 171–176.

Ngumbi, E., Eigenbrode, S. D., Bosque-Pérez, N. A., Ding, H., and Rodriguez, A. (2007). *Myzus persicae* is arrested more by blends than by individual compounds elevated in headspace of PLRV-infected potato. *Journal of Chemical Ecology* 33, 1733–1747.

Nyman, T. and Julkunen-Tiitto, R. (2000). Manipulation of the phenolic chemistry of willows by gall-inducing sawflies. *Proceedings of the National Academy of Sciences* 97, 13184–13187.

Nyman, T., Widmer, A., and Roininen, H. (2000). Evolution of gall morphology and host-plant relationships in willow-feeding sawflies (Hymenoptera: Tenthredinidae). *International Journal of Organic Evolution* 54, 526–533.

O'Shea B, Rebollar-Téllez, E., Werd, R. D., and Hamilton, J. G. C. (2002). Enhanced sandfly attraction to Leishmania infected hosts. *Transactions of the Royal Society of Tropical Medicine and Hygiene* 96, 1–2.

Parker, C. and Riches, C. R. (1993). *Parasitic Weeds of the World: Biology and Control*. CAB International, Wallingford, Oxfordshire.

Penn, D. and Potts, W. (1998). Chemical signals and parasite-mediated sexual selection. *Trends in Ecology & Evolution* 13, 391–396.

Pfunder, M. and Roy, B. A. (2000). Pollinator-mediated interactions between a pathogenic fungus, *Uromyces pisi* (Pucciniaceae), and its host plant, *Euphorbia cyparissias* (Euphorbiaceae). *American Journal of Botany* 87, 480–55.

Pickett, J. A., Wadhams, L. J., Woodcock, C. M., and Hardie, J. (1992). The chemical ecology of aphids. *Annual Review of Entomology* 37, 67–90.

Plant, A. R. (1984). The cause of green islands induced by the Nepticulidae. *Proceedings and Transactions of the British Entomological and Natural History Society* 17, 82–83.

Poulin, R. (2010). Parasite manipulation of host behavior: An update and frequently asked questions. In: *Advances in the Study of Behavior* 41 (ed. Brockmann, H. J.) Academic Press, Burlington.

Price, P. W. and Pschorn-Walcher, H. (1988). Are galling insects better protected against parasitoids than exposed feeders? A test using tenthredinid sawflies. *Ecological Entomology* 13, 195–205.

Price, P. W., Fernandes, G. W., and Waring, G. L. (1987) Adaptive nature of insect galls. *Environmental Entomology* 16, 15–24.

Raguso, R. A. and Roy, B. A. (1998). "Floral" scent production by Puccinia rust fungi that mimic flowers. *Molecular Ecology* 7, 1127–1136.

Raman, A. (1987). On the transfer cell-like nutritive cells of galls induced by thrips (Thysanoptera, Insecta). *Current Science*, 56, 737–738.

Raman, A. and Abrahamson, W. G. (1995). Morphometric relationships and energy allocation in the apical rosette galls of *Solidago altissima* (Asteraceae) induced by *Rhopalomyia solidaginis* (Diptera:Cecidomyiidae). *Environmental Entomology* 24, 635–639.

Raman, A., Schaefer, C. W., and Withers, T. M. (2005). Galls and gall-inducing arthropods: an overview of their biology, ecology and evolution. In: *Biology, Ecology, and Evolution of Gall-Inducing Arthropods* (eds Raman, A., Schaefer, C. W., and Withers, T. M.) Science Publishers, Enfield, CT.

Rathcke, B. J. and Poole, R. W. (1975). Coevolutionary race continues: butterfly larval adaptation to plant trichomes. *Science* 187, 175–176.

Rausher, M. D. (2001). Co-evolution and plant resistance to natural enemies. *Nature* 411, 857–864.

Rohfritsch, O. (1997). Morphological and behavioural adaptations of the gall midge *Lasioptera arundinis* (Schiner) (Diptera, Cecidomyiidae) to collect and transport conidia of its fungal symbiont. *Tijdschrift voor Entomologie* 140, 59–66.

Romanes, G. (1889). Galls. *Nature* 41, 80.

Roy, B. A. (1993). Floral mimicry by a plant pathogen. *Nature* 362, 56–58

Roy, B. A. (1994). The use and abuse of pollinators by fungi. *Trends in Ecology and Evolution* 9, 335–339.

Roy, B. A., and Raguso, R. A. (1997). Olfactory versus visual cues in a floral mimicry system. *Oecologia* 109, 414–426.

Sagers, C. L. (1992). Manipulation of host plant quality: Herbivores keep leaves in the dark. *Functional Ecology* 6, 741–743.

Sandberg, S. L., and Berenbaum, M. R. (1989). Leaf-tying by tortricid larvae as an adaptation for feeding on phototoxic *Hypericum perforatum*. *Journal of Chemical Ecology* 15, 875–885.

Schiestl, F. P., Steinebrunner, F., Schulz, C., von Reuss, S., Francke, W., Weymuth, C., and Leuchtmann, A. (2006). Evolution of "pollinator"-attracting signals in fungi. *Biology Letters* 2, 401–404.

Seibert, T. F. (1993). A nectar-secreting gall wasp and ant mutualism: selection and counterselection shaping gall wasp phenology, fecundity and persistence. *Ecological Entomology* 18, 247–253.

Shivas, R. G. and Hyde, K. D. (1997). Biodiversity of plant pathogenic fungi in the tropics. In: *Biodiversity of Tropical Microfungi*. (ed. Hyde, K. D.) Hong Kong University Press, Hong Kong.

Shorthouse, J. D., Rohfritsch, O. (eds). (1992). *Biology of Insect-induced Galls*. Oxford University Press, New York.

Shykoff, J. K. and Kaltz, O. (1998). Phenotypic changes in host plants diseased by Microbotryum violaceum:

parasite manipulation, side effects, and trade-offs. *International Journal of Plant Sciences* **159**, 236–243.

Shykoff, J. A. and Bucheli, E. (1995). Pollinator visitation patterns, floral rewards and the probability of transmission of *Microbotryum violaceum*, a veneral disease of plants. *The Journal of Ecology* **83**, 189–198.

Shykoff, J. A., Bucheli, E., and Kaltz, O. (1997). Anther smut disease in *Dianthus silvester* (Caryophyllaceae): Natural selection on floral traits. *Evolution* **51**, 383–392.

Steinebrunner, F., Twele, R., Francke, W., Leuchtmann, A., and Schiestl, F. P. (2008a). Role of odour compounds in the attraction of gamete vectors in endophytic Epichloë fungi. *New Phytologist* **178**, 401–411.

Steinebrunner, F., Schiestl, F. P., and Leuchtmann, A. (2008b). Ecological role of volatiles produced by *Epichloë*: differences in antifungal toxicity. *FEMS Microbiology Ecology* **64**, 307–316.

Stern, D. L. (1995). Phylogenetic evidence that aphids, rather than plants, determine gall morphology. *Proceedings of the Royal Society B: Biological Sciences* **260**, 85–89.

Stone, G. N. and Schonrogge, K. (2003). The adaptive significance of insect gall morphology. *Trends in Ecology & Evolution* **18**, 512–522.

Takken, W. and Knols, B. G. (1999). Odor-mediated behavior of Afrotropical malaria mosquitoes. *Annual Review of Entomology* **44,** 131–157.

Taylor, P. J. and Hurd, H. (2001). The influence of host haematocrit on the blood feeding success of *Anopheles stephensi*: Implications for enhanced malaria transmission. *Parasitology* **122**, 491–496.

Thomas F, Adamo S, Moore J (2005). Parasitic manipulation: Where are we and where should we go? *Behavioural Processes* **68**, 185–199.

Thompson G. A. and Goggin, F. L. (2006). Transcriptomics and functional ge- nomics of plant defence induction by phloem-feeding insects. *Journal of Experimental Botany* **57**, 755–766.

Tooker, J. F., Rohr, J. R., Abrahamson, W. G., and De Moraes, C. (2008). Gall insects can avoid and alter indirect plant defenses. *New Phytologist* **178**, 657–671.

Tooker, J. F. and Hanks, L. M. (2006). Tritrophic interactions and reproductive fitness of the prairie perennial *Silphium laciniatum* Gillette (Asteraceae). *Environmental Entomology* **35**, 537–545.

Tooker, J. F. and De Moraes, C. M. (2007). Feeding by Hessian fly [*Mayetiola destructor* (Say)] larvae does not induce plant indirect defences. *Ecological Entomology* **32**, 153–161.

Tscharntke, T. (1994). Tritrophic interactions in gallmaker commu- nities on Phragmites australis: testing ecologi-

cal hypotheses.In: *The Ecology and Evolution of Gall-Forming Insects* (eds Price, P. W., Mattson, W. J., and Baranchikov, Y. N.) USDA.

Vandenabeele, S., Van Der Kelen, K., Dat, J., Gadjev, I., Boonefaes, T., Morsa, S., Rottiers, P., Slooten, L., Van Montagu, M., Zabeau, M., Inze, D., and Van Breusegem, F. (2003). A comprehensive analysis of hydrogen peroxide-induced gene expression in tobacco. *Proceedings of the National Academy of Sciences of the United States of America* **100**, 16113–16118.

Verhage, A., van Wees, Saskia C. M., and Pieterse, Corné M. J. (2010). Plant immunity: it's the hormones talking, but what do they say? *Plant Physiology* **154**, 536–540.

Wagner, D. L. (2005). *Caterpillars of Eastern North America*. Princeton University Press, Princeton, NJ.

Walling, L. L. (2008). Avoiding effective defenses: Strategies employed by phloem-feeding insects. *Plant Physiology* **146**, 859–866.

Walters, D. R., McRoberts, N., and Fitt, B. D. (2008). Are green-islands red herrings? Significance of green-islands in plant interactions with pathogens and pests. *Biological Reviews Cambridge Philosophical Society* **83**, 79–102.

Walters, D. R. and McRoberts, N. (2006). Plants and biotrophs: a pivotal role for cytokinins? *Trends in Plant Science* **11**, 581–586.

Waring, G. L. and Price, P. W. (1989). Parasitoid pressure and the radiation of a gall-forming group (Cecidomyiidae: Asphondylia spp.) on creosote bush (Larrea tridentata). *Oecologia* **79**, 293–299.

Weis, A. E. and Abrahamson, W. G. (1986). Evolution of host-plant manipulation by gall makers: ecological and genetic factors in the *Solidago-Eurosta* system. *American Naturalist* **127**, 681–695.

Whitham, T. G. (1992). Ecology of Pemphigus gall aphids. In: *Biology of Insect-induced Galls* (eds Shorthouse, J. D. and Rohfritsch, O.) Oxford University Press, New York.

Whitham, T. G. (1979). Territorial behavior of Pemphigus gall aphids. *Nature* **279**, 324–325.

Wild, E. H. (1976). Green islands of the Nepticulidae. *The Entomologist's Record and Journal of Variation* **88**, 103.

Zarate, S. I., Kempema, L. A., and Walling, L.L. (2007). Silverleaf whitefly induces salicylic acid defenses and suppresses effectual jasmonic Acid Defenses. *Plant Physiology* **143**, 866–875.

Zwölfer, H. and Arnold-Rinehart, J. (1994). Parasitoids as a driving force in the evolution of the gall size of *Urophora* on *Cardueae* hosts. In: *Plant Galls: Organisms, Interactions, Populations* (ed. Williams, M.A.J.) Clarendon Press, Oxford.

Afterword

Pedro Jordano

Ecological interactions among free-living species are the framework of biodiversity. We cannot understand living species without accounting for the multiplicity of interactions that shape their livings (Thompson 2009). Darwin already recognized in the first edition (1859) of the *Origin* that he was tempted to give: "…one more instance showing how plants and animals, most remote in the scale of nature, are *bound together by a web of complex relations*." Diversified interactions among insects and plants provide a very nice portrayal of this complexity—and this chapter by Mark C. Mescher on the manipulation of plant phenotypes by insects and insect-borne pathogens gives astonishing examples of this diversity.

Interactions among species involve reciprocal effects. These effects constitute the outcomes of the interactions and they can be detrimental, positive, or null; they can be highly asymmetrical in terms of their consequences for the partners (very beneficial for one species, inconsequential for its partner, etc.), or they can involve multiple tiers of complexity, not just a set of reciprocal effects (e.g., tri-trophic interactions of plants, herbivores, and parasitoids). The bulk of ecological literature on plant–animal interactions has emphasized the diversity of interactions patterns, but much less effort has been directed towards analyzing the details of the consequences of those interactions. Central among these consequences are not just the fitness effects deriving from the interaction itself, but also the possibilities that these consequences subtly drive the manipulation of the interaction by one of the species to radically modify the partner's phenotype and thus the scenario of the coevolutionary play. The possibilities for this sort of manipulations are as diverse as the

forms of interaction on which they rely, but two main types of general patterns can be defined, as

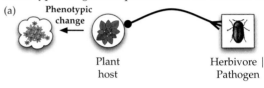

(a)

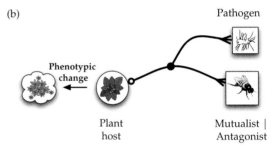

(b)

Figure 5.3 Generalized patterns and scenarios of host manipulation by antagonists and pathogen species. **A**, the interaction itself mediates in a change of the host phenotype directly caused by the herbivore or pathogen (e.g., gall-inducing herbivores; "green islands" of nutrient-enriched tissue induced by symbiotic bacteria associated with leaf-mining caterpillars). These effects are often underpinned by cytokinins induced directly by the herbivore or the bacterial symbiont. **B**, the interaction is capitalized by a "third" actor that manipulates its outcome (e.g., fungal plant pathogens that exploit a pre-existing plant–pollinator mutualism). The induced changes in the host are not driven by the mutualistic species (legitimate pollinator) or the herbivore, but by a pathogen that co-opts the interaction.

Mescher illustrates in his insightful and complete account of host plant manipulation by pathogens (Fig. 5.3).

The first sort of modification underpinned by biological interactions involves the direct modification(s) (dot-ended arrow in Fig. 2.3A) of the plant host phenotype by the antagonist (e.g., a gall-inducing insect). The herbivore drives the modification (manipulation) of a host trait resulting in increased benefits. The examples are numerous and often entail the phenotypic change directly encoded by genes in the herbivore. What are the types of host plant–insect coevolution that might give rise to such amazing details as the natural history of these interactions? Quite frequently, and contrasting with the mega-diversified insect–plant interactions for pollination, host–herbivore–pathogen interactions are much more specific and intimate, with ample possibilities for the fine-tuning of their evolutionary outcomes. This contrasts with coevolutionary patterns in mega-diversified interactions where the selective pressures on the partners' traits are more diluted within a complex network of diversified interactions (Jordano 2010). In this scenario, processes like gene-for-gene coevolution, arms races, or diversifying coevolution help to explain these fascinating outcomes (Thompson 2005). Note that these coevolutionary scenarios do not necessarily imply a strong symmetry of the mutual effects: the pathogen might drive the evolution of the host's traits without reciprocal effects mediated by the host responses.

The second type of modification radically differs from the previous scenario. A third species or group of pathogen or parasitic species "co-opts" the interaction between the plant and, for example a mutualistic pollinator, and entails a modification of the plant host benefiting the pathogen. The pathogen manipulates the interaction (dot-ended arrow in Fig. 5.3B) and alters its outcome (blank dot-ended arrow, Fig. 5.3B). The natural history details of real-world examples of this scenario are truly amazing and involve fascinating cases of mimicry by the pathogens to cheat the mutualism that they co-opt. Specific traits mimicked include anatomical structures (e.g., inducing shape and form change of pseudo-flowers), odors and colors (thus co-opting the sensory channels that mediate in plant–animal mutualistic interactions), and volatiles. Pathogens can also co-opt the interactions between plants and antagonists (herbivores). Given that the number of interactions in a plant–pollinator or a plant–herbivore assemblage is far greater than the number of component species, there are multiple possibilities for co-opting, cheating, and mimicking the interactions themselves. Cheaters of some mutualisms (e.g., nectar robbers) include insect species that legitimately pollinate other plant partners; therefore the range of interaction-mediated host manipulations is extremely broad, ranging from other insects to pathogens like fungi and bacteria.

While the recent literature of host phenotype manipulation by parasites and pathogens has emphasized animal species, where behavioral changes are frequently implied in manipulative actions, the diverse effects on plants have been largely overlooked. However, the enormous diversification of higher plants has pivoted on highly diversified interactions with insects, through herbivory, pollination, and often involving pathogenic relations with microorganisms. These interactions often take place within highly diversified and complex webs involving more than two partners and more than two trophic levels (Lill et al. 2002); and we are still far from understanding how coevolution has shaped their enormous diversity. Cataloguing and describing new species is a major element in the biodiversity research agenda. Yet we also need a deeper and urgent study of the myriad of ecological interactions that shape and support this biodiversity. An additional challenge is to understand the complex physiological mechanisms used by pathogens and parasites in their race to exploit and manipulate the burgeoning biodiversity of plant–animal interactions.

References

Jordano, P. (2010). Coevolution in multispecific interactions among free-living species. *Evolution, Education and Outreach* **3**, 40–46.

Lill, J., Marquis, R., and Ricklefs, R. (2002). Host plants influence parasitism of forest caterpillars. *Nature* **417**, 170–173.

Thompson, J. N. (2005). *The Geographic Mosaic of Coevolution*. University of Chicago Press, Chicago, IL, USA.

Thompson, J. N. (2009). The coevolving web of life. *American Naturalist* **173**, 125–140.

CHAPTER 6

Visual trickery in avian brood parasites

Naomi E. Langmore and Claire N. Spottiswoode

6.1 Introduction

Visual mimicry is a strategy deployed by a wide range of taxa, from spiders that mimic the ants they hunt (Nelson and Jackson 2006), to cross-dressing males in bluegill sunfish (Dominey 1980), and harmless snakes that mimic the vibrant colors of venomous snakes (Harper and Pfenning 2008) (see Chapter 11). Although visual mimicry is widespread amongst invertebrates, and to a lesser extent fish, amphibians, and reptiles (Ruxton et al. 2004), it is generally a rare phenomenon among birds (but see e.g., Dumbacher and Fleischer 2001). However, the avian brood parasites provide a notable exception. Several recent field studies, combined with new techniques for quantifying mimicry and modeling the visual perception of hosts, demonstrate that brood parasites have evolved hitherto unknown forms of visual mimicry as means of manipulating their hosts at every stage of host–parasite interactions.

Avian brood parasites lay their eggs in the nests of other birds, and thereafter abandon their young to the care of the host. The costs of parasitism to the host may be (1) negligible, when incubation of an additional egg is the only care required (Lyon and Eadie 2004); (2) moderate, when the parasitic chick is reared alongside the host young (non-evicting parasites); and (3) severe, when the parasitic chick kills or evicts all the host young from the nest (evicting parasites). In general, with increasing costs of parasitism, host defenses become more sophisticated, in turn selecting for ever more elaborate trickery on the part of the parasite to fool the host

into rearing its young. Thus we generally see the most advanced parasite recognition systems amongst hosts of evicting parasites, which in turn select for highly accurate visual mimicry by parasites to evade host detection.

However, host defenses are not the only selective agent responsible for the evolution of visual mimicry in brood parasites. Mimicry can also arise to facilitate exploitation of parent–offspring channels of communication, ensuring that the parasite chick is able to elicit adequate care from its foster parents (parasite "tuning," Davies 2011). Mimicry for this purpose is equally likely to arise in evicting and non-evicting parasites, and was termed sequential evolution mimicry by Grim (2005), to distinguish it from mimicry that arises through a coevolutionary arms race. However a recent hypothesis suggests that this form of mimicry could also ultimately lead to a coevolutionary arms race between parasites and hosts (see Hauber and Kilner 2007, Section 6.5 below).

Here, we will first consider how visual mimicry evolves as an outcome of a coevolutionary arms race between parasites and their hosts, driven by host defenses against parasitism. We will describe how host defenses have selected for mimicry by brood parasites in all three successive stages of the parasitism process; gaining access to host nests (Section 6.2), the egg stage (Section 6.3), and the chick stage (Section 6.4). Second, we will discuss how visual mimicry can evolve to facilitate exploitation of channels of communication between host parents and offspring (Section 6.5). Finally, in Section 6.6 we will consider the different challenges faced by generalist versus specialist parasites, and the strategies used by generalists to solve the problem of how to mimic a diversity of morphologically distinct hosts.

6. 2 Accessing host nests

The process of laying an egg is fraught with challenges for a parasitic bird. Hosts typically mob parasites violently, and this can provide an effective defense against parasitism (Welbergen and Davies 2009), as well as causing injury (Davies and Brooke 1988b; Welbergen and Davies 2008; Wyllie 1981) or even death (Molnar 1944, Moyer 1980) to the parasite. However, a recent study demonstrated that cuckoos have evolved a clever visual trick to inhibit host aggression, thereby facilitating access to host nests.

Naturalists have long been aware that several cuckoo species bear a striking resemblance to sympatric aggressive or predatory birds (e.g., the drongo cuckoo *Surniculus lugubris* and the drongo *Dicrurus macrocercus*, the large hawk-cuckoo *Cuculus sparverioides* and the besra sparrowhawk *Accipiter virgatus*, the common hawk-cuckoo and the shikra *Accipiter badius*, the common cuckoo *Cuculus canorus* and the Eurasian sparrowhawk *Accipiter nisus*, Fig. 6.1; Johnsgard 1997).

The similarities encompass size, shape, plumage color and pattern, and even flight behavior (Davies and Welbergen 2008). Several hypotheses have been proposed to explain this resemblance, including convergent evolution, inducement of mobbing to facilitate location of potential hosts, protection against hawk attacks, and inhibition of mobbing by hosts (Craib 1994; Davies and Welbergen 2008; Wallace

1889; Wyllie 1981). Resemblance to predatory birds is more common in parasitic than non-parasitic cuckoos, and evolved after the evolution of brood parasitism (Krüger et al. 2007; Payne 1967), favoring hypotheses that suggest this form of mimicry plays some role in brood parasitism. Recently, an experimental test of the functional significance of hawk mimicry by cuckoos was undertaken for the first time (Welbergen and Davies 2011). Plumage manipulations of taxidermic models revealed that the hawk-like barring on the breasts of common cuckoos inhibits approach by their reed warbler *Acrocephalus scirpaceus* hosts, thus reducing harassment of cuckoos during nest parasitism. When the barred underparts were concealed, reed warblers approached the model cuckoo more closely and mobbed it more intensely. This indicates that hawk-like barring is adaptive in the context of brood parasitism and provides rare experimental evidence of Batesian mimicry in birds.

6. 3 The egg stage

The egg stage in the battle between brood parasites and hosts provides some of the most compelling and widely-known examples of visual trickery in nature: a cuckoo egg superbly blending into the clutch of its unsuspecting host is a typical and defining image. Parasitic mimicry of host eggs has arisen independently in six of the seven groups in which interspecific brood parasitism has evolved independently in birds. The seventh is the black-headed duck *Heteronetta atricapilla* of South America, a

Figure 6.1 Many cuckoo species show a physical resemblance to predatory or aggressive birds. Here, the common cuckoo (center, photo: Roy and Marie Battell) bears a superficial resemblance to the peregrine falcon *Falco peregrinus* (left, photo: Jack Wolf) and the Eurasian sparrowhawk *Accipiter nisus* (right, photo: Steve Grimwade).

fascinating exception (Lyon and Eadie 2004) which we consider in more detail below.

The common cuckoo and its hosts provide predictably the longest history of thought on the evolution of visual mimicry in parasitic eggs. The remarkable resemblance between cuckoo eggs of different "strains" (or "gentes," singular "gens"; Newton 1896) and the eggs of their corresponding host species was first commented upon during the eighteenth century, but many early observers attributed it to common elements of diet between host and parasite, rather than any deceptive function (reviewed by Schulze-Hagen et al. 2009). A series of authors, many of them egg collectors, began to speculate about how host–parasite resemblance could serve to reduce egg rejection by hosts (Schulze-Hagen et al. 2009). Systematic egg rejection experiments were pioneered in brown-headed cowbird *Molothrus ater* hosts (e.g., Rothstein 1975; Rothstein 1982), and such an experimental approach applied to common cuckoo hosts elegantly showed that color and pattern mimicry in common cuckoo eggs is under selection from discriminating hosts: host species in which the cuckoo lays a more closely mimetic egg also showed stronger discrimination against foreign eggs (Brooke and Davies 1988; Davies and Brooke 1989a). This provided compelling evidence that egg mimicry is a host-specific adaptation that has arisen through coevolutionary interactions between parasitic exploitation and host counter-defense, and has led to the diversification of cuckoo host-specific gentes.

Subsequent to these classic experimental studies, understanding of avian vision and visual perception has much improved (e.g., Endler and Mielke 2005; Vorobyev and Osorio 1998). The advance of spectrophotometry and visual modeling techniques over the last decade has helped to reveal examples of subtlety in visual trickery by brood parasites that could not have been anticipated by the human eye alone. Early application of spectrophotometry to quantify eggshell reflectance, for example, suggested potential mimicry between red-chested cuckoo *Cuculus solitarius* and host eggs that appear dissimilar to the human eye (Cherry and Bennett 2001), and revealed previously undetected host-specific adaptation to multiple host species in pallid

cuckoos *Cuculus pallidus* (Starling et al. 2006). Spectrophotometry improves on human visual assessment since it is objective and incorporates ultra-violet wavelengths invisible to humans, but does not consider how a signal is processed by a bird's visual system, and hence provides imperfect insight into mimicry from a bird's perspective. Perceptual models of avian vision improve on spectrophotometry alone by processing spectrophotometric data in the light of current understanding of avian color vision, to generate estimates of color differences as they should be perceived by a bird. The last few years have seen a flowering of the application of such models to studies of brood parasitic mimicry in various systems (e.g., Cassey et al. 2008; Avilés 2008; Spottiswoode and Stevens 2010; Langmore et al. 2009; Langmore et al. 2011), and some of their findings will appear later in this chapter.

Avian perceptual modeling has also allowed classic experimental work, which relied on scoring mimicry using human observers, to be revisited in the light of improved understanding of avian vision. The original conclusion has remained unaltered: host species previously shown to have the strongest levels of egg rejection are parasitized by cuckoos laying eggs that are a more mimetic to a bird's eye, too, and this has held true both for egg patterning (Stoddard and Stevens, 2010) and egg color (Stoddard and Stevens, 2011). However, perceptual modeling has provided additional insights that could not have been detected using human assessment alone: for example, the egg color distribution of host species and their corresponding cuckoo gens overlap most in avian perceptual color space in host species that show the strongest rejection behavior (Stoddard and Stevens, 2011). Moreover, as color mimicry improves, so does pattern mimicry, perhaps implying that hosts rely increasingly on pattern cues when mimicry is refined to the point that colors become nearly indistinguishable to a bird's eye (Stoddard and Stevens, 2011).

Host discrimination against foreign eggs based on visual traits has now been demonstrated experimentally using similar approaches in a wide variety of interspecific brood parasitic systems (listed in Table 6.1). Egg rejection experiments are necessary

Table 6.1 Parasitic systems in which egg rejection by hosts has been experimentally investigated, and whether or not parasitic eggs show host mimicry. "Specialist" is used to refer to parasites that specialize either on a single host or a group of very closely-related hosts in the same genus (e.g., *Acanthiza* for shining bronze-cuckoo and *Laniarius* for black cuckoo). "Gentes" here is used to refer to host-specific variation in phenotype, in most cases without any data on whether genetically defined strains of females exist. "Polymorphism" here is used to refer to extreme inter-individual variation in phenotype, following Kilner (2006), rather than classic discrete polymorphisms established to have a simple genetic basis. "Visual" cues here refer to color and patterning, rather than egg size or shape.

Species	Specialist or generalist parasite	Parasite shows gentes	Parasite eggs resemble host's	Host eggs polymorphic	Host egg rejection based on visual cues
Ducks (Anatidae)					
Black-headed duck *Heteronetta atricapilla*	Generalist	No	No	No	Yes[1]
Old World cuckoos (Cuculidae: Cuculinae: Cuculini)					
Horsfield's bronze-cuckoo *Chalcites basalis*	Generalist	No	Yes	No	No[2,3]
Shining bronze-cuckoo *Chalcites lucidus*	Specialist	No	No	No	No[2,4]
Diederik cuckoo *Chrysococcyx caprius*	Generalist	Yes	Yes	Sometimes	Yes[5-8]
Klaas's cuckoo *Chrysococcyx klaas*	Generalist	Yes	Yes	Sometimes	Yes[8]
Common cuckoo *Cuculus canorus*	Generalist	Yes	Sometimes	Sometimes	Sometimes[9-11]
Black cuckoo *Cuculus clamosus*	Specialist	Maybe	Yes	Sometimes	Yes[8]
African cuckoo *Cuculus gularis*	Specialist	No	Yes	Yes	Yes[8,12]
Oriental cuckoo *Cuculus optatus*	Generalist	Yes	Sometimes	No?	Yes[13]
Pallid cuckoo *Cuculus pallidus*	Generalist	Yes	Yes	Yes	Yes[14]
Asian lesser cuckoo *Cuculus poliocephalus*	Generalist	Yes	Sometimes	No?	Yes[13]
Red-chested cuckoo *Cuculus solitarius*	Generalist	Yes	Sometimes	Sometimes	Sometimes[15]
Common koel *Eudynamys scolopaceus*	Specialist	No	Yes	No	Yes[16]
Old World cuckoos (Cuculidae: Cuculinae: Phaenicophaeini)					
Great spotted cuckoo *Clamator glandarius*	Generalist	No	Sometimes	Sometimes	Yes[8,17,18]
Jacobin cuckoo *Clamator jacobinus*	Generalist	No (in Africa)	No	No	No[8,19]

Cowbirds (Icteridae)

Bronzed cowbird *Molothrus aeneus*	Generalist	No	No?	Sometimes[20]
Brown-headed cowbird *Molothrus ater*	Generalist	No	No	Sometimes[21, 22]
Shiny cowbird *Molothrus bonariensis*	Generalist	No	No	Sometimes[23, 24]
Giant cowbird *Molothrus oryzivorus*	Generalist	No	Yes	Yes[25]
Screaming cowbird *Molothrus rufoaxillaris*	Specialist	No	Yes	No[26]

Parasitic Finches (Viduidae)

Cuckoo finch *Anomalospiza imberbis*	Generalist	Yes	Yes	Yes[27]

References

[1] Lyon and Eadie (2004). [2] Brooker and Brooker (1989). [3] Langmore et. al. (2003). [4] Langmore et al. (2005). [5] Victoria (1972). [6] Lahti and Lahti (2002). [7] Jackson (1998). [8] Noble (1995). [9] Brooke and Davies (1988). [10] Yang et al . (2010). [11] Moksnes et al. (1990). [12] Tarboton (1986). [13] Higuchi (1989). [14] Landstrom et al. (2010). [15] Honza et al. (2005). [16] Sumit Sinha and Suhel Quader. Unpublished Data. [17] Soler and Møller (1990). [18] Alvarez et al. (1976). [19] Krüger (2011). [20] Peer et al. (2002). [21] Rothstein (1975). [22] Rothstein (1982). [23] Mermoz (1994) [24] Fraga (1985). [25] Cunningham and Lewis (2005). [26] Fraga (1998). l [27] Spottiswoode and Stevens (2010).

because similarity between host and parasite eggs may arise through mechanisms other than selection (reviewed by Grim 2005), and because selection for mimicry can come from other quarters than host rejection. Table 6.1 shows that there is abundant evidence that egg discrimination is a widespread defense in host species, and has repeatedly selected for parasitic mimicry. To date, the only interspecific parasitic group in which visual mimicry appears to exist (at least in one species), but no experimental investigations have yet been carried out, is the honeyguides (Indicatoridae): the green-backed honeybird *Prodotiscus zambeziae* lays blue or white eggs, as does its white-eye (*Zosterops* spp.) hosts (Vernon 1987). Selection for host-specific mimicry may result in very different evolutionary outcomes for the parasite: while in many brood parasites (including numerous cuckoo species), multiple conspecific strains or "gentes" of host-specialists are maintained within the same species, in other systems host-specific mimicry may have resulted in parasitic speciation; we will return to these alternatives later in Section 6.6.

Could visual trickery at the egg stage have evolved in response to pressures other than host defenses? At least two other possible agents of selection for mimicry have been proposed. The first is cuckoos themselves (Davies and Brooke 1988). This could arise through multiple parasitism, where more than one parasitic individual attempts to lay in the same host nest. When the second-laid parasite would suffer from the presence of the first (either through nestling death, or competition), selection should favor the female parasite herself to selectively destroy mismatched eggs, thus ensuring a clear path for her own offspring. Although there have been several experimental tests of this idea in cuckoos, none have found support for the hypothesis (Davies and Brooke 1988; Davies 1999; Langmore and Kilner 2009). This hypothesis might be more applicable under conditions where multiple parasitism is common, as well as (as is the case in cuckoos) the costs of multiple parasitism to the second-laying female being high.

The second alternative hypothesis is that nest predators may impose selection for mimicry via crypsis, if mismatches in appearance within a clutch

render it more conspicuous (Wallace 1889). While there is some evidence that nest predators sometimes select for egg crypsis in entire clutches of non-parasitic species (reviewed by Kilner 2006), this idea as applied to within-clutch disparity has received no support to date, as several studies have failed to find any effect of mismatched experimental eggs on nest predation rates (Swynnerton 1918; Davies and Brooke 1988; Mason and Rothstein 1987; Davies 1992), nor any interspecific evidence that species with higher levels of natural intra-clutch variability suffer higher predation (Avilés et al. 2006b). Experimental tests of this idea are admittedly difficult in rejecting hosts, because mismatched eggs can be rejected too quickly to allow predators a chance to strike (Davies and Brooke 1988); this underscores the importance of host rejection as a selection pressure many (but clearly not all) parasitic species face during *every* breeding attempt.

Thus the weight of evidence suggests that egg mimicry typically arises from a coevolutionary cycle in which hosts gain a fitness advantage from rejecting mismatched eggs, resulting in selection for parasitic mimicry, which in turn results in selection on hosts to evolve ever more finely-tuned egg discrimination. Egg rejection at the incubation stage, however, is a function not only of the absolute discrimination ability of the host. It is also a function of the degree of visual discrepancy between parasitic and host eggs, which hosts have the potential to manipulate in counter-defense against a mimetic parasite. This idea was first put forward by one of the pioneers of egg rejection experiments, Charles Swynnerton, who while farming in eastern Zimbabwe nearly a century ago wrote with great clarity,

"I doubt whether [mimicry] would always end the matter, for, when a Cuckoo's egg became indistinguishable from its host's, variation in the latter would still afford the means of distinguishing it from the Cuckoo's, and it is even imaginable that a race may in some cases have taken place between the host's eggs and those of the overtaking Cuckoo." (Swynnerton 1918).

Thus, selection should not only favor more discriminating hosts, but also hosts that lay eggs that differ in appearance from parasitic eggs, favoring an increase in inter-clutch variability. Additionally, egg discrimination might be facilitated by reduced variation within individuals: selection from brood parasites should favor a decrease in within-clutch phenotypic variability such that parasites are easier to spot, and to avoid mistakenly rejecting a non-parasitic egg (Davies and Brooke 1989b).

These paired predictions have received several tests in the hosts of the common cuckoo, at the level of species, populations, and individual clutches. At the interspecific and population level there is broad agreement that a history of parasitism by the common cuckoo is associated with subtly higher inter-clutch and, less consistently, lower intra-clutch variability (e.g., Soler and Møller 1996; Øien et al. 1995; Stokke et al. 2002). Within species, a fully experimental study has clearly supported the hypothesis that a decrease in within-clutch variability is associated with improved ability to detect foreign eggs (Moskát et al. 2008), although the numerous correlational studies to date have provided very mixed support (Cherry et al. 2007a; Kilner 2006; Landstrom et al. 2010). It has even been proposed that *increased* within-clutch variability might improve detection of parasitic eggs by allowing hosts to label each of their own eggs with an individual, egg-specific rather than clutch-specific "signature" (Cherry et al. 2007a).

However, hosts of the common cuckoo in Europe all lay relatively invariable eggs compared to some other parasitic systems. By contrast, one common cuckoo host in China, the ashy-throated parrotbill *Paradoxornis alphonsianus*, appears to have evolved discrete polymorphisms in egg appearance, laying immaculate white, blue, or pale blue eggs (Yang et al. 2010). In other parasitic systems, diversity in egg appearance is greater still, involving both color and pattern, and not necessarily forming discrete polymorphisms: these include *Ploceus* weavers parasitized by diederik cuckoos *Chrysococcyx caprius* (Victoria 1972), fork-tailed drongos *Dicrurus adsimilis* parasitized by African cuckoos *Cuculus gularis* (Tarboton 1986, C.N.S. unpubl. data), and *Prinia* and *Cisticola* warblers parasitized by cuckoo finches *Anomalospiza imberbis* (Spottiswoode and Stevens 2010); in the last-mentioned, analyses of egg colors

using visual modeling showed that egg color variation is also continuous when perceived by a bird, rather than falling into distinct morphs (Fig. 6.2). In one host of the diederik cuckoo, the village weaver *Ploceus cucullatus*, there is especially clear evidence that variability in egg appearance has responded to selection from brood parasites: introduced populations released from parasitism pressure show both an increase in within-clutch variability and a decrease in between-clutch variability relative to their ancestral, parasitized populations (Lahti 2005), which compromises the weavers' ability to reject foreign eggs (Lahti 2006). We might speculate that the remarkable phenotypic diversity in the above-mentioned systems is an escalated host defense

Figure 6.2 Eggs of parasitic cuckoo finches (right column) and two host species (left column). The top two rows are from red-faced cisticola nests and the bottom four rows from tawny-flanked prinia nests; note the extensive variation both between and within host species, mimicked by the parasite (photo: Claire Spottiswoode). See also Plate 10.

reflecting a probably much more ancient origin of brood parasitism in the tropics (and in Africa in particular) compared to Europe (Rothstein 1992).

In all the systems mentioned above, the parasite itself is able to reproduce a broad range of host phenotypes; for example, common cuckoos parasitizing ashy-throated parrotbills lay immaculate white or blue eggs (Yang et al. 2010), and cuckoo finches parasitizing tawny-flanked prinias *Prinia subflava* lay eggs with red, blue, or white backgrounds, broadly mimicking those of the prinia but with cruder egg markings that lack the intricate detail of host eggs (Spottiswoode and Stevens 2010). A crucial difference between this form of parasitic variation and host-specific parasitic variation (gentes) is that host-specific imprinting is unavailable as a mechanism through which parasites can target hosts with matching phenotypes. One alternative mechanism is that parasites could learn the appearance of their own eggs and seek out hosts of similar appearance. While common cuckoos have been cautiously suggested to single out reed warbler clutches better matching their own (Cherry et al. 2007b; Avilés et al. 2006a), in field studies it is difficult to exclude fully the possibility that ill-matched cuckoo eggs have been rejected by hosts prior to being found (Avilés et al. 2006a; Swynnerton 1918). Observations from other more phenotypically variable systems would seem to argue against parasites targeting particular individual phenotypes within a host species, at least with any degree of accuracy, despite the advantages this would confer: strikingly mismatched parasitic eggs are common in diederik cuckoos parasitizing masked weavers *Ploceus velatus* (Hunter 1961; Reed 1968), cuckoo finches parasitizing tawny-flanked prinias (Spottiswoode and Stevens 2010), and African cuckoos parasitizing fork-tailed drongos (C.N.S. and J.F.R. Colebrook-Robjent, unpubl. data); in the latter two cases parasites are known to incur heavy losses through host rejection. Thus in these cases parasites seem to lay their eggs haphazardly, relying on chance matches to succeed; this underscores the effectiveness of high degrees of phenotypic variability as a host defense.

Irrespective of whether or not parasites are able to deploy their eggs in a targeted manner when faced with a polymorphic adversary, the highest fitness pay-offs should be gained by parasites that mimic whichever host phenotypes are commonest (Victoria 1972; Lively and Dybdahl 2000). Thus negative frequency-dependent selection might be expected to favor rare host egg phenotypes. Potential coevolutionary outcomes of such host defenses have been modeled theoretically (Takasu 2003; Takasu 2005), and found to depend on the starting distributions of host and parasitic phenotypes within the population, and the mode of inheritance of egg appearance. Possible outcomes are host and parasite egg phenotypes settling to a stable monomorphism; settling to a stable polymorphism; or continuously oscillating owing to hosts being able always to destabilize mimicry by parasites by evolving new phenotypes, in a close analogy of "Red Queen" dynamics (Van Valen 1973). The last-mentioned is expected when egg appearance is inherited autosomally, and when within-population variance in egg phenotype is greater in hosts than in parasites (Takasu 2005).

This raises the intriguing possibility that continuous arms races in egg appearance might result, under conditions that are likely to be realistic for many systems: first, while the mode of inheritance of egg appearance is still poorly understood, some of the few existing case studies have clearly shown an autosomal rather than maternal mode of inheritance (see Fossøy et al. 2010 for a recent review), which we might speculate could itself be the outcome of selection for diverse egg phenotypes. Second, greater variance in host versus parasitic phenotypes (of the appropriate gens) is not implausible and has been shown in at least one parasitic system, the cuckoo finch and its main host the tawny-flanked prinia (Spottiswoode and Stevens, in press). A temporal comparison of cuckoo finch and tawny-flanked prinia eggs over four decades has indeed showed patterns consistent with this prediction: prinia eggs became more diverse in color over time, as did cuckoo finch eggs (Spottiswoode and Stevens, in press). Thus, there is some evidence that visual trickery by brood parasites can result in "chase-away" selection on hosts, in some cases generating extreme levels of phenotypic variation used by hosts as signals of identity.

Effective signals of individual identity across the animal kingdom are predicted to be comprised of multiple uncorrelated characters, since this increases the information content of the signal by maximizing variability amongst individuals (Dale et al. 2001; Beecher 1982). Eggs should be no exception, and recent progress in devising methods to quantify independent aspects of egg appearance, including both color and pattern, has allowed this to be investigated quantitatively. Color, luminance, and five aspects of egg pattern show generally very low levels of correlation with one another in the hosts of the cuckoo finch (Spottiswoode and Stevens 2011). Ideally, this would be compared with background levels of correlation amongst egg traits in species with no history of interactions with brood parasites. Nonetheless, these results do point towards egg phenotypes as complex and multifaceted signals of identity, much as the complicated markings on banknote watermarks render them more difficult to forge by counterfeiters. Taken together, the foregoing examples of counter-defenses involving egg variability argue that visual trickery can also be deployed in return by hosts to their own advantage.

Could parasites use visual trickery in ways other than mimicry alone to aid acceptance of parasitic eggs? This is an interesting question, not least since it may shed light on some of the numerous paradoxical examples of hosts failing to evolve egg discrimination (and hence parasites failing to evolve mimicry) despite the catastrophically high costs of parasitism they incur from blithely accepting parasitic eggs (reviewed by Davies 1999). To date, two non-mimetic visual tricks have been proposed. The first is to evade detection by becoming "invisible" to the host. This tactic is analogous to the "chemical insignificance" of insect brood parasites (Kilner and Langmore 2011), which foil detection based on pheromonal cues by reducing the complexity of their hydrocarbon signature (Lenoir et al. 2001). Several species of *Chalcites* cuckoos appear to evade host detection in a similar way. These cuckoos parasitize hosts that build dark, dome-shaped nests (Langmore et al. 2009b) and lay eggs that are an unusual olive green or bronze color, quite unlike the white speckled eggs of their hosts (Fig. 6.3). Visual modeling revealed that, in the eyes of the

host, the dense, dark pigment reduces the reflectance of the egg to a level similar to that of the nest lining (Langmore et al. 2009b), rendering the eggs cryptic to the host when viewed within the nest. Taken together with the finding that dark egg color is a derived trait in this lineage and occurs only in those species that parasitize dome-nesters exclusively, this strongly suggests that egg color in these *Chalcites* cuckoos has arisen through selection for crypsis (Langmore et al. 2009b). Cryptic eggs appear to be a highly successful tactic for avoiding detection by hosts, because hosts do not reject cryptic cuckoo eggs unless they are laid before the host has commenced laying (N. E. Langmore and G. Maurer unpublished data; Brooker and Brooker 1989; Langmore et al. 2009b; Sato et al. 2010; Tokue and Ueda 2010). Crypsis should have several advantages over mimicry as a form of visual trickery, not least that it should allow parasites to exploit more than one dark-nesting host species without acquiring host-specific adaptations, and to colonize a new host species that has already evolved egg rejection (Brooker et al. 1990; Langmore et al. 2009b). These advantages might suggest that it could be more widespread a parasitic strategy than currently appreciated. Might crypsis help to account for other striking examples of parasitic mismatch in the relatively bright light environment of open-nesting hosts (e.g., common cuckoo vs. dunnock *Prunella modularis*, or red-chested cuckoo vs. cape robin-chat *Cossypha caffra*)? These questions remain to be investigated using the appropriate visual models and, ideally, field experiments.

A second form of non-mimetic visual trickery by parasites is visual superstimulus. Birds in general may have sensory predispositions to certain phenotypes: for example, herring gulls *Larus argentatus* prefer artificial eggs that are larger and greener than their own, and marked with relatively fine and contrasting speckles (Baerends and Drent 1982). Could brood parasites "charm" their hosts by producing eggs (or chicks, e.g., Soler et al. 1995a) with similarly exaggerated traits that are visually attractive to hosts, rather than perfectly mimetic? Three observations from rufous bush-robins *Cercotrichas galactotes* suggest that hosts might be susceptible to this form of trickery: in bush-robins, model eggs that resem-

Figure 6.3 The cryptic egg of a shining bronze-cuckoo *Chalcites lucidus* with two eggs of its host, a yellow-rumped thornbill *Acanthiza chrysorrhoa* (photo: Naomi Langmore). See also Plate 11.

bled neither those of their corresponding gens of cuckoo (which is imperfectly mimetic) nor their own were just as likely to be accepted as perfectly mimetic experimental eggs, whereas eggs with partial mimicry were usually rejected (Alvarez 1999). This suggests that visual traits of the non-mimetic experimental eggs (which were either whiter, or had more contrasting markings) aided their acceptance. Moreover, identically painted model eggs were more likely to be accepted if they were larger than either bush-robin eggs, or cuckoo eggs of the bush-robin gens (Alvarez 2000). Similar results with respect to both color and size have also been found in European magpies *Pica pica* parasitized by the great spotted cuckoo *Clamator glandarius* (Alvarez et al. 1976), and in both cuckoo species real cuckoo eggs differ from host eggs in accordance with this apparent host preference, being larger, paler, and with more contrasting markings. New methods for quantifying independent components of egg appearance (including both pattern and color) should make future analyses of this kind even more powerful. While the hypothesis that blue-green egg colors are attractive to birds as sexual signals remains controversial (Moreno and Osorno 2003; reviewed by Reynolds et al. 2009), it does suggest one avenue through which brood parasites might exploit pre-existing host preferences to their own advantage.

Finally, egg mimicry should evolve when parasites impose strong fitness costs on hosts, and thus selection for egg recognition, and we began this section by mentioning an exception. The one brood parasitic group (although it comprises just one species) in which egg mimicry has not evolved, the black-headed duck, would at first sight seem to be perfectly consistent with this prediction: owing to its highly precocial young it imposes negligible fitness costs on hosts, and lays unmarked whitish eggs that contrast strikingly with the mottled eggs of its favorite hosts, two species of coot (*Fulica* spp.) (Weller 1968). But a paradox arises because coots do regularly discriminate strongly against mismatched eggs, resulting in high fitness costs for the duck. In this instance egg rejection behavior has been suggested to be the product of defenses against intraspecific brood parasitism between coot females, in which the duck now appears to find itself trapped and unable to evolve its way out owing to strong selection by hosts against even partially mimetic eggs (Lyon and Eadie 2004). An intriguing alternative may be that the parasitic duck egg's lack of mimicry serves to signal to the hosts "I am a duck egg," and therefore harmless in comparison to a coot chick requiring parental care and not worth the potential costs of ejecting (N. B. Davies, pers. comm.).

6.4 The nestling stage

Brood parasitic nestlings are generally more renowned for their lack of resemblance to host young than for mimicry. In contrast to the exquisite mimicry of host eggs by many brood parasites, brood parasitic chicks typically differ from host young in traits such as size, mouth color, skin color in naked chicks and plumage color in feathered chicks. The failure of many brood parasites to mimic host young appears to reflect the relative rarity of host defenses against parasitism at the nestling stage (Davies 2000). Most hosts attack an adult brood parasite near the nest, and many hosts reject foreign eggs in their nests, but rejection of brood parasitic chicks was, until recently, unknown.

The first evidence of cuckoo nestling rejection by hosts came from a study of the superb fairy-wren

Malurus cyaneus, which is parasitized by Horsfield's bronze-cuckoo *Chalcites basalis* (Langmore et al. 2003). Over a third of host parents abandon the young cuckoo a few days after hatching, leaving it to die in the nest. More recently, cuckoo nestling rejection has also been documented in other bronze-cuckoo hosts; cuckoo chick rejection by eviction from the nest was discovered in two hosts of the little bronze-cuckoo *C. minutillus*; the large-billed gerygone *Gerygone magnirostris* (Sato et al. 2010); and the mangrove gerygone *G. laevigaster* (Tokue and Ueda 2010); and there is also some indication of shining cuckoo *Chalcites lucidus* chick discrimination in the grey warbler *G. igata* (McLean and Maloney 1998). Just as cuckoo egg rejection by hosts has selected for egg mimicry in cuckoos, we might expect that cuckoo chick rejection by these hosts would select for chick mimicry in cuckoos. A recent study confirms just that; the chicks of three species of bronze-cuckoos are striking visual mimics of their morphologically diverse hosts (Fig. 6.4; Langmore et al. 2011). This study uses recent models of avian visual processing to demonstrate that the cuckoo nestlings resemble host young, not only through human eyes, but also in the eyes of the hosts themselves. The nestling cuckoos mimic their hosts in the color of their skin and rictal flange (the broad, colorful border of the nestling mouth), as well as in the incidence of natal down. One cuckoo species even expresses a derived trait that mimics that of their hosts; nestlings of the little bronze-cuckoo (*C. m. minutillus*) display multi-barbed "fluffy" natal down (Langmore et al. 2011), which is typical of nestling passerines but unique amongst cuckoos (Payne 1977).

The bronze-cuckoos appear to track variation in host nestling morphology with remarkable accuracy, even down to the level of subspecies (Table 6.2). For example, northern Australian hosts of the little bronze-cuckoo typically display dark skin and white natal down at the nestling stage, which is mimicked by the cuckoo (Fig. 6.4 Langmore et al. 2011; Sato et al. 2010; Tokue and Ueda 2010). However, the more southerly host, the white-throated gerygone *G. albogularis*, displays pink skin and yellow down. This host is parasitized by a subspecies of the little bronze-cuckoo, *C. m. barnardi*, which also displays pink skin and yellow down

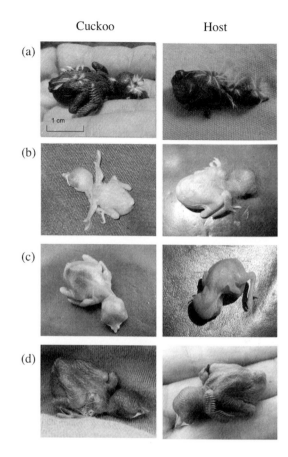

| Cuckoo | Host |

Figure 6.4 Visual mimicry of host chicks by nestling *Chalcites* cuckoos. a) Little bronze-cuckoo *C. minutillus* and large-billed gerygone *Gerygone magnirostris*; b) Shining bronze-cuckoo *C. lucidus* and yellow-rumped thornbill *Acanthiza chrysorrhoa*; c) Horsfield's bronze-cuckoo and superb fairy-wren *Malurus cyaneus*; d) Horsfield's bronze-cuckoo and purple-crowned fairy-wren *M. coronatus* (from Langmore, NE, Stevens, M, Maurer, G, Heinsohn, R, Hall, ML, Peters, A, Kilner, RM (2011) Visual mimicry of host nestlings by cuckoos. Proceedings of the Royal Society: Biological Sciences, 278: 2455–2463. Photos: Naomi Langmore, Golo Maurer, Michelle Hall). See also Plate 12.

(McGill and Goddard 1979, Table 6.2). The morphological differences between the different subspecies of these bronze-cuckoo species are markedly more pronounced as nestlings than as adults.

Experiments suggest that the high fidelity of bronze-cuckoo nestling mimicry is adaptive and has arisen through coevolution with their hosts. The superb fairy-wren is a primary host of Horsfield's bronze-cuckoo and an infrequent host of the shining bronze-cuckoo. Horsfield's bronze-cuckoo chicks (pink and grey skin; white flange) are

Table 6.2 Subspecies of *Chalcites* cuckoos mimic their morphologically diverse hosts.

Cuckoo nestling	Host nestling
Little bronze-cuckoo *C. m. barnardi*[1]	**White-throated gerygone** *Gerygone olivacea*[1, 2]
Pink skin	Pink skin
? rictal flange	White rictal flange
Pale yellowish down	Pale yellowish down
Little bronze-cuckoo *C. m. minutillus*[3]	**Large-billed gerygone** *Gerygone magnirostris*[3]
Black skin	Black skin
Greenish-white rictal flange	Greenish-white rictal flange
Multi-barbed white down	Multi-barbed white down
Shining bronze-cuckoo *C. l. plagosus*[3]	**Yellow-rumped thornbill** *Acanthiza chrysorrhoa*[3]
Yellow skin	Yellow skin
Yellow rictal flange	Yellow rictal flange
Sparse or no down	Fine grey down
Shining cuckoo *C. l. lucidus*[4,5]	**Grey warbler** *Gerygone igata*[5,6]
Pink and grey skin	Pink or grey skin
White rictal flange, becomes yellow	White rictal flange, becomes yellow
Sparse white hair-like down	Abundant white down

References: [1] McGill and Goddard 1979, [2] Pers. obs. [3] Langmore et al. 2011; [4] Gill 1982; [5] Gill 1998; [6] Anderson, M. pers. comm.

more similar than shining bronze-cuckoo chicks (yellow, or rarely black skin; yellow flange) to their fairy-wren hosts (pink skin; white flange) (Fig. 6.4 Langmore et al. 2011). Cross-fostering experiments, in which cuckoos were transferred into superb fairy-wren nests at the egg stage, revealed that shining bronze-cuckoo chicks were rejected by superb fairy-wrens at a significantly higher rate (100%) than Horsfield's bronze-cuckoo chicks (40%, Langmore et al. 2003).

A close resemblance has also been reported between two non-evicting brood parasites and their hosts. Great spotted cuckoo nestlings appear (to human eyes, at least) to be similar to host young (Lack 1968). This resemblance has been attributed to low variation in altricial nestling appearance (Grim 2005; Soler et al. 1995b), but it is also plausible that it has evolved in response to host defenses. Cross-fostering experiments revealed that host magpies reject some nestlings of other species that are placed in their nests, as well as rejecting some of their own young when their skin or plumage was dyed (Alvarez et al. 1976). Mimicry of host young also occurs in screaming cowbirds *Molothrus*

rufoaxillaris, a specialist parasite of the bay-winged cowbird *M. badius*. Hosts do not discriminate against non-mimetic nestlings, but they do show discrimination against non-mimetic fledglings (Fraga 1998). Correspondingly, the parasite does not mimic host nestling skin color (Fraga 1998; but see Lichtenstein, 2001), but shows precise mimicry of host plumage as a dependent fledgling (Fraga 1998). A similarity between parasite and host young has also been reported in several other host–parasite systems (reviewed in Davies and Brooke 1988b; Grim 2006), but it is unknown whether these have evolved as a result of host–parasite interactions or through some other process (e.g. common ancestry, predator avoidance; reviewed in Grim 2005).

The precise mimicry shown by some young cuckoos and cowbirds is at odds with the perception that brood parasite chicks generally fail to mimic their hosts (Rothstein and Robinson 1998). Why is chick mimicry necessary to fool some hosts but not others? Lotem (1993) proposed a convincing explanation for the failure of many hosts to discriminate parasite chicks through learning. He demonstrated that the high cost of misimprinting on a cuckoo nestling and subsequently rejecting all future host offspring should prevent the evolution of learned chick recognition. Similarly, among hosts of non-evicting parasites learned recognition of brood parasite chicks is only likely to evolve under a relatively rare suite of conditions: when host chick survival in parasitized nests is low, rates of parasitism are high and clutch size is large (Lawes and Marthews 2003). Further, some non-evicting parasites carry a lower cost to hosts than evicting parasites, and may not depress host reproductive success sufficiently to outweigh the costs of nestling recognition errors.

Despite the solid theoretical basis for dismissing the possibility of learned nestling recognition, a subsequent study demonstrated that learning can indeed facilitate discrimination of parasite chicks, if used in conjunction with other mechanisms for discrimination that decrease the probability of misimprinting. Superb fairy-wrens optimize chick rejection decisions by integrating learned recognition cues and external cues (Langmore et al. 2009a). Females only reject chicks that are alone in the nest, and only when adult cuckoos are present in the area, thereby greatly reducing the possibility of making a recognition error. Further, experienced breeders are both more likely both to reject a lone cuckoo chick and accept a lone fairy-wren chick than naïve females, indicating that rejection decisions improve with breeding experience.

An alternative explanation for the lack of chick discrimination in many hosts is that the success of earlier lines of defense against parasitism (e.g., mobbing, egg rejection) may result in diminishing returns from a later line of defense (Britton et al. 2007; Dawkins and Krebs 1979; Grim 2006; Planque et al. 2002). For example, if mobbing successfully foils most attempts at parasitism, a parasite egg or chick in the nest may be such a rare event that the benefits of evolving rejection behavior are outweighed by rejection costs (e.g., Krüger 2011). This could explain the variability between hosts in the elaboration of defense portfolios, and thereby explain the corresponding variability between brood parasites in the evolution of mimicry to circumvent host defenses.

6.5 Visual trickery to elicit parental care

Brood parasitic nestlings face a twofold challenge in the nests of their hosts; not only must they circumvent any host recognition systems (Section 6.4 above), but they must also provide the appropriate signals to stimulate provisioning by their hosts. To do so requires exploiting the parent–offspring communication system of the host species (Kilner and Davies 1999). The begging display of nestling birds typically comprises a combination of visual and vocal signals, including begging calls, posturing, and brightly colored gapes. Parents integrate these signals to adjust their provisioning rate to the hunger level of their brood (Kilner and Davies 1999).

In order to stimulate provisioning by the host, mimicry of host begging signals may be necessary. For example, the brood parasitic *Vidua* finches of Africa, which are reared alongside the host young, mimic the elaborate gape patterns of their estrildid finch hosts. Most *Vidua* species are specific to a single host species, and precisely match the distinctive gape pattern of that host (Fig. 6.5). The resemblance does not appear to have evolved in response to host defenses, because hosts fail to reject young with abnormal gape markings (Schuetz 2005a; Schuetz 2005b). However, when gape markings were obscured experimentally, nestlings suffered a reduced provisioning rate, suggesting that the markings provide a necessary stimulus to elicit provisioning by host parents (Schuetz 2005b). This experiment provides an elegant explanation for why gape mimicry has evolved, but it does not explain why gape patterns have diversified between host species. Recently, an intriguing hypothesis was put forward to explain this conundrum. Hauber and Kilner (2007) suggested that competition between parasite and host young for parental resources might lead to exaggeration of those aspects of the signal that most effectively exploit host parents. *Vidua* chicks are likely to experience stronger selection for exaggerated signals than host young, because they are unrelated to the other chicks in the nest and are therefore under selection to behave more selfishly (Lichtenstein 2001). However, as parasite chicks become more efficient

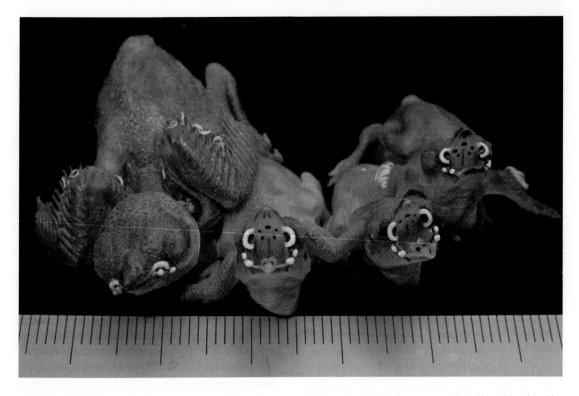

Figure 6.5 Parasitic pin-tailed whydah *Vidua macroura* nestlings (the two outer chicks) are visual mimics of common waxbill *Estrilda astrild* nestlings (the two center chicks, photo: Justin Schuetz). See also Plate 13.

at monopolizing parental care, host young will likewise experience selection for exaggerated signals to allow them to compete effectively with parasites. Although the exaggeration of signals is likely to be constrained to remain below a certain threshold to ensure acceptance by the host (Holen et al. 2001), this scenario could ultimately result in a paradoxical situation in which host young evolve mimicry of parasite young (Hauber and Kilner 2007).

Brood parasites that are larger than their hosts face an additional challenge in acquiring sufficient resources from their host parents. Most cuckoo species are larger than their hosts, and some cuckoo chicks can grow to more than ten times the size of their foster parents (Davies 2000). Therefore a cuckoo chick requires more food than a single host chick, yet it can display only one gape and the begging call of one individual. How then do parasitic chicks ensure that host parents provide them with sufficient food? Studies on two species of cuckoo reveal ingenious solutions to this problem.

The common cuckoo is reared alone in the nest, yet the rate at which the host parents feed the cuckoo chick is similar to the rate at which they provision a brood of four of their own chicks (Kilner and Davies 1999). The cuckoo chick solicits such a high rate of care partly by exploiting the tendency of its reed warbler hosts to vary provisioning rate in relation to the total area of nestling gapes presented during a nest visit. By presenting a relatively large gape to its foster parents, the cuckoo can solicit a greater rate of care than a single host chick. However, this alone does not elicit sufficiently high provisioning rates, and the cuckoo supplements its larger gape by producing an extremely rapid begging call, which simulates the sounds produced by four host nestlings (Kilner and Davies 1999).

Figure 6.6 The Horsfield's hawk-cuckoo reveals a bare wing patch during its begging display to simulate multiple begging mouths and thereby drive its foster parents to greater provisioning efforts. In this photo, the host is thoroughly fooled, and tries to place food in the wing patch (photo: Keisuke Ueda).

The Horsfield's hawk-cuckoo *Cuculus fugax* of south-east Asia has come up with a different solution to the challenge of eliciting sufficient care from its hosts. During the begging display, the cuckoo chick raises its wing to reveal a naked skin patch under the wing, which mimics the appearance of a nestling gape (Fig. 6.6; Tanaka and Ueda 2005). The wing patch is displayed at a greater rate when the provisioning rate of the foster parents is slow. The hosts are apparently fooled by the wing patch, because they sometimes try to place food in the wing patch rather than the gape (Fig. 6.6), and because experimental darkening of the wing patch results in a reduced provisioning rate (Tanaka and Ueda 2005). These two studies cleverly illustrate how cuckoos have evolved the ability to solicit sufficient care by tuning into the sensory predispositions of their hosts.

6. 6 Mimicry in generalist versus specialist parasites

Mimicry of hosts is a more challenging feat for a generalist parasite than a specialist. Generalists parasitize several different host species, each of which may display distinct egg and chick morphologies. Some types of signals can be modified after parasitism to resemble the appropriate host species. For example, nestling Horsfield's bronze-cuckoos can modify their begging calls to mimic those of the host species that rears them (Langmore et al. 2008).

Similarly, insect brood parasites may acquire the colony-specific hydrocarbon signature of their host after parasitism (Kilner and Langmore 2011; Lenoir et al. 2001). However, visual signals generally lack the phenotypic plasticity necessary for modification after parasitism. Instead, generalist brood parasites have come up with two alternative solutions to the problem of mimicking morphologically diverse hosts.

First, brood parasites may evolve host-specific gentes, each of which produces progeny that matches that of its favored host. For example, as discussed earlier (Section 6.3) several cuckoo species (as well as, independently, the cuckoo finch) have evolved gentes, each of which lays an egg that matches that of its host. Host-specific races can be maintained as a single species where there are cross-matings of males with females of different gentes, as has been described in the common cuckoo (Gibbs et al. 2000) and greater honeyguide (Spottiswoode et al. 2011), or they can diverge into separate species where there is assortative mating between males and females of the same gens. The latter process appears to account for the remarkably rapid speciation of *Vidua* finches, which not only resemble their specialist hosts with respect to chick mouth markings, but are exceptional in that both sexes imprint on their respective hosts' songs (males mimic them and females are attracted to them); thus instant host-specific reproductive isolation and speciation can result (Sorenson et al. 2003). In common cuckoos, evidence for a degree of assortative mating according to host origin has also been found on a local scale (Fossøy et al. 2010), which if widespread and consistent may represent the beginnings of such a process. Likewise, we can speculate that diversifying selection driven by host discrimination of cuckoo chicks (and consequent selection for cuckoo mimicry of host chicks) could explain the diversity of species and subspecies of *Chalcites* cuckoos (Langmore et al. 2011). Thus selection on cuckoos for mimicry of host eggs and chicks could explain recent evidence that species richness is higher in parasitic than non-parasitic cuckoos (Krüger et al. 2009).

Second, generalist parasites may adopt a morphology that is intermediate between that of several hosts ("imperfect mimicry," Sherratt 2002). This strategy appears to have been adopted by the common cuckoo in addition to evolving gentes. Each gens lays an egg that matches its favored host, but the eggs display more "average" markings than host eggs, perhaps to facilitate parasitism of subsidiary hosts (Stoddard and Stevens 2010). Similarly, the Horsfield's bronze-cuckoo, which is the most generalist of the Australian bronze-cuckoos, produces nestlings that are intermediate in color between their various host species (Langmore et al. 2011). The nestlings display pink skin on the upper back, and grey skin on the head and lower back, thereby achieving a moderate resemblance to both light and dark skinned host species (Langmore et al. 2011). Imperfect mimicry may carry a cost to the parasite, because generalist Horsfield's bronze-cuckoo nestlings in superb fairy-wren nests tend to suffer a higher rejection rate (38%, Langmore et al. 2003) than the specialist shining bronze-cuckoo in yellow-rumped thornbill nests (0%, Langmore unpublished data) or the specialist little bronze-cuckoo in mangrove gerygone (18%, Tokue and Ueda 2010) or large-billed gerygone nests (36%, Sato et al. 2010).

6.7 Conclusions

The dazzling array of visual tricks displayed by brood parasites to manipulate their hosts has evolved through two processes: either as a coevolutionary response to host defenses against brood parasites, or to "tune in" to pre-existing host provisioning strategies. Davies (2011) terms these two processes "trickery" and "tuning." The most common form of visual trick selected by both processes is mimicry, which has evolved at all three stages of the parasitic process: gaining access to host nests, the egg stage, and the chick stage (Welbergen and Davies, 2011). Aggressive mimicry, in which the predatory or parasitic species mimics a harmless or inviting species in order to gain access to resources (Ruxton et al. 2004), is the most common form of mimicry in brood parasites, but Batesian mimicry, in which a defenseless species mimics a harmful one, also occurs in the case of hawk mimicry by common cuckoos (Welbergen and Davies 2011). The special case of parasite "tuning" into host provisioning strategies has given rise to additional visual tricks in the form of supernormal stimuli, such as large or false gapes. In general, trickery and tuning have different impacts on host populations. Trickery evolves in response to host defenses and usually provokes reciprocal genetic change in host populations, as they evolve counter-adaptations to cuckoo tricks (e.g., Lahti 2005; Langmore et al. 2003; Spottiswoode and Stevens, in press; Yang et al. 2010). By contrast, tuning allows the parasite to exploit pre-existing host provisioning strategies and does not typically lead to reciprocal adaptations in hosts (but see Hauber and Kilner 2007). The recent technological advances that allow quantification of mimicry through the eyes of the host have provided us with fascinating insights into the exquisite subtlety of visual trickery in brood parasites. They also pave the way for future research into the relatively less well-understood reciprocal adaptations of host populations in response to the visual tricks of brood parasites.

Acknowledgements

We thank Nick Davies and Justin Welbergen for helpful comments on the manuscript, and Roy and Marie Battell, Steve Grimwade, Michelle Hall, Golo Maurer, Justin Schuetz, Keisuke Ueda, and Jack Wolf for providing photographs. NEL was supported by an Australian Research Council Australian Research Fellowship. CNS was supported by a

Royal Society Dorothy Hodgkin Fellowship and the DST/NRF Centre of Excellence at the Percy FitzPatrick Institute, University of Cape Town.

References

Alvarez, F. (1999). Attractive non-mimetic stimuli in Cuckoo *Cuculus canorus* eggs. *Ibis* **141**, 142–144.

Alvarez, F. (2000). Response to Common Cuckoo *Cuculus canorus* model egg size by a parasitized population of Rufous Bush Chat *Cercotrichas galactotes*. *Ibis* **142**, 683–686.

Alvarez, F., Arias De Reyna, L., and Segura, M. (1976). Experimental brood parasitism of the magpie (*Pica pica*). *Animal Behaviour* **24**, 907–916.

Avilés, J. M. (2008). Egg colour mimicry in the common cuckoo *Cuculus canorus* as revealed by modelling host retinal function. *Proceedings of the Royal Society of London B* **275**, 2345–2352.

Avilés, J. M., Stokke, B. G., Moksnes, A., Røskaft, E., Asmul, M., and Møller, A. P. (2006a). Rapid increase in cuckoo egg matching in a recently parasitized reed warbler population. *Journal of Evolutionary Biology* **19**, 1901–1910.

Avilés, J. M., Stokke, B. G., Moksnes, A., Røskaft, E., and Møller, A. P. (2006b). Nest predation and the evolution of egg appearance in passerine birds in Europe and North America. *Evolutionary Ecology Research* **8**, 493–513.

Baerends, G. P. and Drent, R. H. (1982). The herring gull and its egg. Part II. The responsiveness to egg features. *Behaviour* **82**, 1–416.

Beecher, M. D. (1982). Signature systems and kin recognition. *American Zoologist*, **22**, 477–490.

Britton, N. F., Planque, R., and Franks, N. R. (2007). Evolution of defence portfolios in exploiter-victim systems. *Bulletin of Mathematical Biology* **69**, 957–988.

Brooke, M. de L. and Davies, N. B. (1988). Egg mimicry by cuckoos *Cuculus canorus* in relation to discrimination by hosts. *Nature* **335**, 630–632.

Brooker, M. G. and Brooker, L. C. (1989). The comparative breeding behaviour of two sympatric cuckoos, Horsfield's bronze-cuckoo *Chrysococcyx basalis* and the shining bronze-cuckoo *C. lucidus*, in Western Australia: a new model for the evolution of egg morphology and host specificity in avian brood parasites. *Ibis* **131**, 528–547.

Brooker, L. C., Brooker, M. G., and Brooker, A. M. H. (1990). An alternative population/genetics model for the evolution of egg mimesis and egg crypsis in cuckoos. *Journal of Theoretical Biology* **146**, 123–143.

Cassey, P., Honza, M., Grim, T., and Hauber, M. E. (2008). The modelling of avian visual perception predicts behavioural rejection responses to foreign egg colours. *Biology Letters* **4**, 515–517.

Cherry, M. I. and Bennett, A. T. D. (2001). Egg colour matching in an African cuckoo, as revealed by ultraviolet-visible reflectance spectrophotometry. *Proceedings of the Royal Society of London B* **268**, 565–572.

Cherry, M. I., Bennett, A. T. D., and Moskát, C. (2007a). Host intra-clutch variation, cuckoo egg matching and egg rejection by great reed warblers. *Naturwissenschaften* **94**, 441–447.

Cherry, M. I., Bennett, A. T. D., and Moskát, C. (2007b). Do cuckoos choose nests of great reed warblers on the basis of host egg appearance? *Journal of Evolutionary Biology* **20**, 1218–1222.

Craib, J. (1994). Why do common cuckoos resemble raptors? *British Birds* **87**, 78–79.

Dale, J., Lank, D. B., and Reeve, H. K. (2001). Signaling individual identity versus quality: a model and case studies with ruffs, queleas, and house finches. *American Naturalist* **158**, 75–86.

Davies, N. B. (1992). *Dunnock Behaviour and Social Evolution.* Oxford University Press, Oxford.

Davies, N. B. (1999). Cuckoos and cowbirds versus hosts: co-evolutionary lag and equilibrium. *Ostrich* 70, 71–79.

Davies, N. B. (2011). Cuckoo adaptations: trickery and tuning. *Journal of Zoology* 284, 1–14.

Davies, N. B. and Brooke, M.de L. (1988). Cuckoos versus reed warblers: adaptations and counteradaptations. *Animal Behaviour* **36**, 262–284.

Davies, N. B. and Brooke, M. de L. (1989a). An experimental study of co-evolution between the cuckoo, *Cuculus canorus*, and its hosts. I. Host egg discrimination. *Journal of Animal Ecology* **58**, 207–224.

Davies, N. B. and Brooke, M. de L. (1989b). An experimental study of co-evolution between the cuckoo, *Cuculus canorus*, and its hosts. II. Host egg markings, chick discrimination and general discussion. *Journal of Animal Ecology* **58**, 225–236.

Davies, N. B. and Welbergen, J. A. (2008). Cuckoo-hawk mimicry? An experimental test. *Proceedings of the Royal Society of London B* **275**, 1817–1822.

Dawkins, R. and Krebs, J. R. (1979). Arms races between and within species. *Proceedings of the Royal Society of London B* **205**, 489–511.

Dominey, W. J. (1980). Female mimicry in male bluegill sunfish—a genetic polymorphism? *Nature* **284**, 546–548.

Dumbacher, J. P. and Fleischer, R. C. (2001). Phylogenetic evidence for colour pattern convergence in toxic pitohuis: Müllerian mimicry in birds? *Proceedings of the Royal Society of London B* **268**, 1971–1976.

Endler, J. A. and Mielke, P. W. J. (2005). Comparing color patterns as birds see them. *Biological Journal of the Linnean Society* **86**, 405–431.

Fossøy, F., Antonov, A., Moksnes, A., Røskaft, E., Vikan, J. R., Møller, A. P., Shykoff, J. A., and Stokke, B. G. (2010). Genetic differentiation among sympatric cuckoo host races: males matter. *Proceedings of the Royal Society of London B* **278**, 1639–1645.

Fraga, R. M. (1985). Host-parasite interactions between Chalk-browed Mockingbirds and Shiny Cowbirds. *Ornithological Monographs* **36**, 829–844.

Fraga, R. M. (1998). Interactions of the parasitic screaming and shiny cowbirds (Molothrus rufoaxillaris and M. bonariensis) with a shared host, the bay-winged cowbird (M. badius). In: *Parasitic Birds and their Hosts* (eds Rothstein, S.I. and Robinson, S.K.) Oxford University Press, Oxford.

Gibbs, H. L., Sorenson, M. D., Marchetti, K., Brooke, M. D., Davies, N. B., and Nakamura, H. (2000). Genetic evidence for female host-specific races of the common cuckoo. *Nature* **407**, 183–186.

Gill, B. J. (1982). Notes on the shining cuckoo (*Chrysococcyx lucidus*) in New Zealand. *Notornis* **29**, 215–227.

Gill, B. J. (1988). Behaviour and ecology of the shining cuckoo, *Chrysococcyx lucidus*. In: *Parasitic Birds and their Hosts* (eds Rothstein, S. I. and Robinson, S. K.) Oxford University Press, Oxford.

Grim, T. (2005). Mimicry vs. similarity: which resemblances between brood parasites and their hosts are mimetic and which are not? *Biological Journal of the Linnean Society* **84**, 69–78.

Grim, T. (2006). The evolution of nestling discrimination by hosts of parasitic birds: why is rejection so rare? *Evolutionary Ecology Research* **8**, 785–802.

Harper, G. R. and Pfenning, D. W. (2008). Selection overrides gene flow to break down maladaptive mimicry. *Nature* **451**, 1103–1106.

Hauber, M. E. and Kilner, R. M. (2007). Coevolution, communication and host-chick mimicry in parasitic finches: who mimics whom? *Behavioural Ecology and Sociobiology* **61**, 497–503.

Higuchi, H. (1989). Responses of the bush warbler *Cettia diphone* to artificial eggs of Cuculus cuckoos in Japan. *Ibis* **131**, 94–98.

Holen, O. H., Saetre, G.-P., Slagsvold T., and Stenseth, N. C. (2001). Parasites and supernormal manipulation. *Proceedings of the Royal Society of London B* **268**, 2551–2558.

Honza, M., Kuiper, S. M., and Cherry, M. I. (2005). Behaviour of African turdid hosts towards experimental parasitism with artificial red-chested cuckoo *Cuculus solitarius* eggs. *Journal of Avian Biology* **36**, 514–522.

Hunter, H. C. (1961). Parasitism of the Masked Weaver *Ploceus velatus arundinaceus*. *Ostrich* **32**, 55–63.

Jackson, W. M. (1998). Egg discrimination and egg-color variability in the northern masked weaver: the importance of conspecific versus interspecific parasitism. In: In: *Parasitic Birds and their Hosts* (eds Rothstein, S. I. and Robinson, S. K.). Oxford University Press, Oxford.

Johnsgard, P. A. (1997). *The Avian Brood Parasites: Deception at the Nest,* Oxford University Press, Oxford.

Kilner, R. M. (2006). The evolution of egg colour and patterning in birds. *Biological Reviews* **81**, 383–406.

Kilner, R. M. and Davies, N. B. (1999). How selfish is a cuckoo chick? *Animal Behaviour* **58**, 797–808.

Kilner, R. M. and Langmore, N. E. (2011). Cuckoos versus hosts in insects and birds: adaptations, counter-adaptations and outcomes. *Biological Reviews* **86**, 836–852.

Krüger, O. (2011). Brood parasitism selects for no defence in a cuckoo host. *Proceedings of the Royal Society of London B* **278**, 2777–2783.

Krüger, O., Davies, N. B., and Sorenson, M. D. (2007). The evolution of sexual dimorphism in parasitic cuckoos: sexual selection or coevolution? *Proceedings of the Royal Society of London B* **274**, 1553–1560.

Krüger, O., Sorenson, M. D., and Davies, N. B. (2009). Does coevolution promote species richness in parasitic cuckoos? *Proceedings of the Royal Society of London B* **276**, 3871–3879.

Lack, D. (1968). *Ecological Adaptations for Breeding in Birds.* Methuen & Co. Ltd, London.

Lahti, D. C. (2005). Evolution of bird eggs in the absence of cuckoo parasitism. *Proceedings of the National Academy of Sciences USA* **102**, 18057–18062.

Lahti, D. C. (2006). Persistence of egg recognition in the absence of cuckoo brood parasitism: pattern and mechanism. *Evolution* **60**, 157–168.

Lahti, D. C.and Lahti, A. R. (2002). How precise is egg discrimination in weaverbirds? *Animal Behaviour* **63**, 1135–1142.

Landstrom, M. T., Heinsohn, R., and Langmore, N. E. (2010). Clutch variation and egg rejection in three hosts of the pallid cuckoo, *Cuculus pallidus*. *Behaviour* **147**, 19–36.

Langmore, N. E., Cockburn, A., Russell, A. F., and Kilner, R. M. (2009a). Flexible cuckoo chick-rejection rules in the superb fairy-wren. *Behavioral Ecology* **20**, 978–984.

Langmore, N. E., Hunt, S., and Kilner, R. M. (2003). Escalation of a coevolutionary arms race through host rejection of brood parasitic young. *Nature* **422**, 157–160.

Langmore, N. E. and Kilner, R. M. (2009). Why do Horsfield's bronze-cuckoo *Chalcites basalis* eggs mimic those of their hosts? *Behavioural Ecology and Sociobiology* **63**, 1127–1131.

Langmore, N. E., Kilner, R. M., Butchard, S. H. M., Maurer, G., Davies, N. B., Cockburn, A., Macgregor, N. A., Peters, A., Magrath, M. J. L., and Dowling, D. (2005). The evolution of egg rejection by cuckoo hosts in Australia and Europe. *Behavioural Ecology* **16**, 686–692.

Langmore, N. E., Maurer, G., Adcock G. J., and Kilner, R. M. (2008). Socially acquired host-specific mimicry and the evolution of host races in Horsfield's bronze-cuckoo Chalcites basalis. *Evolution* **62**, 1689–1699.

Langmore, N. E., Stevens, M., Heinsohn, R., Hall, M. L., Peters, A., and Kilner, R. M. (2011). Visual mimicry of host nestlings by cuckoos. *Proceedings of the Royal Society of London B* **278**, 2455–2463.

Langmore, N. E., Stevens, M., Maurer, G., and Kilner, R. M. (2009b). Are dark cuckoo eggs cryptic in host nests? *Animal Behaviour* **78**, 461–468.

Lawes, M. J. and Marthews, T. R. (2003). When will rejection of parasite nestlings by hosts of nonevicting avian brood parasites be favored? A misimprinting-equilibrium model. *Behavioural Ecology* **14,** 757–770.

Lenoir, A., D'Ettoir, P., and Errard, C. (2001). Chemical ecology and social parasitism in ants. *Annual Review of Entomology* **46**, 573–599.

Lichtenstein, G. (2001). Low success of shiny cowbird chicks parasitizing rufous-bellied thrushes: chick-chick competition or parental discrimination? *Animal Behaviour* **61**, 401–413.

Lively, C. M. and Dybdahl, M. F. (2000). Parasite adaptation to locally common host genotypes. *Nature* **405**, 679–681.

Lotem, A. (1993). Learning to recognize nestlings is maladaptive for cuckoo Cuculus canorus hosts. *Nature* **362**, 743–745.

Lyon, B. E. and Eadie, J. M. (2004). An obligate brood parasite trapped in the intraspecific arms race of its hosts. *Nature* **432**, 390–393.

Mason, P. and Rothstein, S. I. (1987). Crypsis versus mimicry and the color of shiny cowbird eggs. *American Naturalist* **130**, 161–167.

McGill, I. G. and Goddard, M. T. (1979). The little bronze cuckoo in New South Wales. *Australian Birds* **14**, 23–24.

Mclean, I. G. and Maloney, R. F. (1998). Brood parasitism: recognition, and response: the options. In: *Parasitic Birds and their Hosts*. (Rothstein, S. I. and Robinson, S. K. eds) Oxford University Press, Oxford.

Mermoz, M. E. and Reboreda, J. C. (1994). Brood parasitism of the shiny cowbird, *Molothrus bonariensis*, on the brown-and-yellow marshbird, *Pseudoleistes virescens*. *Condor* **96**, 716–721.

Moksnes, A., Røskaft, E., Braa, A. T., Korsnes, L., Lampe, H. L., and Pedersen, H. C. (1990). Behavioural responses of potential hosts towards artificial cuckoo eggs and dummies. *Behaviour* **166**, 64–89.

Molnar, B. (1944). The cuckoo in the Hungarian plain. *Aquila* **51**, 100–112.

Moreno, J. and Osorno, J. L. (2003). Avian egg colour and sexual selection: does eggshell pigmentation reflect female condition and genetic quality? *Ecology Letters* **6**, 803–806.

Moskát, C., Avilés, J. M., Ban, M., Hargitai, R., and Zolei, A. (2008). Experimental support for the use of egg uniformity in parasite egg discrimination by cuckoo hosts. *Behavioural Ecology and Sociobiology* **62**, 1885–1890.

Moyer, D. C. (1980). On lesser honeyguide and black-collared barbet. *Zambian Ornithological Society Newsletter* **10**, 159.

Nelson, K. J. and Jackson, R. R. (2006). Compound mimicry and trading predators by the males of sexually dimorphic Batesian mimics. *Proceedings of the Royal Society of London B* **273**, 367–372.

Newton, A. (1896). *A Dictionary of Birds*. Adam & Charles Black, London.

Noble, D. G. (1995). *Coevolution and ecology of seven sympatric cuckoo species and their hosts in Namibia* (PhD thesis), University of Cambridge, Cambridge.

Øien, I. J., Moksnes, A., and Røskaft, E. (1995). Evolution of variation in egg color and marking pattern in European passerines: adaptations in a coevolutionary arms race with the cuckoo, *Cuculus canorus*. *Behavioural Ecology* **6**, 166–174.

Payne, R. B. (1967). Interspecific communication signals in parasitic birds. *American Naturalist* **101**, 363–375.

Payne, R. B. (1977). The ecology of brood parasitism in birds. *Annual Review of Ecology and Systematics* **8**, 1–28.

Peer, B. D., Ellison, K. S., and Sealy, S. G. (2002). Intermediate frequencies of egg ejection by Northern Mockingbirds (Mimus polyglottos) sympatric with two cowbird species. *Auk* **119**, 855–858.

Planque, R., Britton, N. F., Franks, N. R., and Peletier, M. A. (2002). The adaptiveness of defence strategies against cuckoo parasitism. *Bulletin of Mathematical Biology* **64**, 1045–1068.

Reed, R. A. (1968). Studies of the Diederik Cuckoo *Chrysococcyx caprius* in the Transvaal. *Ibis* **110**, 321–331.

Reynolds, S. J., Martin, G. R., and Cassey, P. (2009). Is sexual selection blurring the functional significance of eggshell coloration hypotheses? *Animal Behaviour* **78**, 209–215.

Rothstein, S. I. (1975). Mechanisms of avian egg-recognition: do birds know their own eggs? *Animal Behaviour* **23**, 268–278.

Rothstein, S. I. (1982). Mechanisms of avian egg recognition: which egg parameters elicit responses by rejecter species? *Behavioural Ecology and Sociobiology* **11**, 229–239.

Rothstein, S. I. (1992). Brood parasitism, the importance of experiments and host defences of avifaunas on different continents. *Proc. VII Pan-Afr. Orn. Congr.*, 521–535.

Rothstein, S. I. and Robinson, S. K. (1998). *Parasitic Birds and their Hosts: Studies in Coevolution.* Oxford University Press, Oxford.

Ruxton, G. D., Sherratt, T. N., and Speed, M. P. (2004). *Avoiding Attack: The Evolutionary Ecology of Crypsis, Warning Signals and Mimicry.* Oxford University Press, Oxford.

Sato, N. J., Tokue, K., Noske, R. A., Mikami, O. K., and Ueda, K. (2010). Evicting cuckoo nestlings from the nest: a new anti-parasite behaviour. *Biology Letters* **6**, 67–69.

Schuetz, J. G. (2005a). Low survival of parasite chicks may result from their imperfect adaptation to hosts rather than expression of defenses against parasitism. *Evolution* **59**, 2017–2024.

Schuetz, J. G. (2005b). Reduced growth but not survival of chicks with altered gape patterns: implications for the evolution of nestling similarity in a parasitic finch. *Animal Behaviour* **70**, 839–848.

Schulze-Hagen, K., Stokke, B. G., and Birkhead, T. R. (2009). Reproductive biology of the European Cuckoo *Cuculus canorus*: early insights, persistent errors and the acquisition of knowledge. *Journal of Ornithology* **150**, 1–16.

Sherratt, T. N. (2002). The evolution of imperfect mimicry. *Behavioural Ecology* **13**, 821–826.

Smith, N. G. (1968). The advantage of being parasitized. *Nature* **219**, 690–694.

Soler, M., Martinez, J. G., Soler, J. J., and Møller, A. P. (1995a). Preferential allocation of food by magpies *Pica pica* to great spotted cuckoo *Clamator glandarius* chicks. *Behavioural Ecology and Sociobiology* **37**, 7–13.

Soler, M. and Møller, A. P. (1990). Duration of sympatry and coevolution between the great spotted cuckoo and its magpie host. *Nature* **343**, 748–750.

Soler, J. J. and Møller, A. P. (1996). A comparative analysis of the evolution of variation in appearance of eggs of European passerines in relation to brood parasitism. *Behavioural Ecology* **7**, 89–94.

Soler, M., Soler, J. J., Martinez, J. G., and Møller, A. P. (1995b). Chick recognition and acceptance: a weakness in magpies exploited by the parasitic great spotted cuckoo. *Behavioural Ecology and Sociobiology* **37**, 243–248.

Sorenson, M. D., Sefc, K. M., and Payne, R. B. (2003). Speciation by host switch in brood parasitic indigobirds. *Nature* **424**, 928–931.

Spottiswoode, C. N. and Stevens, M. (2010). Visual modeling shows that avian host parents use multiple visual cues in rejecting parasitic eggs. *Proceedings of the National Academy of Sciences USA* **107**, 8672–8676.

Spottiswoode, C. N. and Stevens, M. (2011). How to evade a coevolving brood parasite: egg discrimination versus egg variability as host defences. *Proceedings of the Royal Society of London B* **278**, 3566–3573.

Spottiswoode, C. N. & Stevens, M. (2012) Host-parasite arms races and rapid changes in bird egg appearance. *American Naturalist*.

Spottiswoode, C. N., Stryjewski, K. F., Quader, S., Colebrook-Robjent, J. F.R., and Sorenson, M. D. (2011). Ancient host-specificity within a single species of brood parasitic bird. *Proceedings of the National Academy of Sciences of the USA* **108**, 17738–17742.

Starling, M., Heinsohn, R., Cockburn, A., and Langmore, N. E. (2006). Cryptic gentes revealed in pallid cuckoos *Cuculus pallidus* using reflectance spectrophotometry. *Proceedings of the Royal Society of London B* **273**, 1929–1934.

Stoddard, M. C. and Stevens, M. (2010). Pattern mimicry of host eggs by the common cuckoo, as seen through a bird's eye. *Proceedings of the Royal Society of London B* **277**, 1387–1393.

Stoddard, M. C. and Stevens, M. (2011). Avian vision and the evolution of egg color mimicry in the common cuckoo. *Evolution* **65**, 2004–2013.

Stokke, B. G., Moksnes, A. & Røskaft, E. (2002). Obligate brood parasites as selective agents for the evolution of egg appearance in passerine birds. *Evolution* **56**, 199–205.

Swynnerton, C. F. M. (1918). Rejections by birds of eggs unlike their own: with remarks on some of the cuckoo problems. *Ibis* **6**, 127–154.

Takasu, F. (2003). Co-evolutionary dynamics of egg appearance in avian brood parasitism. *Evolutionary Ecology Research* **5**, 345–362.

Takasu, F. (2005). A theoretical consideration on co-evolutionary interactions between avian brood parasites and their hosts. *Ornithological Science* **4**, 65–7.

Tanaka, K. D. and Ueda, K. (2005). Horsfield's hawk-cuckoo nestlings simulate multiple gapes for begging. *Science* **308**, 653–653.

Tarboton, W. (1986). African cuckoo: the agony and ecstasy of being a parasite. *Bokmakierie* **38**, 109–111.

Tokue, K. and Ueda K. (2010). Mangrove gerygones reject and eject little bronze-cuckoo hatchlings from parasitized nests. *Ibis* **152**, 835–839.

Van Valen, L. (1973). A new evolutionary law. *Evolutionary Theory* **1**, 1–30.

Plate 1 The nematode *Gammarinema gammari*. This nematode uses *G. insensibilis* as a habitat and source of nutrition, not as a vector for transmission to birds. There is evidence that this nematode prefers non-manipulated gammarids and seems also able to reverse the manipulation induced by *M. papillorobustus*, changing the gammarid's altered behavior back to a normal behavior (Thomas et al. 2002b) © Pascal Goetgheluck. See also Fig. 2.4, page 24.

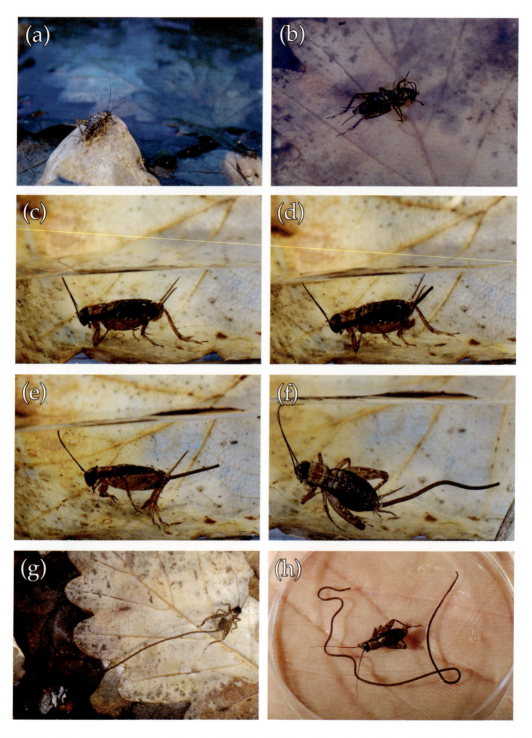

Plate 2 Water-seeking behavior in crickets (*Nemobius syslvestris*) harboring the hairworm *Paragordius tricuspidatus*. This behavioral sequence is only the second step within the host manipulation process. © Pascal Goetgheluck. See also Fig. 2.5, page 24.

Plate 3 (a) Crazy gammarids actively swimming toward the surface following a mechanical disturbance in the water. (b) Photophilic behavior of crazy gammarids in the absence of a mechanical disturbance. © Pascal Goetgheluck. See also Fig. 2.7, page 28.

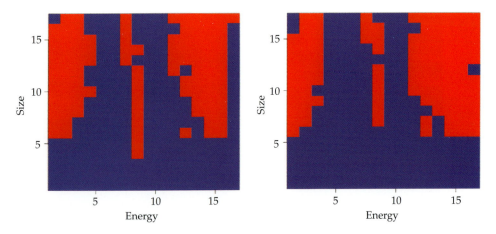

Plate 4 Optimal activities as a function of energy (x axis) and body size (y axis) for a healthy "caterpillar" when feeding on plants with two different within-plant distributions of nutrient quality: in 4.2a, the distribution of resources is left skewed toward low values and in 4.2b, the distribution of resources is left skewed toward high values. Three activities are possible: Lightest shade is for leave the plant, medium shade is to move within the plant, and darkest shade is for a remain-at-the-site decision. For the set of values shown, plant emigration (lightest shade) is never an optimal decision. See also Fig. 4.2, page 60.

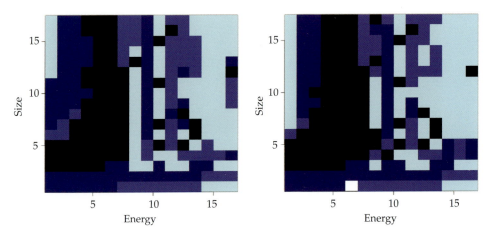

Plate 5 Optimal allocation of energy to growth as a function of energy (x axis) and body size (y axis) for a healthy "caterpillar" when feeding on plants with two different within-plant distributions of nutrient quality: in 4.3a, the distribution of resources is left skewed toward low values and in 4.3b, the distribution of resources is left skewed toward high values. The darker the cell the greater the proportion of energy that is allocated to growth. See also Fig. 4.3, page 61.

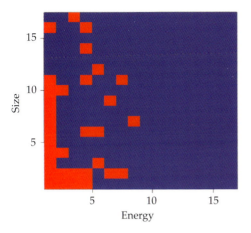

Plate 6 Optimal activities as a function of energy (x axis) and body size (y axis) for a parasitized "caterpillar", in control of its body, when feeding on a plant where the distribution of nutrients is left skewed toward low values. Lightest shade is for leave the plant, medium shade is to move within the plant, and darkest shade is for a remain-at-the-site decision. See also Fig. 4.4, page 63.

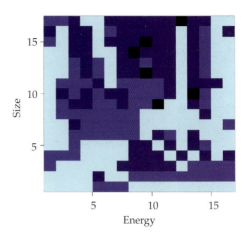

Plate 7 Optimal allocation of energy to growth as a function of energy (x axis) and body size (y axis) for a parasitized "caterpillar" in control of its own body, when feeding on a plant where the distribution of nutrients is left skewed toward low values. The darker the cell the greater the proportion of energy that is allocated to growth. See also Fig. 4.5, page 63.

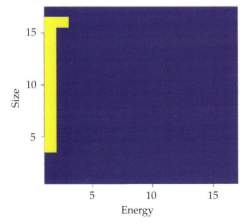

Plate 8 Optimal activities for a "caterpillar" as a function of its energy (x axis) and body size (y axis), where the parasite controls its host body, when the host is feeding on a plant where the distribution of nutrients is left skewed toward low values. Lightest shade is for leave the plant, medium shade is to move within the plant and darkest shade is for a remain-at-the-site decision. See also Fig. 4.6, page 63.

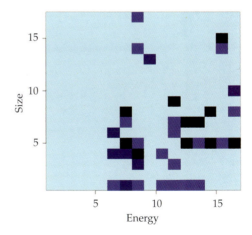

Plate 9 Optimal extraction rates as a function of host energy (x axis) and body size (y axis) for a parasite in control of its host's body when found on plants where the distribution of nutrients is left skewed toward low values. The darker the cell the greater the extraction rate. See also Fig. 4.7, page 65.

Plate 10 Eggs of parasitic cuckoo finches (right column) and two host species (left column). The top two rows are from red-faced cisticola nests and the bottom four rows from tawny-flanked prinia nests; note the extensive variation both between and within host species, mimicked by the parasite (Photo by Claire Spottiswoode). See also Fig. 6.2, page 101.

Plate 11 The cryptic egg of a shining bronze-cuckoo *Chalcites lucidus* with two eggs of its host, a yellow-rumped thornbill *Acanthiza chryssorhoa* (Photo by Naomi Langmore). See also Fig. 6.3, Page 104.

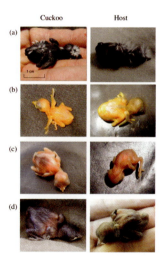

Plate 12 Visual mimicry of host chicks by nestling Chalcites cuckoos. a) Little bronze-cuckoo *C. minutillus* and large-billed gerygone *Gerygone magnirostris*; b) Shining bronze-cuckoo *C. lucidus* and yellow-rumped thornbill *Acanthiza chrysorrhoa*; c) Horsfield's bronze-cuckoo and superb fairy-wren *Malurus cyaneus*; d) Horsfield's bronze-cuckoo and purple-crowned fairy-wren *M. coronatus* (from Langmore, NE, Stevens, M, Maurer, G, Heinsohn, R, Hall, ML, Peters, A, Kilner, RM (2011) Visual mimicry of host nestlings by cuckoos. Proceedings of the Royal Society: Biological Sciences, 278: 2455–2463. (Photos by Naomi Langmore, Golo Maurer, Michelle Hall). See also Fig. 6.4, page 105.

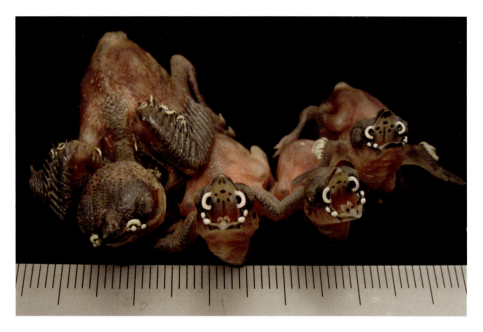

Plate 13   Parasitic pin-tailed whydah *Vidua macroura* nestlings (the two outer chicks) are visual mimics of common waxbill *Estrilda astrild* nestlings (the two center chicks, Photo by Justin Schuetz). See also Fig. 6.5, page 108.

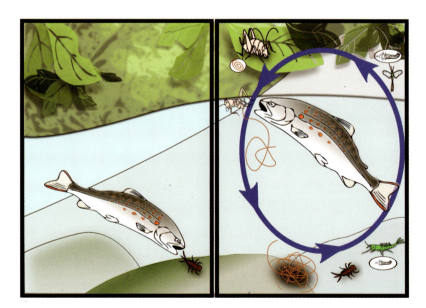

Plate 14   Hypothesized effects of a nematomorph worm on a Japanese stream ecosystem. In the left panel, the parasite is absent and charr forage on a sparse prey base of benthic arthropods. In the right panel, adult nematomorph worms mate in the stream. Their larvae infect larval insects and leave the stream as the insects mature and disperse into the forest. Crickets scavenging on dead insects ingest the larvae, and become the second host. As the worm matures in the cricket, it drives its host from the forest into the stream, attracting predation by charr. Such crickets provide 60% of the charr's energetic intake. Satiated by crickets, the charr eat fewer benthic insects, resulting in a higher production of adult insects that leave the stream for the forest. (Artwork by K. D. Lafferty). See also Fig. 9.1, page 161.

Plate 15 Hypothesized effects of a tapeworm on the Isle of Royale ecosystem. In the left panel, the parasite is absent, and wolves have a hard time persisting on moose. Moose, unchecked by predators, over-graze the forest. Right panel, the adult worms of *Echinococcus granulosus* live in the intestines of wolves where they cause little pathology. Tapeworm eggs contaminate the soil and are incidentally ingested by moose. Larval tapeworms form hydatid cysts in the lungs of moose. Wolves can more easily prey on infected moose, reducing the abundance of the moose population and allowing re-vegetation of the island. (Artwork by K. D. Lafferty). See also Fig. 9.2, page 163.

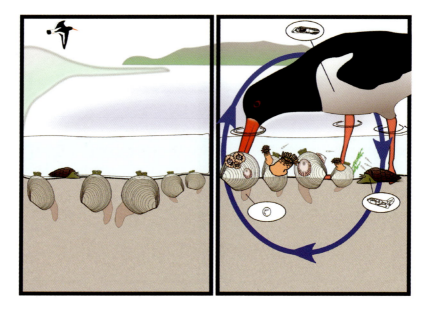

Plate 16 Hypothesized effects of a trematode on a New Zealand mudflat ecosystem. In the left panel, cockles burrow into the sediment with only their siphons protruding. In the right panel, adult trematodes live in the intestine of shorebirds. Birds defecate trematode eggs onto the mudflat where they infect an estuarine snail as the first intermediate host. Trematode cercariae emerge from the snail and then seek out a cockle as a second intermediate host, forming a cyst in the foot. Infected cockles have impaired digging abilities, making them easier prey for birds. Raised above the mud surface, the cockles provide hard substrate for a community of invertebrates and algae. (Artwork by K. D. Lafferty). See also Fig. 9.3, page 164.

Vernon, C. J. (1987). On the Eastern Green-backed Honeyguide. *Honeyguide* **33**, 6–12.

Victoria, J. K. (1972). Clutch characteristics and egg discriminative ability of the African village weaverbird *Ploceus cucullatus*. *Ibis* **114**, 367–376.

Vorobyev, M. and Osorio, D. (1998). Receptor noise as a determinant of colour thresholds. *Proceedings of the Royal Society of London B* **265**, 351–358.

Wallace, A. R. (1889). *Darwinism: An Exposition of the Theory of Natural Selection With Some Of Its Applications.* Macmillan, London.

Welbergen, J. A. and Davies, N. B. (2008). Reed warblers discriminate cuckoos from sparrowhawks with graded alarm signals that attract mates and neighbours. *Animal Behaviour* **76**, 811–822.

Welbergen, J. A. and Davies, N. B. (2009). Strategic variation in mobbing as a front line of defense against brood parasitism. *Current Biology* **19**, 235–240.

Welbergen, J. A. and Davies, N. B. (2011). A parasite in wolf's clothing: hawk mimicry reduces mobbing of cuckoos by hosts. *Behavioural Ecology* **22**, 574–579.

Weller, M. W. (1968). The breeding biology of the parasitic black-headed duck. *Living Bird* **7**, 169–208.

Wyllie, I. (1981). *The Cuckoo.* Batsford, London.

Yang, C., Liang, W., Cai, Y., Shi, S., Takasu, F., Møller, A. P., Antonov, A., Fossøy, F., Moksnes, A., Røskaft, E., and Stokke, B. G. (2010). Coevolution in action: disruptive selection on egg colour in an avian brood parasite and its host. *PLoS One* **5**, e10816.

Afterword

Scott V. Edwards

The tricks that parasites play on their hosts to gain access to reproductive opportunities are among the most elaborate and sinister in the world of behavior. As an evolutionary biologist, I find that instances of visual mimicry—of insects or frogs mimicking leaves, for example—often stretch my credulity in the power of natural selection to hone and refine the mimicry to the smallest of details. I wonder how it could be that mimicry in nature is often so perfect—was the advantage of a slightly better mimic really so great as to render an imperfect mimic less fit? As a population geneticist, I like to imagine that genetic drift—an ever present force in all but the largest populations, and most likely a permanent fixture of the genetics of all vertebrates (along other forces such as linkage of genes to one another on chromosomes), must place constraints on the efficacy of natural selection to increase the match between mimic and host. After all, population genetics and behavior can sometimes be uneasy bedfellows—the one frequently eschewing simple deterministic scenarios involving linear paths to adaptation unfettered by genetic constraints, and the other reveling in them. As an ornithologist, however, the often unbelievable yet sometimes contradictory and surprising outcomes of host–parasite arms races come into better focus. Birds are, after all, excellent model systems for the study and experimental manipulation of behavior in general and nest parasitism in particular. Langmore and Spottiswoode's chapter on visual manipulation of hosts by avian brood parasites nicely illustrates the amazing diversity of tricks used by parasites to gain access to nests and parental behaviors of their hosts, the varied responses of avian hosts to these parasites, and yet how much we still don't understand about these intricate interactions. In particular, I argue, we need to understand the genetics of avian brood parasite adaptations, and their realization over broad phylogenetic, temporal, and geographic landscapes, in order to better understand the dynamics and contingencies underlying these remarkable adaptations.

The chapter opens with a good example of how avian parasites utilize different strategies to achieve access to host nests. (Actually, I very much dislike the anthropomorphic term "strategy" in biology but I will here and elsewhere let my guard down.) Despite the fact that most brood parasites try to mimic "harmless or inviting" species so as to gain access to nests (p. 110). Langmore and Spottiswoode point out that avian parasites such as cuckoos frequently possess the plumage, morphology, and flight behavior of predatory species such as hawks and falcons, an hypothesis from natural history observation that now has experimental support (Welbergen and Davies 2011). One wonders why the incidence of Batesian mimicry—when a (physically) harmless species mimics one that could cause harm—is so high in the cuckoos. It is intriguing that the clade of cuckoos—non-passerine birds with medium to large body sizes compared to many passerines—have sometimes converged on predatory birds with similar body sizes. Perhaps the morphological distance that needs to be traversed by cuckoos in order to mimic hawks is less than in other clades? Or are there a suite of mutations deep in the cuckoo clade that makes such visual mimicry easier to achieve, even convergently across multiple lineages?

As Langmore and Spottiswoode point out, cuckoos are among the best studied avian brood parasites, providing excellent examples of egg mimicry and within-species egg diversification, and recently providing the first well-studied example of host mimicry at the chick stage (Langmore et al. 2011). For example, in the classic scenario for many cuckoo species, selection for egg mimicry by parasites is driven by a history of host-rejection of conspicuous or outstanding eggs in the clutch. But some host species, such as superb fairy-wren (*Malurus cyaneus*) appear to be unable to identify and reject conspicuous eggs when placed in the nest experimentally. As Langmore and Spottswoode show, the multiple and often unexpected hypotheses on the origins of egg mimicry constitute a sort of two-sided mirror through which nearly every adaptation of brood parasites and their hosts can be viewed.

A useful perspective on the speed and mechanisms by which egg mimicry arises in brood parasites comes from molecular methods. Several species of cuckoo exhibit host races, or "gentes," differentiated forms that lay distinct egg types that closely match eggs of their hosts. Such forms are hypothesized to be differentiated at genes underlying egg color and possibly other traits associated with parasitism. Indeed, such gentes have been shown to exhibit mild genetic differentiation in maternally inherited mtDNA in some populations of cuckoos and in indigobirds, the latter among closely related parasitic species (reviewed in Sorenson and Payne 2002). Still, the extent of differentiation in mtDNA is not large and does not approach the pattern of fixed genetic differences that the egg colors themselves imply. Phylogeographic patterns such as these suggest that the evolution of cuckoo gentes has proceeded rapidly and that no stable associations between parasite and host exist in some species, including Horsfield's bronze-cuckoo (Joseph et al. 2002). This conclusion in turn suggests that the intricate host–parasite interactions and multiple rounds of coevolution hypothesized by Langmore and Spottiswoode to occur in many avian brood parasite systems have arisen very rapidly and recently, and may not per-

sist for long periods before being obliterated by environmentally contingent demographic or geographic shifts in host or parasite. Given this potentially short time frame, one wonders whether learning or some type of cultural evolution could facilitate the rapid spread of parasite-rejection by some hosts. Reviewing the evidence and models, Langmore and Spottiswoode suggest that a role for learning in the evolution of parasite rejection is likely small and would work only under a restricted set of evolutionary parameters, including low host-chick survival. Still, the precise fitness loss experienced by hosts under parasitism can be a complex function of many demographic parameters (Servedio and Hauber 2006), and further studies on the role of learning in host rejection are clearly needed.

With its strictly maternal inheritance in birds and other vertebrates, mtDNA is a useful proxy for the genes underlying egg color, especially if the avian W chromosome, with which mtDNA is effectively linked in birds, is involved. Still, with its lack of frequent recombination for much of its length in birds, the W chromosome may not be the whole story behind variation in egg coloration of brood parasites and genetic interactions with autosomal loci have been hypothesized (Sorenson and Payne 2002). Recently, a polymorphic population of cuckoo gentes was shown to exhibit mild but significant differentiation at autosomal microsatellites (Fossoy et al. 2011). Given the bias toward low estimates of differentiation (such as F_{st}) for highly polymorphic microsatellites as compared to sequence-based markers (Charlesworth 1998; Brito and Edwards 2009), detecting such differentiation among such closely related populations is indeed a feat, and bodes well for future genomic studies aimed at finding the genes underlying egg shell coloration and other adaptations of avian brood parasites (Yuan and Lu 2007).

And to be sure, the mapping, sequencing, and evolutionary study of these still hypothetical brood parasite and host response genes is of great relevance to the study of avian brood parasitism. As with any behavioral or morphological adaptation, understanding the genetic basis can clarify a lot about how and when the adaptation arose in a particular lineage, how it coevolves with other traits and how it is maintained and potentially lost over time: think of the marvelous work on the genetics of social behavior in *Dictostylium* (Sucgang et al. 2011). I never did adhere to a strict interpretation of Paul Sherman's (1988) famous four-fold distillation of the sources of adaptations, involving evolutionary origins, ontogenetic processes, mechanisms, and functional consequences, because I view these four levels as so inextricably linked—potentially genetically—as to make their clean separation nearly impossible, particularly when one tries to quantify these aspects of adaptation. This is why understanding the genomics of brood parasitism will be helpful for interpreting host and parasite behavior, however distant that goal is. By elucidating the genetic mechanisms underpinning the many brood parasitic adaptations and host responses discussed by Langmore and Spottiswoode, they will help us make sense of—indeed, quantify—their evolutionary history, constraints, and contingencies, thereby making all the more realistic our evolutionary scenarios of these wildly diverse traits.

References

Brito, P., and Edwards, S. V. (2009). Multilocus phylogeography and phylogenetics using sequence-based markers. *Genetica* **135**, 439–455.

Charlesworth, B. (1998). Measures of divergence between populations and the effect of forces that reduce variability. *Molecular Biology and Evolution* **15**, 538–543.

Fossoy, F., Antonov, A., Moksnes, A., Roskaft, E., Vikan, J. R., Moller, A. P., Shykoff J. A., and Stokke, B. G. (2011). Genetic differentiation among sympatric cuckoo host races: males matter. *Proceedings of the Royal Society B-Biological Sciences* **278**,1639–1645.

Joseph, L., Wilke, T., and Alpers, D. (2002). Reconciling genetic expectations from host specificity with historical population dynamics in an avian brood parasite, Horsfield's Bronze-Cuckoo Chalcites basalis of Australia. *Molecular Ecology* **11**, 829–837.

Langmore, N. E., and Kilner, R. M. (2010). The coevolutionary arms race between Horsfield's Bronze-Cuckoos and Superb Fairy-wrens. *Emu* **110**, 32–38.

Langmore, N. E., Hunt, S., and Kilner, R. M. (2003). Escalation of a coevolutionary arms race through host rejection of brood parasitic young. *Nature* **422**, 157–160.

Langmore, N. E., Srevens, M., Heinsohn, R., Hall, M. L., Peters, A., and Kilner, R. M. (2011). Visual mimicry of host nestlings by cuckoos. *Proceedings of the Royal Society of London B* **278**, 2455–2463.

Servedio, M. R. and Hauber, M. E. (2006). To eject or to abandon? Life history traits of hosts and parasites interact to influence the fitness payoffs of alternative anti-parasite strategies. *Journal of Evolutionary Biology* **19**, 1585–1594.

Sherman, P. W. (1988). The levels of analysis. *Animal Behaviour* **36**, 616–619.

Sorenson, M. D. and Payne, R. B. (2002). Molecular genetic perspectives on avian brood parasitism. *Integrative and Comparative Biology* **42**, 388–400.

Sucgang, R., Kuo, A., Tian, X., Salerno, W., Parikh, A., Feasley, C., Dalin, E., Tu, H., Huang, E., Barry, K., Lindquist, E., Shapiro, H., Bruce, D., Schmutz, J., Salamov, A., Fey, P., Gaudet, P., Anjard, C., Babu, M.M., Basu, S., Bushmanova, Y., van der Wel, H., Katoh-Kurasawa, M., Dinh, C., Coutinho, P., Saito, T., Elias, M., Schaap, P., Kay, R., Henrissat, B., Eichinger, L., Rivero, F., Putnam, N., West, C., Loomis, W., Chisholm, R., Shaulsky, G., Strassmann, J., Queller, D., Kuspa, A., and Grigoriev, I. 2011. Comparative genomics of the social amoebae *Dictyostelium discoideum* and *Dictyostelium purpureum*. *Genome Biology* 12:R20.

Welbergen, J. A. and Davies N. B. (2011). A parasite in wolf's clothing: hawk mimicry reduces mobbing of cuckoos by hosts. *Behavioral Ecology* **22**, 574–579.

Yuan, Q.-Y. and Lu. L.-Z. (2007). Progresses in inheritance of genes for avian eggshell color. *Yichuan* **29**, 265–268.

CHAPTER 7

Endosymbiotic microbes as adaptive manipulators of arthropod behavior and natural driving sources of host speciation

Wolfgang J. Miller and Daniela Schneider

7.1 Introduction

The German botanists Heinrich Anton de Bary and Albert Bernhard independently developed the concept of symbiosis in 1868, and it was de Bary who finally coined the term "symbiosis" in 1879. According to his definition, symbiosis includes a complex of associations between two or more dissimilar organisms that live together, at least during part of their lifetime, ranging from parasitic (detrimental) to commensalistic (neutral) and mutualistic (beneficial) relationships (de Bary 1879). In this chapter, we will adhere to Buchner's original "global" definition of symbiosis, since recent symbiosis research shows that, even within short evolutionary time frames, symbionts can transform from parasitic to mutualistic lifestyles and *vice versa* (reviewed in Riegler and O'Neill 2007).

Symbiont-induced manipulation can occur among several phenotypic dimensions, ranging from morphological to physiological and behavioral alterations of the host, representing the complex outcome of relationships that yield transmission benefits for the symbiont. Hence, such beneficial side effects of host–symbiont interactions can become adaptations that help symbionts increase their own transmission success throughout the host population. In this respect, we follow Poulin's definition of adaptive manipulation as the following: all cases of symbiont-triggered manipulations are judged adaptive, if (1) a symbiont-induced change

in host behavior leads to improved transmission of the symbiont, (2) there is a genetic basis to this effect, and (3) this is fortuitous or not (Poulin 2010). In this chapter we will review recent cases from literature and from our own laboratory concerning microbial symbionts capable of triggering adaptive behavioral changes in their respective hosts with emphases on the multilevel manipulator *Wolbachia*.

7.2 *Wolbachia*: the multidimensional manipulator of arthropods

Within the large group of symbiotic bacteria, one genus within the α-proteobacteria, *Wolbachia*, attracts particular attention. *Wolbachia* were first described as intracellular *Rickettsia*-like organisms (RLOs) in the gonad cells of the mosquito *Culex pipiens* (Hertig and Wolbach 1924) and named *Wolbachia pipientis* (Hertig 1936). Since their first description in the early twentieth century, *Wolbachia* have been found in a uniquely wide range of host species, mostly belonging to the arthropod phylum. *Wolbachia* infect up to two thirds of all insect species (Jeyaprakash and Hoy 2000; Hilgenboecker et al. 2008), as well as a variety of mites, spiders, scorpions, and terrestrial crustaceans (Rowley et al. 2004; Bordenstein and Rosengaus 2005; Baldo et al. 2007, 2008; Wiwatanaratanabutr 2009). In addition to arthropods, *Wolbachia* are also present in filarial nematodes (see Taylor et al. 2010 for recent review) and have also been detected in a plant-associated nematode (Haegeman et al. 2009). In arthropods, *Wolbachia* are obligate endosymbionts that are mainly transmitted transovarially with the cytoplasm from infected females to their offspring, but can also spread into distantly related hosts via horizontal transmission, presumably mediated by vectors such as parasitoid wasps or mites (Werren et al. 2008).

Over the last decades, *Wolbachia* have attracted major foci in arthropod research and evince the largest growth in publications compared to any other bacterial arthropod symbiont. The reasons for this are many: *Wolbachia* are extremely common, have a wide host range, and feature manifold ways of influencing host fitness and behavior (Table 7.1). Furthermore arthropod hosts tend to accumulate different strains of a defined symbiont group in their cells, perhaps as a strategy to extend their genomes with gene networks that they could not otherwise exploit so quickly (Schneider et al. 2011).

Besides neutrality (commensalism), symbiotic relationships can be either parasitic or mutualistic, depending on whether one or both organisms benefit from the interaction. Parasitism occurs, when one partner benefits at the expense of the other in order to secure its own survival through exploiting the host for diverse resources, such as water, food, habitat, and so on. Nature offers a huge variety of parasitism but classical cases include interrelations between vertebrate hosts and diverse invertebrate hosts such as parasitic viruses, bacteria, protozoa, worms, and insects. In contrast, mutualism is a form of symbiotic interaction where each organism derives a fitness benefit; for example protection from predators and provision of nutritional resources. Mutualistic relationships exist in all kingdoms of life, including biological interactions between humans and their microbial gut flora, without which they would not be able to digest food properly (Blaut and Clavel 2007; Garrett et al. 2010).

Among *Wolbachia*, a group of highly manipulative symbionts, both forms of biological interactions were shown to be operative since these bacteria feature a broad range of phenotypic traits in their hosts, extending from parasitism to mutualism (reviewed in Riegler and O'Neill 2007; Cook and McGraw 2010). Numerous studies have demonstrated that *Wolbachia*-based arthropod manipulation is multidimensional, ranging from reproductive phenotypes (see reviews by Stouthamer et al. 1999; Saridaki and Bourtziz 2010, and references therein) to non-reproductive forms affecting host physiol-

Table 7.1   Multidimensionality of arthropod manipulation by *Wolbachia*

Wolbachia-induced phenotype	Affected	First description
Cytoplasmic incompatibility (CI)	Reproduction	Yen & Bar 1971; Hoffmann et al. 1986
Parthenogenesis	Reproduction	Stouthamer et al. 1990
Male killing (MK)	Reproduction	Hurst et al. 1999
Feminization	Reproduction	Rousset et al. 1992
Interference with host iron metabolism	Metabolism	Kremer et al.; Brownlie et al. 2009
Manipulation of host imprinting	Sex determination	Negri et al. 2009
Host life shortening effect	Immunity	McMeniman et al. 2009
Resistance to RNA viruses	Immunity	Hedges et al.; Teixeira et al. 2008
Immunity against filarian nematodes	Immunity	Kambris et al. 2009
Resistance to infection with Dengue fever, Chikungunya, Plasmodium	Immunity	Moreira et al. 2009
Inhibition of Plasmodium development	Immunity	Kambris et al. 2010
Assortative mating & oviposition	Sexual behaviour	Vala et al. 2004
Pre- & post-zygotic isolation	Host speciation	Miller et al. 2010

ogy (Skorokhod et al. 1999; Brownlie et al. 2009; Ikeya et al. 2009; Kremer et al. 2009), immunity (Min and Benzer 1997; Fytrou et al. 2006; Panteleev et al. 2007; Hedges et al. 2008; Teixeira et al. 2008; Kambris et al. 2009; McMeniman et al. 2009; Moreira et al. 2009), and behavior (Vala et al. 2004; Koukou et al. 2006; Miller et al. 2010, and see Section 7.3 within this chapter).

7.2.1 Reproductive parasitism triggered by *Wolbachia*

The most commonly described *Wolbachia*-induced phenotype assigned to reproductive parasitism is cytoplasmic incompatibility (CI), resulting in embryonic lethality among offspring. Already in the 1970s, Yen and Barr (1971) had elucidated *Wolbachia*'s pivotal roles in causing CI between populations of the mosquito *Culex pipiens*. Induction of CI is a highly efficient tool of the endosymbiont to promote and secure its own transgenerational transmission and thus promote rapid spreading and persistence within a population. Cross-mating that is disadvantageous for *Wolbachia* is that leading to the production of uninfected offspring. Such incompatible crossings can occur in two different modes, that is uni- or bidirectionally. Unidirectional CI arises whenever *Wolbachia*-infected males mate with uninfected females (Fig. 7.1 A). Mating in all other directions is fully compatible. In contrast, bidirectional CI arises when individuals carry different strains of *Wolbachia* that are incompatible with each other (Fig. 7.1 B; recently reviewed by Merçot and Poinsot 2009). However, both modes of CI result in embryonic mortality and dramatically reduced numbers of offspring (Hoffmann et al. 1986; Tram and Sullivan 2002; Vavre et al. 2000, 2002; Bordenstein and Werren 2007).

Although the molecular mechanism underlying CI is not yet fully understood, recent biological

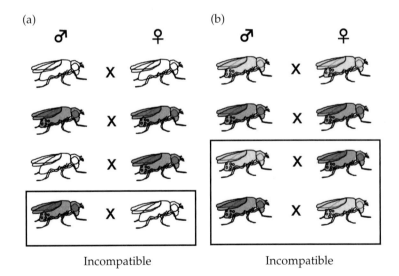

(a)   (b)

♂   ♀   ♂   ♀

Incompatible   Incompatible

Figure 7.1 Scheme of cytoplasmic incompatibility (CI). *Wolbachia*-uninfected flies are presented in white; flies infected with A-*Wolbachia* in dark gray and flies infected with B-*Wolbachia* in light gray. Incompatible crossings are indicated by black boxes. (A) Uni directional CI refers to an incompatible cross-mating between A-*Wolbachia*-infected males and *Wolbachia*-uninfected females resulting in high mortality in early embryos. The following cross-directions are compatible and do not trigger CI: (i) female and male are both uninfected; (ii) females are infected (A-*Wolbachia*) and males are uninfected, and (iii) both female and male are infected with A-*Wolbachia*. (B) Bi directional CI results when A-*Wolbachia*-infected males inseminate females carrying a different *Wolbachia* strain (B-*Wolbachia*). Crossings between partners harboring the same *Wolbachia* strain, that is, A x A or B x B are fully compatible.

studies have significantly contributed to our current understanding of this impressive *Wolbachia*-induced host manipulation process. Cytological analyses suggest that *Wolbachia* might induce CI via altering host-cell cycle timing. Infected male pronuclei are delayed when entering first mitotic divisions, resulting in aneuploid nuclei and death in early stages of embryonic development of uninfected egg cells (Tram and Sullivan 2002; Tram et al. 2003; Ferree and Sullivan 2006; Tram et al. 2006).

Besides CI, *Wolbachia* utilize male killing in order to influence the reproduction of their hosts. The strictly maternal transmission mode of *Wolbachia* via the egg cytoplasm puts males at a clear disadvantage compared to females. Hence, host sex ratios are severely distorted in favor of females by bacterial-induced killing of male embryos. Although male killing is another efficient way of securing symbiont persistence in a host population, it is not the most prominent of all *Wolbachia*-induced reproductive phenotypes. Male killing *Wolbachia* strains are rare (Hornett et al. 2006; Sheeley and McAllister 2009) and few have been described among three different orders of arthropods: coleoptera (Majerus et al. 2000), lepidoptera (Jiggins et al. 1998), and diptera (Hurst et al. 2000). Interestingly, a counter-defense mechanism of the host against *Wolbachia*-induced male killing has been recently described. The nymphalid butterfly *Hypolimnas bolina* is capable of suppressing this phenotype completely in southeast Asian populations, leading to a rapid expansion of the resistance gene throughout populations of these butterflies in the Pacific region (Hornett et al. 2006, 2008; Jaenike 2007; Charlat et al. 2009). However, the potential to induce male killing is not restricted to *Wolbachia*. A diverse range of bacteria such as *Rickettsia*, *Flavobacteria*, *Spiroplasma*, and *Arsenophonus* have been associated with the described phenotype (for review see Hurst and Jiggins 2000).

The third described reproductive phenotype triggered by *Wolbachia* is thelytokous parthenogenesis. In general, parthenogenesis refers to a form of asexual reproduction where growth and development of embryos takes place without fertilization by male individuals. Among insects, sex-determination systems span diploidy (diploid males and females), haplodiploidy (haploid males and diploid females), and thelytoky, resulting in production of only diploid female offspring without males (reviewed in Normark 2003). Since *Wolbachia* depend on female transovarial transmission, it is in their interest to dramatically shift the overall sex ratio within a population towards females by parthenogenesis induction. Cases of *Wolbachia*-induced parthenogenetical reproduction have been reported in mites and Hymenoptera (Huigens and Stouthamer 2003). Besides *Wolbachia*, other bacteria such as *Cardinium hertigii* (Giorgini et al. 2009) and *Rickettsia* (Hagimori et al. 2006) have also been shown to be capable of inducing thelytoky.

Numerous theoretical models have predicted that CI-inducing transovarially transmitted microbial symbionts can play an important role as drivers of host speciation processes (Caspari and Watson 1959; Weeks et al. 2002; Bordenstein 2003; Telschow et al. 2005; Jaenike et al. 2006; Bordenstein and Werren 2007). A key parameter in speciation is the establishment of reproductive isolation between populations of the same species in order to prevent wasteful gene flow and the production of potentially inferior hybrids (Otte and Endler 1989). It was recently proposed that under specific and environmental conditions *Wolbachia* can promote host speciation in nature (Telschow et al. 2005; Telschow et al. 2007), but only a few empirical model systems from nature are available. Two examples strongly suggesting that *Wolbachia* might have played a role in fostering host speciation in their evolutionary past come from parasitoid wasps of the *Nasonia* sibling species group (Breeuwer and Werren 1990; Bordenstein et al. 2001) and mushroom feeding *Drosophila* species (Shoemaker et al. 1999; Jaenike et al. 2006). However, a very recent case supporting *Wolbachia*'s impact on promoting incipient host speciation in nature via the induction of both strong bidirectional CI in hybrids plus the manipulation of assortative mating behavior is reported from the evolutionary young *D. paulistorum* species complex (Miller et al. 2010, and see Section 7.3.2).

7.2.2 *Wolbachia*'s repertoire of inducing non-reproductive, adaptive phenotypes

Besides their impressive potential for triggering reproductive parasitism in the many arthropods mentioned above, recent studies have uncovered *Wolbachia*'s impact on adaptive non-reproductive phenotypes affecting host protection, immunity, and olfactory cued locomotion. Such changes further improve the fitness of the host, and thereby the transmission success of the symbiont via vertical transmission throughout subsequent host generations. Thus it is in the symbiont's interest to upgrade host defense traits such as camouflage and host innate immune systems against predators and parasites such as viruses, bacteria, protozoa, or even nematodes. Insects in nature are usually infected by multiple phylogenetically unrelated symbionts that may have opposing interests in their use of the host. In some cases, symbionts have been shown to sabotage the manipulation exerted by other parasites, reverting the phenotype of an infected host to its "normal" uninfected state (Rigaud and Haine 2005).

As demonstrated in recent studies, *Wolbachia* are involved in more host-dominated natural histories than was earlier anticipated. *Wolbachia* may generally interfere with iron metabolism and protect host cells from oxidative stress and apoptosis in *D. melanogaster* and parasitic wasps, respectively (Brownlie et al. 2009; Kremer et al. 2009). Furthermore, the insulin/IGF-like signaling (IIS) pathway, which has pleiotrophic functions in multi-cellular organisms (Skorokhod et al. 1999), is influenced by this omni-present symbiont in *Drosophila*. *Wolbachia* potentially increase IIS and thereby exert influence on host aging, such that removal of the bacteria via antibiotic treatment results in reduced life spans (Ikeya et al. 2009).

Furthermore, accumulating evidence in recent symbiosis literature impressively demonstrates the capacity of *Wolbachia* in providing host resistance to a variety of parasites ranging from viruses to protists and filarian nematodes (Table 7.1). *Wolbachia* protect *Drosophila melanogaster* from infections with RNA viruses (Hedges et al. 2008; Teixeira et al. 2008), whereas similar studies on the *Wolbachia*-induced protection from entomopathogenic fungi

in *D. melanogaster* are still inconclusive (Fytrou et al. 2006; Panteleev et al. 2007). Analyses of *Wolbachia*-mediated protective effects against microbes have recently been extended from *Drosophila* host systems to the mosquito *Aedes aegypti*. In this host, a *Wolbachia* strain, originating from *D. melanogaster*, interferes with the establishment of viral infections, as well as *Plasmodium* (Moreira et al. 2009; Kambris et al. 2010) and filarial nematodes (Kambris et al. 2009). The *Wolbachia* strain of *D. melanogaster* was originally transferred into *Ae. aegypti* because of its virulent life span reduction in the original host (Min and Benzer, 1997). As expected from the phenotype in its original host, *Wolbachia*-infected *Ae. aegypti* have shorter lives (McMeniman et al. 2009). It has been suggested that symbiont-triggered enhanced immunity against parasites is costly, at the expense of host longevity (McMenimen et al. 2009; Moreira et al. 2009; Kambris et al. 2009, 2010). *Wolbachia*-infected *Ae. aegypti* are less likely to transmit viral diseases such as dengue, requiring specific incubation periods in mosquito vectors before transmission to humans can occur. By interfering with vector biology, this virulent *Wolbachia* strain provides a novel strategy for fighting vector-borne diseases, such as dengue, malaria, and filariasis (Sinkins 2004; McMeniman et al. 2009; Moreira et al. 2009; Kambris et al. 2009, 2010). However, future studies will be of utmost interest in order to decipher the molecular mechanisms and metabolic pathways by which *Wolbachia* interfere with the host immune system and thereby boost their own protection and propagation success.

Furthermore, natural *Wolbachia* infection also alters olfactory-cued locomotion in *D. melanogaster* and *D. simulans* by increasing the basal activity level of both host insects as well as their responsiveness to food cues (Peng et al. 2008). Sequencing of the *D. melanogaster Wolbachia* strain *w*Mel genome has revealed that the symbiont is well suited to direct uptake of amino acids from its environment and is incapable of synthesizing several metabolic intermediates (Wu et al. 2004). Alternatively, but not exclusively, the accumulation of *Wolbachia* in brain tissues of both *Drosophila* host species (Albertson et al. 2009) strongly suggests that the symbiont may actively manipulate the olfactory-cued locomotion of its host in an adaptive manner, in order to increase its own nutritional status and thereby its own transmission success.

7.3 Symbiont-directed adaptive manipulation of host sexual behavior

Besides their impressive repertoire of manipulating sex ratios, fecundity, and physiology of arthropods in an adaptive manner, some *Wolbachia* have the potential to affect host sexual behavior at multiple levels, and to trigger sexual isolation. Consequently, behaviorally adaptive symbioses can emerge that foster host speciation over evolutionary periods. In this part, we will review recent findings from adaptive symbiotic systems affecting sexual behavior in arthropods with emphasis on *Wolbachia*.

7.3.1 Feminization—the transformation of genetic males into functional females

Some *Wolbachia* strains in isopods, lepidoptera, and hemiptera can cause an imbalance in host sex ratios by converting genetic males into functional females, and thereby accelerating endosymbiont distribution throughout their host range. In these hosts, *Wolbachia*-induced feminization of genetic males can be complete or almost complete by mimicking female phenotypes at multidimensional levels, that is, morphologically, reproductively, and even behaviorally. Feminization can be seen as a classic case of symbiont-directed adaptive manipulation of host behavior, improving their transmission successes.

In terrestrial isopods such as woodlice, microbial endosymbionts cause feminization in most host species by inducing revision of genotypic males into functional females (Martin et al. 1973); and these endosymbionts were later identified as members of the genus *Wolbachia* (Rousset et al. 1992). As a result of their strict maternal transmission mode, *Wolbachia*-infected mothers produce highly female-biased progeny due to the feminizing potential of the endosymbiont (Juchault et al. 1993; Bouchon et al. 1998; Rigaud 1997; Rigaud et al. 1999; Moreau and Rigaud 2000, 2003). In the woodlouse *Armadillidium vulgare*,

for example, the genetic basis of sex determination is female heterogamety (ZZ in males and ZW in females), common in lepidopteran insects. Genetic males of *A. vulgare* possess an endocrine organ, the androgenic gland, from which a polypeptide factor, androgenic gland hormone (AGH), is secreted and in the absence of AGH males transform into females (Legrand et al. 1987). Upon *Wolbachia* infection, this parasite proliferates in the androgenic gland, leading to androgenic gland hypertrophy and inhibited AGH secretion function, and therefore genetic males develop as females, the default development in isopodes (Rigaud 1997; Vanderkerckhove 2003). This sex ratio distortion increases *Wolbachia*'s frequency in host populations rapidly, but it also has potential consequences for host biology that might, in turn, have consequences for parasite prevalence, such as extinction of the host population and consequently also the parasite population (Hamilton 1967). This genetic conflict over sex determination, triggered by feminizing endosymbionts, can result in rapid evolution of the sex determination system (Juchault et al. 1994), such as the accumulation of suppressing host genes that act as male-determining factors or resistance genes abolishing *Wolbachia*-directed feminization in some populations.

Indeed, feminizing *Wolbachia* are often found at relatively low frequencies in natural host populations, suggesting that, besides the potential presence of suppressor or resistance genes, sexual selection against *Wolbachia*-infected neo-females (feminized males) could maintain a polymorphism of the infection in populations at low frequencies. In mating choice assays where woodlice males had the choice between infected and uninfected females, Moreau and colleagues have shown that infected neo-females have lower mating rates and receive less sperm relative to uninfected females, and that males interact significantly more frequently with uninfected females and made more mating attempts, implying reduced attractiveness of infected neo-females (Moreau et al. 2001). Active discrimination of males against neo-females could be due to differences in pheromone profiles between uninfected females and neo-females, suggesting that *Wolbachia*-induced transformation of genetic males is incomplete. Furthermore, a female behavioral difference was observed in response to male mating attempts: infected neo-females more often exhibited behaviors that stop the mating sequence (Moreau et al. 2001). Since the sex of true woodlice females is naturally determined by the presence of the heterogamic W chromosomes (ZW), the authors suggest that the W chromosome might carry some genes necessary for proper female mating behavior and pheromone expression, not present in *Wolbachia*-transformed neo-females (ZZ). Symbiont-directed feminization of male isopods is incomplete, most likely due to weak sexual selection against neo-females permitting the infection to persist at low prevalence in natural populations without endangering host and symbiont extinction by the sex ratio distortion.

In contrast to terrestrial isopods, sex determination in diptera and lepidoptera and many other insects is not defined by sex hormones such as masculinizing hormones produced by androgenic glands, but it is determined genetically by sex chromosomes in a cell-autonomous manner. Once established after fertilization, the sex of each cell is maintained during later development through a gene expression cascade of sex determination genes, such as *Sex lethal*, *transformer*, *doublesex*, and others, in which sex-specific splicing plays a pivotal role (recently reviewed in Sanchez 2008). In these host systems, however, reproductive parasites such as *Wolbachia*, which follow their strategy of feminizing genetic males in favor of their own vertical transmission, need to orchestrate and redirect the host sex determination pathway at a very early embryonic stage in a cell-autonomous manner. One of the most prominent examples of parasite-directed feminization comes from lepidoptera, belonging to the genus *Eurema*, where *Wolbachia* completely transform genetic males into functional females (Hiroki et al. 2002, 2004).

In some Japanese populations of the butterfly *Eurema hecabe*, two different types of individuals are found relative to their infection status with *Wolbachia* (Hiroki et al. 2002, 2004). Genetic males doubly infected with distinct strains of *Wolbachia* (*w*HecCI and *w*HecFem) exhibit the complete female phenotype. These neo-females are morphologically, reproductively, and behaviorally func-

tional females that only produce infected daughters with the male genotype. Individuals singly infected with *w*HecCI, however, produce normal males and females at a ratio of nearly 1:1. When adults of double-infected feminized strains were treated with antibiotics, their progeny consisted exclusively of males, whereas antibiotic treatment of double-infected feminized larvae gave rise to adults expressing intermediate sexual traits in their wings, reproductive organs, and genitalia (Narita et al. 2007). However, the expression of the sexually intermediate traits is dependent on the timing of antibiotic treatment, since intersexual phenotypes were significantly more frequent in the insects treated from first instar than in insects treated from fourth instar. This data strongly suggests that feminizing *Wolbachia* are necessary to continuously act on genetic males during larval development for the maintenance of the complete feminizer phenotype in adults and to ensure proper sexual behavior and reproduction. The authors speculate that feminizing *Wolbachia* in *E. hecabe* may interact persistently with some female- or male-specific molecular pathways that are located downstream of the sex determination system responsible for the complete expression of female-specific phenotypes in *Wolbachia*-infected neo-females.

Another case of feminizing *Wolbachia* in lepidoptera was initially described in two moth species, *Ostrinia scapulalis* and *Ostrinia furnacalis* (Kageyama et al. 2002, 2003), but in subsequent studies, it was shown that *Wolbachia* actually kill genetic males after an unsuccessful "attempt" at feminization (Kageyama and Traut 2004; Sakamoto et al. 2007). Although the *Ostrina–Wolbachia* symbiosis does not serve as an example for adaptive manipulation of host behavior, this system beautifully demonstrates the flexibility and adaptability of symbiont-directed strategies that can evolve in the course of an evolutionary host–parasite arms race: first to make an attempt to transform genetic males into functional females appropriate for vertical parasite spreading and, if this strategy fails because of counteractions by the host or incomplete feminization, come up with a back-up strategy such as selective killing of genetic males. As mentioned earlier, male-killing *Wolbachia* are widespread in arthropods and provide a potent reproductive phenotype that favors infected females by ensuring sufficient nutritional resources during their larval development at the expense of their brothers, that are regarded from the perspective of the parasite as useless for symbiont transmission.

Realization of a specific sex in insects can also be manipulated by environmental factors promoting transgenerational epigenetic changes, such as feminizing *Wolbachia* that interfere with host-directed imprinting (Negri et al. 2009). Since *Wolbachia* are maternally transmitted, cytoplasmatic endosymbionts in host germ cells they can be easily transferred between populations, but can also cross species borders via horizontal transmission. Hence this symbiont can be regarded as an extra-nuclear, external, or even environmental factor. For many *Wolbachia*-host systems it is known that a critical threshold of *Wolbachia* density has to be reached in order to express reproductive parasitism and that the symbiont *per se* is highly sensitive to external signals such as: heat, stress, larval crowding density, and the availability of natural antibiotics (Boyle et al. 1993; Hoffmann et al. 1990; Hoffmann et al. 1996; Clancy and Hoffmann 1998; Wiwatanaratanabutr and Kittayapong 2006; Negri et al. 2009; Wiwatanaratanabutr and Kittayapong 2009). A recent case in literature provides initial hints as to potential molecular mechanisms by which the symbiont juggles insect development and sexual orientation, that is by means of altering DNA-methylation profiles of host genomes (Negri et al. 2009).

In the grass-dwelling European leafhopper *Zyginidia pullula* maternally transmitted *Wolbachia* can feminize genetic males, leaving them as intersexes, that, with the exception of male chitinous structures present in the last abdominal segments, display all the phenotypic features typical of female *Z. pullula* (Negri et al. 2006). Importantly, the entire manifestation of the feminizer phenotype, with functional ovaries, occurs only if *Wolbachia* titer exceeds a critical threshold, since lower infection densities in males causes incomplete feminization with non-functional dystrophic ovaries and, in some rare cases, even externally feminized males have been found with an ovipositor associated with male

gonads (Negri et al. 2009). The authors could show that the expressivity of the feminizer phenotype in *Z. pullula* is directly correlated to their sex-specific DNA methylation patterns. In contrast to most lepidoptera that possess a ZZ/ZW sex chromosome system, with females as the heterogamic sex (Hiroki et al. 2002), the leafhopper *Z. pullula* has an XX/X0 sex-determination system, with XX females and X0 males. In the latter case, Negri and colleagues demonstrated that in *Z. pullula*, nuclear DNAs of males and females have diagnostic sex-specific methylation profiles. Whereas mildly feminized intersexes with retained testes show low *Wolbachia* densities and male-specific methylation patterns, fully transformed females with functional ovaries exert high *Wolbachia* densities and female-specific methylation patterns, suggesting that *Wolbachia* could feminize *Z. pullula* males by disrupting male-specific DNA imprinting.

Moreover, some of the fully-transformed intersexes were found with their spermathecae full of sperm, indicating that they had successfully mated with males under natural conditions (Negri et al. 2009). This finding suggests that such intersexes can function, at least in part, as biologically complete females accepted by males. However, detailed behavioral studies in leafhoppers will be necessary to determine the full complexity of this intriguing host–parasite interaction beyond sex-determination and development in order to answer the question as to whether or not such intersexes express fully-fledged female sexual behavior.

The examples mentioned above impressively demonstrate *Wolbachia*'s potential and plasticity to manipulate host sex development and the expression of sex-specific behavior in a variety of ways, although the physiological pathways and molecular players still remain to be elucidated.

However, the capacity of transforming genetic males into females is not restricted to endosymbiotic *Wolbachia*, since other parasites with similar propagation strategies have been reported in arthropods (recently reviewed in Narita et al. 2010). In some amphipods, such as *Gammarus duebeni* and *Orchestia gammarellus*, males can be partially or completely feminized by spore-forming unicellular eukaryotes, such as the microsporidian parasites *Octosporea effeminans* and *Nosema granulosis*, and the haplosporidian parasite *Paramartelia orchestiae*, leading to intersexual development (Bulnheim and Vavra 1968; Bulnheim 1977, 1978, Kelly et al. 2004; Ginsburger-Vogel et al. 1980). Not only microorganisms, but also nematodes, trematodes, and turbellarians can have significant effects on gonad development and/or secondary sexual characters of various arthropods (reviewed by Baudoin 1975). It has been demonstrated that the infection by the mermithid nematode *Gasteromermis* sp. is known to feminize morphologically and behaviorally genetic males of the mayfly host *Baetis bicaudatus* (Vance 1996).

7.3.2 Manipulating sexual mating behavior

Among *Wolbachia*-induced reproductive manipulations in arthropods, cytoplasmic incompatibility (CI) is the most commonly documented (see Section 11.2.1). This phenotype leads to high embryonic mortality, thus "sterilization" of uninfected females that have mated with infected males confers a fitness benefit on infected females. Hence, as soon as the *Wolbachia* infection is above the critical threshold frequency in a panmictic population, CI reduces the fitness of uninfected females below that of infected females and, consequently, the proportion of infected hosts increases (Stevens and Wade 1990). Thus, CI is a classic case of reproductive parasitism that benefits the symbiont, but not the host.

However, hosts could gain a reproductive advantage by avoiding CI through the development of mate recognition systems that allow them to distinguish the infection status of a potential mate. Thus, under panmictic conditions, the advantage that CI confers to *Wolbachia* will be ameliorated as hosts selectively choose compatible mates. Under such conditions, any host trait that suppresses CI will be positively selected (Turelli 1994; Vala et al. 2002).

The first case in literature suggesting that hosts can counteract CI, by the evolution of assortative mating behavior, came from spider mites (Vala et al. 2004). In this system, *Wolbachia*-directed unidirectional CI can be avoided by uninfected females at the pre-mating level. The authors have demonstrated that spider-mite females exhibit both precopulatory and ovipositional behaviors that increase the odds of successful compatible mating. As shown by mating choice experiments, uninfected females preferably mate with uninfected males, and in doing so, directly reduce opportunities for *Wolbachia*-induced unidirectional CI expression (Vala et al. 2004). The authors speculate that the specificity of the behavioral response of uninfected females, that is to perceive the presence of *Wolbachia* and to avoid mating with infected males, seems to be triggered by olfactory stimuli. Contrary to the situation in woodlice, where *Wolbachia* can feminize genetic males and under choice conditions uninfected males discriminate against infected neo-females (Moreau et al. 2001, and see Section 7.3.1); in spider mites it is the female that actively avoids infected mates carrying CI-inducing *Wolbachia*. Importantly, in both cases endosymbiotic *Wolbachia* capable of causing different reproductive phenotypes (CI and feminization respectively) have fostered the evolution of adaptive assortative mating behavior independently in their respective arthropod hosts.

Another recent example of the influence of *Wolbachia* on assortative mating behavior comes from the *Drosophila* model system. Koukou and colleagues demonstrated that in some selected laboratory populations of *D. melanogaster*, the artificial loss of *Wolbachia* infection induced by antibiotic treatment can exert significant effects on the assortative mating behavior of female flies (Koukou et al. 2006). The mechanism and biological significance in nature, however, remains obscure since this study is based on long-term cage conditions and artificial selectional constraints have acted on the lines assayed in this study.

Other adaptive behavioral host strategies that may have evolved in nature in order to avoid or reduce detrimental effects of CI-inducing *Wolbachia* affect the sexual mating behavior of infected males and uninfected females. As proposed by Zeh and Zeh (1996, 1997) genetic incompatibilities such as CI could promote the evolution of male promiscuity and female polyandry. In the two *Drosophila* species *D. melanogaster* and *D. simulans*, it was found that males infected with CI-inducing *Wolbachia* show significantly higher mating rates and enhanced male promiscuity compared to uninfected males (de Crespigny et al. 2006). In addition to inducing CI, *Wolbachia* can impose significant physiological costs on their hosts. Infected *D. simulans* males, for example, produce approximately 40% fewer sperm than uninfected males, thus, reducing their fertility (Snook et al. 2000). On the other hand, it has been shown that courtship experiences, progressive age, and repeated mating of CI-infected males leads to a decline in levels of CI (Karr et al. 1998; Reynolds and Hoffmann 2002). Based on their observations, de Crespigny and collaborators suppose that the promotion of higher levels of promiscuity of *Wolbachia*-infected males is an adaptive behavioral strategy that has evolved in response to counteract the high fitness cost of CI-inducing *Wolbachia*. Besides promoting promiscuity in infected males, uninfected females might also have evolved adaptive behavioral strategies that may alter their mating behavior in order to escape fecundity costs after mating with infected males carrying CI-inducing *Wolbachia*. One potential escape strategy would be the promotion of polyandry in uninfected females, that is, multiple matings with more males. Depending on the frequency of the *Wolbachia* infection in a population, increased polyandric female behavior could enhance the chance of mating with a compatible mate. A similar adaptive behavioral change was recently documented by Wedell and collaborators in natural populations of *D. pseudoobscura* in response to the spreading of "selfish" sex-distortion elements unrelated to *Wolbachia* (Price et al. 2008, 2010).

However, significant effects of symbiotic *Wolbachia* on female assortative mating in a natural model system have recently been reported in the neotropical superspecies *Drosophila paulistorum* (Miller et al. 2010). This species complex, a system *in statu nascendi*, comprises six incipient species or semispecies (Amazonian, Andean-Brazilian, Centroamerican, Interior, Orinocan, and Transitional) distributed sympatrically throughout the neotropics (Dobzhansky and Spassky 1959). All six semispecies of *D. paulistorum* were shown to carry conspecific and closely related, but incompatible, *Wolbachia* variants (Miller et al. 2010; illustrated in Fig. 7.2). Infection in this system has evolved towards a tight mutualistic host–symbiont relationship, since attempts

to completely clear hosts from *Wolbachia* have failed and no true bacteria-free strain has ever been generated (Ehrman 1968; Kernaghan and Ehrman 1970; Ehrman and Kernaghan 1971, 1972; Ehrman and Daniels 1975; Ehrman et al. 1995). Indeed, a strict equilibrium between natural host and symbiont density is maintained in this model system, and disruption of this state has dramatic consequences. The default status of *Wolbachia* infection in *D. paulistorum* is benign, representing aforementioned mutualistic host–symbiont interrelation. However, the symbiont has preserved its capacity to switch to a malign state, thus becoming a true parasite that is pathogenic to its host (Fig. 7.2). This happens whenever the equilibrium between host and symbiont is perturbated by means of altering symbiont density and titer levels (Miller et al. 2010).

In *D. paulistorum*, three intrinsic isolating mechanisms are operative: sexual isolation acting before copulation via mating behavior, hybrid sterility, plus hybrid inviability acting after copulation (Ehrman and Kernaghan 1972). Females from one of the six *D. paulistorum* semispecies express strong assortative mating behavior against males from alternate semispecies, meaning that they do not normally mate with them. This behavior, referred to as strong positive assortative mating (likes mate with likes) is illustrated in Fig. 7.3 A.

The classical tool for testing such behavior experimentally involves mating choice assays. In these assays, females are offered proper and improper mates, in this case, males from two different semispecies. In order to distinguish between members from each semispecies, the flies are marked by for example wing clipping. Mating between homogamic (within one semispecies) and heterogamic individuals (among two different semispecies) is then recorded, allowing for estimation of female assortative mating (Ehrman 1961, 1965). Assortative mating plays an important role in pre-mating isolation because it prevents pairing between improper mates. A recent study revealed the prominent impact of symbiotic *Wolbachia* on manipulating assortative mating in the *D. paulistorum* species complex. As demonstrated by extensive behavioral and molecular assays, *Wolbachia* of all six *D. paulistorum* semispecies are capable of triggering

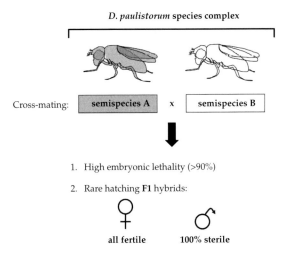

D. paulistorum species complex

Cross-mating: | semispecies A | x | semispecies B |

1. High embryonic lethality (>90%)

2. Rare hatching **F1** hybrids:

♀   ♂

all fertile   100% sterile

Figure 7.2 Dual effects of *Wolbachia* on the induction of post-mating isolation in the *D. paulistorum* species complex. Forced mating in the laboratory between both semispecies A and B causes in hybrids two reproductive phenotypes, that is high embryonic lethality (bidirectional CI), plus complete male sterility. Mild antibiotic treatments, however, are sufficient to partly rescue both reproductive phenotypes in this species complex (Ehrman 1961; Miller et al. 2010).

semispecies is sufficient to trigger complete abandonment of female assortative mating (Fig. 7.3). Hence, females that normally accept only homogamic mates, do accept "improper" males from another semispecies with which they produce only sterile sons and grandsons upon symbiont depletion (Miller et al. 2010).

The study mentioned above elucidates influences of a transovarially transmitted symbiont on female mate choice in *Drosophila* via the triggering of two key parameters of speciation: pre- and post-mating isolation. In general, certain prerequisites, focusing on both host and symbiont, have to be considered for symbiont-driven speciation to occur. These prerequisites are (1) the host system should be tightly associated with a fixed symbiont and (2) this symbiont must be capable of triggering strong postzygotic bidirectional incompatibilities in hybrids, plus, pre-zygotic isolation in order to avoid wasteful gene flow via hybrid formation. The *Wolbachia*– *D. paulistorum* symbiosis system, which perfectly fulfills the aforementioned criteria, serves an excellent model for studying the global impact of endosymbiotic bacteria on sexual behavior and consequentially on symbiont-driven host speciation in insects. Since endosymbionts are universal entities of all living organisms, we suppose that in the near future more cases of "infectious" speciation will be reported in our literature when the impact of symbionts in experimentally accessible, natural

pre-mating isolation via selective mating behavior, that is, by avoiding heterogamic mates harboring another, incompatible *Wolbachia* variant (Miller et al. 2010). A series of behavioral experiments demonstrated that partial depletion of *Wolbachia* via mild antibiotic treatment that significantly reduces natural symbiont titer levels in the *D. paulistorum*

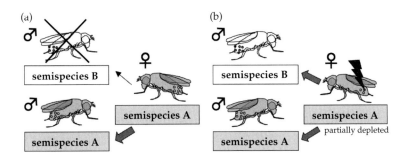

Figure 7.3 Effect of partial symbiont-depletion on pre-mating behavior of *D. paulistorum*. (A) Expression of strong sexual selection in the *D. paulistorum* species complex. In mating choice assays, females from semispecies A, infected with *Wolbachia* A (indicated by grey dots in the fly abdomen) actively avoid males of the heterogamic semispecies B, infected with incompatible *Wolbachia* B (indicated by black dots), but accept mates from the homogamic semispecies A. (B) Loss of female assortative mating behavior in *D. paulistorum* after artificial symbiont-depletion. Upon mild antibiotic treatment, semispecies A females no longer recognize "improper" mates of semispecies B, infected with incompatible *Wolbachia* B, and accept them randomly. Black flash symbolizes antibiotic treatment.

incipient speciation systems is more carefully reevaluated. We expect that such studies will support a general applicability of the results obtained from the *Wolbachia–D. paulistorum* model system and will expand our understanding of the mechanisms and dynamics of speciation.

Finally, *Wolbachia* are not a unique class of microbial symbionts capable of manipulating sexual behavior in arthropods, but additional, solid cases are still rare even anecdotally in literature. The subelytral mite, *Chrysomelobia labidomerae*, parasitizes the milkweed leaf beetle, *Labidomera clivicollis*, as a sexually transmitted ectosymbiont that is dispersed throughout populations when the host copulates. These mites are common parasites of *L. clivicollis* developing the elytra, and, at mature stage, adult females move to the abdomen of the beetle, where they feed on hemolymph (Abbot and Dill 2001). The infection of this contagious parasite is costly to the host since it significantly reduces the survival of nutritionally stressed beetles. However, in this symbiotic system, parasitic mites affect host behavior by manipulating male–male competition, thereby increasing the reproductive success of infected males. In experiments in which test males were paired with copulating beetles, it was shown that parasitized males can out-compete unparasitized rivals by displacing them and taking over the females. This "winner" phenotype of infected males is most likely caused by a heightened competitive vigor, potentially triggered by the parasite (Abbot and Dill 2001). A similar enhancement in reproductive efforts and resulting increased mating speeds and successes was observed in *Drosophila* males that were infected with life-shortening parasitic mites in the laboratory (Polak and Starmer 1998). However, both studies failed to provide compelling experimental evidence that the increased expenditure of reproductive effort results from parasitic manipulation of host behavior to increase parasite fitness in order to promote the parasite's transmission between hosts (Polak and Starmer 1998; Abbot and Dill 2001). Hence higher sexual competitiveness of infected males compared to uninfected controls was interpreted as a trade-off strategy of the hosts to compensate for reduced host fitness by increasing their reproductive efforts. Therefore, the

evidence that contact-transmitted parasites manipulate host sexual behavior in favor of their own increased transmission success is not very convincing (Poulin 2010, but see Chapter 2).

However, in the corn earworm moth *Helicoverpa zea*, it has been demonstrated that a parasitic virus can change female sexual behavior by intensifying their male calling frequencies and overproducing sex pheromones in order to increase the reproductive success of the virus in insect populations (Burand et al. 2005). In this host–symbiont interface, virus-infected females are hyper attractive to male butterflies. Since the virus is sexually transmitted during copulation, alteration of the host physiology and sexual behavior most likely serves to directly facilitate virus transmission in these insect populations. The virus named Hz-2V was found to replicate exclusively in the male and female reproductive tissues and oviducts of *H. zea*, a major agricultural pest of many crops (Rallis and Burand 2002). Most infected females are sterile and termed agonadal, due to over-replication of the virus in the female oviduct, the accumulation of virus particles in the cervix bursa, and the formation of a virus-filled vesicle covering the tip of the vulva of such females. These virus "plugs" are highly contaguous to healthy males upon contacting agonadal females, so they then transmit the virus to other females during subsequent mating (Burand et al. 2004). Recently, it has been demonstrated that presence of the virus "plug" in sterile females affects the frequency of male-calling behavior and pheromone production too. While control females stop calling after mating, most of the infected females resume calling males even after being in contact with males, since the virus "plug" found at the tip of the abdomen of agonadal females acts like a mating plug, preventing the transfer of pheromonestatic peptide (PSP) and other inhibitors of calling to virus-infected females (Burand et al. 2005).

Furthermore, in flight-tunnel experiments where males were given a choice between infected and uninfected females, it was shown that males flew to infected females almost twice as frequently as they did to control females (Burand et al. 2005). This increase in the attractiveness of infected females results from overproduction of pheromones, since agonadal females produce six to seven fold more pheromones than control females. The authors conclude that the overproduction of pheromones makes infected females more attractive to potential mates which possibly contributes to the spread of the virus among populations through the contamination of males that attempt to mate with these females (Burand et al. 2005). Hence, the parasite most likely has evolved such that its transmission has become linked with host reproduction physiology and sexual behavior facilitating spreading the virus in insect populations.

However, complete sterilization of infected females might work as an effective short-term spreading strategy of the virus, but in a longer-term perspective, the induction of such dramatic phenotypes might endanger the fitness of both the host and the parasite. Hence, the virus has developed a second route of infection with a more benign phenotype. Some infected females are fertile, asymptomatic carriers of the virus that can vertically transmit the infection to their progeny inside eggs. Such asymptomatic females can be regarded as an ancestral reservoir that facilitates maternal transmission of the virus but, upon over-replication in reproductive tissues on the expense of female fecundity, these agonadal females can spread the virus horizontally to uninfected males during attempted copulation. It remains to be elucidated, however, whether the vertical transmission mode of the Hz-2V virus is the derived or ancestral state of this host–symbiont interaction.

Indeed, in all duplex horizontal/vertical transmission instances, we do not have priorities assigned. Our guess, however, is that horizontal forms evolved first, later secured by vertical transmission though horizontal transmission was never abandoned. This incorporates an "easier" evolutionary flow, from extrinsic to intrinsic, from free-living to captured, from extracellular to intracellular, from contagion to heredity. And heredity is ultimate.

7.4 Conclusions

In contrast to earlier studies, a growing body of cases from recent literature impressively demonstrates that a diverse set of symbionts, ranging from viruses to bacteria and worms can orchestrate host biology in an adaptive manner, primarily in order to improve their own transmission successes. In such cases, behaviorally adaptive symbionts manipulate host reproductive behavior by modulating sex-determination pathways, sex hormone expression, and even mating strategies such as ovipositional and assortative mating behavior, as well as promiscuity. Although the exact physiological pathways and molecular mechanisms remain to be determined, in many cases symbionts follow a common strategy, that is, they obviously juggle the expression of mate-recognition signals such as pheromones and/or the perception and interpretation of courtship signals. In response, affected host populations have developed counter-defense strategies that allow endangered individuals to diagnose the infection status of a potential mate. In evolutionary long-term interactions, however, some of these symbiotic interactions are no longer purely antagonistic but have transformed into compensatory relationships wherein both symbiotic partners become "addicted" to each other and thus mutualistic symbionts can reach fixation in populations. Due to the high mutation rates manifest in microbial symbionts, adaptive interactions affecting sexual behavior can evolve speedily within relatively short periods, especially in isolated host populations. Upon their migration, however, such symbiont-adapted populations are no longer sexually compatible with members of other populations that carry, albeit closely related, incompatible variants of the symbiont. In this scenario adaptive symbiotic interactions can easily promote sexual selection, fostering incipient speciation in nature.

Acknowledgements

We thank Traude Kehrer and Christoph Kneisz for excellent technical support and fly work; we thank Lee Ehrman for very helpful comments on this manuscript. WJM and DS were partly supported by the research grant P22634-B17 from the Austrian Science Fund and the COST Action FA0701.

References

Abbot, P. and Dill, L. M. (2001). Sexually-transmitted parasites and sexual selection in the milkweed leaf beetle, *Labidomera clivicollis*. Oikos **92**, 91–100.

Albertson, R., Casper-Lindley, C., Cao, J., Tram, U., and Sullivan, W. (2009). Symmetric and asymmetric mitotic segregation patterns influence *Wolbachia* distribution in host somatic tissue. *Journal of Cell Science* **122**, 4570–4583.

Baldo, L., Ayoub, N. A., Hayashi, C. Y., Russell, J. A., and Werren, J. H. (2008). Insight into the routes of *Wolbachia* invasion: high levels of horizontal transfer in the spider genus Agelenopsis revealed by *Wolbachia* strain and mitochondrial DNA diversity. *Molecular Ecology* **17**, 557–569.

Baldo, L., Prendini, L., Corthals, A., and Werren, J. H. (2007). *Wolbachia* are present in southern african scorpions and cluster with supergroup F. *Current. Microbiology* **55**, 367–373.

Baudoin, M. (1975). Host castration as a parasitic strategy. *Evolution* **29**, 335–352.

Blaut, M. and Clavel, T. (2007). Metabolic diversity of the intestinal microbiota: implications for health and disease. *The Journal of Nutrition* **137**, 751S-755S.

Bordenstein, S. R. (2003) Symbiosis and the origin of species. In: *Insect Symbiosis* (eds Bourtzis K. and Miller T.) pp 283–304. CRC Press, New York.

Bordenstein, S. R., O'Hara, F. P., and Werren, J. H. (2001). *Wolbachia*-induced incompatibility precedes other hybrid incompatibilities in Nasonia. *Nature* **409**, 707–710.

Bordenstein, S. and Rosengaus, R. B. (2005). Discovery of a novel *Wolbachia* super group in Isoptera. *Current Microbiology* **51**, 393–398.

Bordenstein, S. R. and Werren, J. H. (2007). Bidirectional incompatibility among divergent *Wolbachia* and incompatibility level differences among closely related *Wolbachia* in Nasonia. *Heredity* **99**, 278–287.

Bouchon, D., Rigaud, T., and Juchault, P. (1998). Evidence for widespread *Wolbachia* infection in isopod crustaceans: molecular identification and host feminization. *Proceedings of the Royal Society B: Biological Sciences* **265**, 1081–1090.

Boyle, L., O'Neill, S. L., Robertson, H. M., and Karr, T. L. (1993). Interspecific and intraspecific horizontal transfer of *Wolbachia* in Drosophila. *Science* **260**, 1796–1799.

Breeuwer, J. A. and Werren, J. H. (1990). Microorganisms associated with chromosome destruction and reproductive isolation between two insect species. *Nature* **346**, 558–560.

Brownlie, J. C., Cass, B. N., Riegler, M., Witsenburg, J. J., Iturbe-Ormaetxe, I., McGraw, E. A., O'Neill, S. L. (2009). Evidence for metabolic provisioning by a common invertebrate endosymbiont, *Wolbachia pipientis*, during periods of nutritional stress. *PLoS Pathogens* **5**, e1000368.

Bulnheim, H. P. (1977). Geschlechtsumstimmung bei *Gammarus duebeni* (Crustaceana, Amphipoda) unterdem Einfluss hormonaler und parasitärer Faktoren. *Biologisches Zeitblatt* **96**, 61–78.

Bulnheim, H. P. (1978). Interaction between genetic, external and parasitic factors in the sex determination of the crustacean amphipod *Gammarus duebeni*. *Helgoländer wissenschaftliche Meeresuntersuchungen* **31**, 1–33.

Bulnheim, H. P. and Vávra, J. (1968). Infection by the microsporidian *Octosporea effeminans* sp. n., and its sex determining influence in the amphipod *Gammarus duebeni*. *Journal of Parasitology* **54**, 241–248.

Burand, J. P., Rallis, C. P., and Tan, W. (2004). Horizontal transmission of Hz-2V by virus infected *Helicoverpa zea* moths. *Journal of Invertebrate Pathology* **85**, 128–131.

Burand, J. P., Tan, W., Kim, W., Nojima S., and Roelofs, W. (2005). Altered mating behavior and pheromone production in female *Helicoverpa zea* moths infected with the insect virus Hz-2V. *Journal of Insect Science* **5**, 6.

Caspari, E. and Watson, G. (1959) On the evolutionary importance of cytoplasmic sterility in mosquitoes. *Evolution* **13**, 568–570.

Charlat, S., Duplouy, A., Hornett, E. Dyson, E. A., Davies, N., Roderick, G. K., Wedell, N., Hurst, G. D. (2009). The joint evolutionary histories of *Wolbachia* and mitochondria in *Hypolimnas bolina*. *BMC Evolutionary Biology* **9**, 64.

Clancy, D. J. and Hoffmann, A. A. (1998). Environmental effects on cytoplasmic incompatibility and bacterial load in *Wolbachia*-infected *Drosophila simulans*. *Entomologia Experimentalis et Applicata* **86**, 13–24.

Cook, P. E. and McGraw, E. A. (2010). *Wolbachia pipientis*: an expanding bag of tricks to explore for disease control. *Trends in Parasitology* **26**, 373–375.

de Bary, H. A. (1879). Die Erscheinung der Symbiose. In: *Vortrag auf der Versammlung der Naturforscher und Ärzte zu Kassel*, pp. 1–30. Karl J. Trübner Verlag, Strasbourg.

de Crespigny, F. E., Pitt, T. D., and Wedell, N. (2006). Increased male mating rate in *Drosophila* is associated with *Wolbachia* infection. *Journal of Evolutionary Biology* **19**, 1964–1972.

Dobzhansky, Th. and Spassky, B. (1959). *Drosophila paulistorum*, a cluster of species *in statu nascendi*. *Proceedings of the National Academy of Sciences* **45**, 419–428.

Ehrman, L. (1961). The genetics of sexual isolation in *Drosophila paulistorum*. *Genetics* **46**, 1025–1038.

Ehrman, L. (1965). Direct observation of sexual isolation between allopatric and between sympatric strains of different *Drosophila paulistorum* races. *Evolution* **19**, 459–464.

Ehrman, L. (1968). Antibiotics and infections hybrid sterility in *Drosophila paulistorum*. *Molecular and General Genetics* **103**, 218–222.

Ehrman, L. and Daniels, S. (1975). Pole cells of *Drosphila paulistorum*: Embryologic differentiation with symbionts. *Australian Journal of Biological Sciences* **28**, 133–144.

Ehrman, L. and Kernaghan, R. P. (1971). The microorganismical basis of the infectious hybrid sterility in *Drosophila paulistorum. Heredity* **62**, 66–71.

Ehrman, L. and Kernaghan, R. P. (1972). Infectious heredity in *Drosophila paulistorum*. In *Pathogenic Mycoplasmas*, pp. 227–250. Ciba Foundation Symposium. Associated Scientific Publishers, Amsterdam.

Ehrman, L., Perelle, I., and Factor, J. (1995). Endosymbiotic infectivity in *D. paulistorum* semispecies. In: *Genetics of Natural Populations: The continuing influence of Theodosius Dobzhansky* (ed. Levine) pp. 241–261. Columbia University Press, New York.

Ferree, P. M. and Sullivan, W. (2006). A genetic test of the role of the maternal pronucleus in *Wolbachia*-induced cytoplasmic incompatibility in *Drosophila melanogaster. Genetics* **173**, 839–847.

Fytrou, A., Schofield, P. G., Kraaijeveld, A. R., and Hubbard, S. F. (2006). *Wolbachia* infection suppresses both host defence and parasitoid counter-defence. *Proceedings of the Royal Society of London. Series B, Biological sciences* **273**, 791–796.

Garrett, W. S, Gordon, J. S., and Glimcher, L. H. (2010). Homeostasis and inflammation in the intestine. *Cell* **140**, 859–870.

Ginsburger-Vogel, T., Carre-Lecuyer, M. C., and Fried-Montaufier M. C. (1980). Transmission expérimentale de la thélygenie liee a l'intersexualité chez *Orchestia gammarellus* (Pallas); analyse des génotypes sexuels dans la descendance des femelles normales transformées en femelles thélygenes. *Archives de Zoologie Expérimentale et Générale* **122**, 261–270.

Giorgini, M., Monti, M. M., Caprio, E., Stouthamer, R., and Hunter, M. S. (2009). Feminization and the collapse of haplodiploidy in an asexual parasitoid wasp harboring the bacterial symbiont *Cardinium. Heredity* **102**, 365–371.

Haegeman, A., Vanholme, B., Jacob, J., Vandekerckhove, T. T., Claeys, M., Borgonie, G., Gheysen, G. (2009). An endosymbiotic bacterium in a plant-parasitic nematode: member of a new *Wolbachia* supergroup. *International Journal of Parasitology* **39**, 1045–1054.

Hagimori, T., Abe, Y., Date, S., and Miura, K. (2006). The first finding of a *Rickettsia* bacterium associated with parthenogenesis induction among insects. *Current Microbiology* **52**, 97–101.

Hamilton, W. D. (1967). Extraordinary sex ratios. A sex-ratio theory for sex linkage and inbreeding has new implications in cytogenetics and entomology. *Science* **156**, 477–488.

Hedges, L. M., Brownlie, J. C., O'Neill, S. L., and Johnson, K. N. 2008. *Wolbachia* and virus protection in insects. *Science* **322**, 702.

Hertig, M. (1936). The *Rickettsia, Wolbachia pipientis* and associated inclusions of the mosquito, *Culex pipiens. Parasitology* **28**, 453–490

Hertig, M. and Wolbach, S. B. (1924). Studies on *Rickettsia*-like microorganisms in insects. *Journal of Medical Research* **44**, 329–374.

Hilgenboecker, K., Hammerstein, P., Schlattmann, P., Telschow, A., and Werren, J. H. (2008). How many species are infected with *Wolbachia*?—A statistical analysis of current data. *FEMS Microbiology Letters* **281**, 215–220.

Hiroki, M., Kato, Y., Kamito, T., and Miura, K. (2002). Feminization of genetic males by a symbiotic bacterium in a butterfly, *Eurema hecabe* (Lepidoptera: Pieridae). *Naturwissenschaften* **89**, 167–170.

Hiroki, M., Tagami, Y., Miura, K., and Kato, Y. (2004). Multiple infection with *Wolbachia* inducing different reproductive manipulations in the butterfly *Eurema hecabe. Proceedings of the Royal Society B: Biological Sciences* **271**, 1751–1755.

Hoffmann, A. A., Clancy D., and Duncan, J. (1996). Naturally-occurring *Wolbachia* infection in *Drosophila simulans* that does not cause cytoplasmic incompatibility. *Heredity* **76**, 1–8.

Hoffmann, A. A., Turelli, M., and Harshman, L. G. (1990). Factors affecting the distribution of cytoplasmic incompatibility in *Drosophila simulans. Genetics* **126**, 933–948.

Hoffmann, A. A., Turelli, M., and Simmons, G. M. (1986). Unidirectional incompatibility between populations of *Drosophila simulans. Evolution* **40**, 692–701.

Hornett, E. A., Charlat, S., Duplouy, A. M. Davies, N., Roderick, G. K., Wedell, Hurst, G. D. (2006). Evolution of male-killer suppression in a natural population. *PLoS Biology* **4**, e283.

Hornett, E. A., Duplouy, A. M., Davies, N., Roderick, G. K., Wedell, N., Hurst, G. D., Charlat, S. (2008). You can't keep a good parasite down: evolution of a male-killer suppressor uncovers cytoplasmic incompatibility. *Evolution* **62**, 1258–1263.

Huigens, M. E. and Stouthamer, R. (2003). Parthenogenesis associated with *Wolbachia*. In *Insect Symbiosis* (eds K. Bourtzis and T. A. Miller) pp.247–266. CRC, Boca Raton, FL.

Hurst, G. D. D., Graf von der Schulenburg, J. H., Majerus, T. M., Bertrand, D., Zakharov, I. A., Baungaard, J., Völkl, W., Stouthamer, R., Majerus, M. E. (1999). Invasion of one insect species, *Adalia bipunctata*, by two different male-killing bacteria. *Insect Molecular Biology* **8**, 133–139.

Hurst, G. D. and Jiggins, F. M. (2000). Male-killing bacteria in insects: mechanisms, incidence, and implications. *Emerging Infectious Diseases* **6**, 329–336.

Hurst, G. D., Johnson, A. P., Schulenburg, J. H., and Fuyama, Y. (2000). Male-killing *Wolbachia* in *Drosophila*: a temperature-sensitive trait with a threshold bacterial density. *Genetics* **156**, 699–709.

Ikeya, T., Broughton, S., Alic, N., Grandison, R., and Partridge, L. (2009). The endosymbiont *Wolbachia* increases insulin/IGF-like signalling in *Drosophila*. *Proceedings of the Royal Society B: Biological Sciences* **276**, 3799–3807.

Jaenike, J. (2007). Fighting back against male-killers. *Trends in Ecology and Evolution* **22**, 167–169.

Jaenike, J., Dyer, K. A., Cornish, C., and Minhas, M. S. (2006). Asymmetrical reinforcement and *Wolbachia* infection in Drosophila. *PLoS Biology* **4**, e325.

Jeyaprakash, A. and Hoy, M. A. (2000). Long PCR improves *Wolbachia* DNA amplification: *wsp* sequences found in 76% of sixty-three arthropod species. *Insect Molecular Biology* **9**, 393–405.

Jiggins, F. M., Hurst, G. D. D., and Majerus, M. E. N. (1998). Sex ratio distortion in *Acraea encedon* (Lepidoptera: Nymphalidae) is caused by a male-killing bacterium. *Heredity* **81**, 87–91.

Juchault, P., Frelon, M., Bouchon, D., and Rigaud, T. (1994). New evidence for feminizing bacteria in terrestrial isopods: evolutionary implications. *Comptes Rendus de l'Académie des sciences Paris III* **317**, 225–230.

Juchault, P., Rigaud, T., and Mocquard, J. P. (1993). Evolution of sex determination and sex ratio variability in wild populations of *Armadillidium vulgare* (Latr) (Crustacea, Isopoda): a case study in conflict resolution. *Acta Oecologia* **14**, 547–562.

Kageyama, D., Nishimura, G., Hoshizaki, S., and Ishikawa, Y. (2002). Feminizing *Wolbachia* in an insect, *Ostrinia furnacalis* (Lepidoptera: Crambidae). *Heredity* **88**, 444–449.

Kageyama, D., Nishimura, G., Hoshizaki, S., and Ishikawa, Y. (2003). Two kinds of sex ratio distorters in a moth, *Ostrinia scapulalis*. *Genome* **46**, 974–982.

Kageyama, D. and Traut, W. (2004). Opposite sex-specific effects of *Wolbachia* and interference with sex determination of its host *Ostrinia scapulalis*. *Proceedings of the Royal Society B: Biological Sciences* **271**, 251–258.

Kambris Z., Blagborough, A. M., Pinto, S. B., Blagrove, M. S., Godfray, H. C., Sinden, R. E., Sinkins, S. P. (2010). *Wolbachia* stimulates immune gene expression and inhibits plasmodium development in *Anopheles gambiae*. *PLoS Pathogens* **6**, e1001143.

Kambris, Z., Cook, P. E., Phuc, H. K., and Sinkins, S. P. (2009). Immune activation by life-shortening *Wolbachia* and reduced filarial competence in mosquitoes. *Science* **326**, 134–136.

Karr, T. L., Yang, W., and Feder, M. E. (1998). Overcoming cytoplasmic incompatibility in *Drosophila*. *Proceedings of the Royal Society B: Biological Sciences* **265**, 391–395.

Kelly, A., Hatcher, M. J., and Dunn, A. M. (2004). Intersexuality in the amphipod *Gammarus duebeni* results from incomplete feminisation by the vertically transmitted parasitic sex ratio distorter *Nosema granulosis*. *Evolutionary Ecology* **18**, 121–132.

Kernaghan, R. P. and Ehrman, L. (1970). Antimycoplasmal antibiotics and hybrid sterility in *Drosophlia paulistorum*. *Science* **169**, 63–64.

Kremer, N., Voronin, D., Charif, D., Mavingui, P., Mollereau, B., and Vavre, F. (2009). *Wolbachia* interferes with ferritin expression and iron metabolism in insects. *PLoS Pathogens* **5**, e1000630.

Koukou, K., Pavlikaki, H., Kilias, G., Werren, J. H., Bourtzus, K., and Alahiotis, S. N. (2006). Influence of antibiotic treatment and *Wolbachia* curing on sexual isolation among *Drosophila melanogaster* cage populations. *Evolution* **60**, 87–96.

Legrand, J. J, Legrand-Hamelin, E., and Juchault, P. (1987). Sex determination in Crustacea. *Biological Reviews* **62**, 439–470.

Majerus, M. E., Hinrich, J., Schulenburg, G. V., and Zakharov, I. A. (2000). Multiple causes of male-killing in a single sample of the two-spot ladybird, *Adalia bipunctata* (Coleoptera: coccinellidae) from Moscow. *Heredity* **84**, 605–609.

Martin, G., Juchault, P., and Legrand, J. J. (1973). Mise en évidence d'un micro-organisme intracytoplasmique symbiote de l'Oniscoïde *Armadillidium vulgare* Latr. dont la présence accompagne l'intersexualité ou la féminisation totale des mâles génétiques de la lignée thély-gène. *Comptes Rendus Acad de l'Académie des Sciences Paris III* **276**, 2313–2316.

McMeniman, C. J., Lane, R. V., Cass, B. N., Fong A. W., Sidhu M., Wang Y. F., O'Neill, S. L. (2009). Stable introduction of a life-shortening *Wolbachia* infection into the mosquito *Aedes aegypti*. *Science* **323**, 141–144.

Merçot, H. and Poinsot, D. (2009). Infection by *Wolbachia*: from passengers to residents. *Comptes Rendus Biologies* **332**, 284–297.

Miller, W. J., Ehrman, L., and Schneider, D. (2010) Infectious speciation revisited: Impact of symbiont-depletion on female fitness and mating behavior of *Drosophila paulistorum*. *PLoS Pathogens* **6**, e1001214.

Min, K. T. and Benzer, S. (1997). *Wolbachia*, normally a symbiont of *Drosophila*, can be virulent, causing degeneration and early death. *Proceedings of the National*

Academy of Sciences of the United States of America **94**, 10792–10796.

Moreau, J., Bertin, A., Caubet, Y., and Rigaud, T. (2001). Sexual selection in an isopod with *Wolbachia*-induced sex reversal: males prefer real females. *Journal of Evolutionary Biology* **14**, 388–394.

Moreau, J. and Rigaud, T. (2000). Operational sex ratio in terrestrial isopods: interaction between potential rate of reproduction and *Wolbachia*-induced sex ratio distortion. *Oikos* **92**, 477–484.

Moreau, J. and Rigaud, T. (2003). Variable male potential rate of reproduction: high male mating capacity as an adaptation to parasite-induced excess of females? *Proceedings of the Royal Society B: Biological Sciences* **270**, 1535–1540.

Moreira, L. A., Iturbe-Ormaetxe, I., Jeffery, J. A., Lu, G., Pyke, A. T., Hedges L. M., Rocha, B. C., Hall-Mendelin, S., Day, A., Riegler, M., Hugo, L. E., Johnson, K. N., Kay, B. H., McGraw, E. A., van den Hurk, A. F., Ryan, P. A., O'Neill, S. L. (2009). A *Wolbachia* Symbiont in *Aedes aegypti* Limits Infection with Dengue, Chikungunya, and *Plasmodium*. *Cell* **139**, 1268–1278.

Narita, S., Kageyama, D., Nomura, M., and Fukatsu, T. (2007). Unexpected mechanism of symbiont-induced reversal of insect sex: feminizing *Wolbachia* continuously acts on the butterfly *Eurema hecabe* during larval development. *Applied and Environmental Microbiology* **73**, 4332–4341.

Narita, S., Pereira, R. A. S., Kjellberg, F., and Kageyama, D. (2010). Gynandromorphs and intersexes: potential to understand the mechanism of sex determination in arthropods. *Terrestrial Arthropod Reviews* **3**, 63–96.

Negri, I., Franchini, A., Gonella, E., Daffonchio, D., Mazzoglio, P. J., Mandrioli, M., Alma, A. (2009). Unravelling the *Wolbachia* evolutionary role: the reprogramming of the host genomic imprinting. *Proceedings of the Royal Society B: Biological Sciences* **276**, 2485–2491.

Negri, I., Pellecchia, M., Mazzoglio, P. J., Patetta, A., and Alma, A. (2006). Feminizing *Wolbachia* in *Zyginidia pullula* (Insecta, Hemiptera), a leafhopper with an XX/X0 sex-determination system. *Proceedings of the Royal Society B: Biological Sciences* **273**, 2409–2416.

Normark, B. B. (2003). The evolution of alternative genetic systems in insects. *Annual Reviews in Entomology* **48**, 397–423.

Otte, D. and Endler, J. E. (1989). Speciation and its consequences. Sinauer Associates, Sunderland, MA.

Panteleev, DIu., Goryacheva, I. I., Andrianov, B. V., Reznik, N. L., Lazebnyǐ, O. E., and Kulikov, A. M. (2007). The endosymbiotic bacterium *Wolbachia* enhances the nonspecific resistance to insect pathogens and alters behavior of *Drosophila melanogaster*. *Genetika* **43**, 277–280.

Peng, Y., Nielsen, J. E., Cunningham, J. P., and McGraw, E. A. (2008). *Wolbachia* infection alters olfactory-cued locomotion in *Drosophila spp*. *Applied and Environmental Microbiology* **74**, 3943–3948.

Polak, M. and Starmer, W. T. (1998). Parasite-induced risk of mortality elevates reproductive effort in male Drosophila. *Proceedings of the Royal Society of London Series B, Biological Sciences* **265**, 2197–2201.

Poulin, R. (2010). Parasite manipulation of host behavior: an update and frequently asked questions. *Advances in the Study of Behavior* **41**, 151–186.

Price, T. A., Hodgson, D. J., Lewis, Z., Hurst, G. D., and Wedell, N. (2008). Selfish genetic elements promote polyandry in a fly. *Science* **322**, 1241–1243.

Price, T. A., Hurst, G. D., and Wedell, N. (2010). Polyandry prevents extinction. *Current Biology* **20**, 471–475.

Rallis, C. P. and Burand, J. P. (2002). Pathology and ultrastructure of the insect virus, Hz-2V, infecting agonadal female corn earworms, *Helicoverpa zea*. *Journal of Invertebrate Pathology* **81**, 33–44.

Reynolds, K. T. and Hoffmann, A. A. (2002). Male age, host effects and the weak expression or non-expression of cytoplasmic incompatibility in *Drosophila* strains infected by maternally transmitted *Wolbachia*. *Genetical Research* **80**, 79–87.

Riegler, M. and O'Neill, S. L. (2007). Evolutionary dynamics of insect symbiont associations. *Trends in Ecology and Evolution* **22**, 625–627.

Rigaud, T. (1997). Inherited microorganisms and sex determination of arthropod hosts. In: *Influential passengers (eds* O'Neill, S. L., Hoffmann, A. A., and Werren, J. H.) pp. 81–101. Oxford University Press, Oxford.

Rigaud, T. and Haine, E. R. (2005). Conflicting parasites as confounding factors in parasitic manipulation studies. *Behavioural Processes* **68**, 259–262.

Rigaud, T., Moreau, J., and Juchault, P. (1999). *Wolbachia* infection in the terrestrial isopod *oniscus asellus*: sex ratio distortion and effect on fecundity. *Heredity* **83**, 469–475.

Rousset, F., Bouchon, D., Pintureau, B., Juchault, P., and Solignac, M. (1992). *Wolbachia* endosymbionts responsible for various alterations of sexuality in arthropods. *Proceedings of the Royal Society B: Biological Sciences* **250**, 91–98.

Rowley, S. M., Raven, R. J., and McGraw. (2004). *Wolbachia pipientis* in Australian spiders. *Current Microbiology* **49**, 208–214.

Sakamoto, H., Kageyama, D., Hoshizaki, S., and Ishikawa, Y. (2007). Sex-specific death in the Asian corn borer moth (*Ostrinia furnacalis*) infected with *Wolbachia* occurs across larval development. *Genome* **50**, 645–652.

Sanchez, L. (2008). Sex-determining mechanisms in insects. *The International Journal of Developmental Biology* **52**, 837–856.

Saridaki, A. and Bourtzis, K. (2010). *Wolbachia*: more than just a bug in insects genitals. *Current Opinion in Microbiology* **13**, 67–72.

Schneider, D., Miller, W. J., and Riegler, M. (2011). Arthropods Shopping for *Wolbachia*. In: *Manipulative Tenants: Bacteria Associated with Arthropods* (eds Zchori-Fein, E. and Bourtzis, K.) pp. 149–74. CRC Press, New York.

Sheeley, S. L. and McAllister, B. F. (2009). Mobile male-killer: similar *Wolbachia* strains kill males of divergent *Drosophila* hosts. *Heredity* **102**, 286–892.

Shoemaker, D. D., Katju, V., and Jaenike, J. (1999). *Wolbachia* and the evolution of reproductive isolation between Drosophila recens and *Drosophila subquinaria*. *Evolution* **53**, 1157–1164.

Sinkins, S. P. (2004). *Wolbachia* and cytoplasmic incompatibility in mosquitoes. *Insect Biochemistry and Molecular Biology* **34**, 723–729.

Skorokhod, A., Gamulin, V., Gundacker, D., Kavsan, V., Muller, I. M., and Muller, W. E. (1999). Origin of insulin receptor-like tyrosine kinases in marine sponges. *Biological Bulletin* **197**, 198–206.

Snook, R. R., Cleland, S. Y., Wolfner, M. F., and Karr, T. L. (2000). Offsetting effects of *Wolbachia* infection and heat shock on sperm production in *Drosophila simulans*: Analyses of fecundity, fertility and accessory gland proteins. *Genetics* **155**, 167–178.

Stevens, L. and Wade, M. J. (1990). Cytoplasmically inherited reproductive incompatibility in Tribolium flour beetles: the rate of spread and effect on population size. *Genetics* **124**, 367–372.

Stouthamer, R., Breeuwer, J. A., and Hurst, G. D. (1999). *Wolbachia pipientis*: microbial manipulator of arthropod reproduction. *Annual Review of Microbiology* **53**, 71–102.

Stouthamer, R., Luck, R. F., and Hamilton, W. D. 1990. Antibiotics cause parthenogenetic *Trichogramma* (Hymenoptera/Trichogrammatidae) to revert to sex. *Proceedings of the National Academy of Sciences of the United States of America* **87**, 2424–2427.

Taylor, M. J., Hoerauf, A., and Bockarie, M. (2010). Lymphatic filariasis and onchocerciasis. *The Lancet* **376**, 1175–1185.

Teixeira, L., Ferreira, A., and Ashburner, M. (2008). The bacterial symbiont *Wolbachia* induces resistance to RNA viral infections in *Drosophila melanogaster*. *PLoS Biology* **6**: e2.

Telschow, A., Flor, M., Kobayashi, Y., Hammerstein, P., and Werren, J. H. (2007) *Wolbachia*-induced unidirectional cytoplasmic incompatibility and speciation: mainland-island model. *PLoS One* **2**: e701.

Telschow, A., Hammerstein, P., and Werren, J. H. (2005) The effect of *Wolbachia* versus genetic incompatibilities on reinforcement and speciation. *Evolution* **59**, 1607–1619.

Tram, U., Ferree, P. M., and Sullivan, W. (2003). Identification of *Wolbachia*-host interacting factors through cytological analysis. *Microbes Infection* **5**, 999–1011.

Tram, U., Fredrick, K., Werren, J. H., and Sullivan, W. (2006). Paternal chromosome segregation during the first mitotic division determines *Wolbachia*-induced cytoplasmic incompatibility phenotype. *Journal of Cell Science* **119**, 3655–3663.

Tram, U. and Sullivan, W. (2002). Role of delayed nuclear envelope breakdown and mitosis in *Wolbachia*-induced cytoplasmic incompatibility. *Science* **296**, 1124–1126.

Turelli, M. (1994). Evolution of incompatibility-inducing microbes and their hosts. *Evolution* **48**, 1500–1513.

Vala, F., Egas, M., Breeuwer, J. A.J., and Sabelis, M. W. (2004). *Wolbachia* affects oviposition and mating behaviour of its spider mite host. *Journal of Evolutionary Biology* **17**, 692–700.

Vala, F., Weeks, A., Claessen, D., Breeuwer, J. A. J., and Sabelis, M. W. (2002). Within- and between-population variation for *Wolbachia* induced reproductive incompatibility in a haplodiploid mite. *Evolution* **56**, 1331–1339.

Vance, S. A. (1996). Morphological and behavioural sex reversal in mermithid-infected mayflies. *Proceedings of the Royal Society of London, Series B* **263**, 907–912.

Vandekerckhove, T., Watteyne, S., Bonne, W., Vanacker, D., Devaere, S., Rumes, B., Maelfait, J. P., Gillis, M., Swings, J. G., Braig, H. R., Mertens, J. (2003). Evolutionary trends in feminization and intersexuality in woodlice (Crustacean, Isopoda) infected with *Wolbachia pipientis* (α-proteobacteria). *Belgian Journal of Zoology* **133**, 61–69.

Vavre, F., Fleury, F., Varaldi, J., Fouillet, P., and Boulétreau, M. (2000). Evidence for female mortality in *Wolbachia*-mediated cytoplasmic incompatibility in haplodiploid insects, epidemiologic and evolutionary consequences. *Evolution* **54**, 191–200.

Vavre, F., Fleury, F., Varaldi, J., Fouillet, P., and Boulétreau, M. (2002). Infection polymorphism and cytoplasmic incompatibility in *Hymenoptera-Wolbachia* associations. *Heredity* **88**, 361–365.

Weeks, A. R., Reynolds K. T. and Hoffmann, A. A. (2002). *Wolbachia* dynamics and host effects: what has (and has not) been demonstrated? *Trends in Ecology and Evolution* **17**, 257–262.

Werren, J. H., Baldo, L., and Clark, M. E. (2008). *Wolbachia*: master manipulator of invertebrate biology. Nature Reviews Microbiology **6**, 741–751.

Wiwatanaratanabutr, S. and Kittayapong, P. (2006). Effects of temephos and temperature on *Wolbachia* load and life history traits of *Aedes albopictus*. *Medical and Veterinary Entomology* **20**, 300–307.

Wiwatanaratanabutr, I. and Kittayapong, P. (2009). Effects of crowding and temperature on *Wolbachia* infection density among life cycle stages of *Aedes albopictus*. *Journal of Invertebrate Pathology* **102**, 220–224.

Wu, M., Sun, L. V., Vamathevan, J., Riegler, M., Deboy, R., Brownlie, J. McGraw, E. A., Martin, W., Esser, C., Ahmadinejad, N., Wiegand, C., Madupu, R., Beanan, M. J., Brinkac, L. M., Daugherty S. C., Durkin, A. S., Kolonay1, J. F., Nelson, W. C., Mohamoud, Y., Lee, P., Berry, K., Young, M. B., Utterback, T., Weidman, J., Nierman, W. C., Paulsen, I. T., Nelson, K. E., Tettelin, H., O'Neill, S. L., and Eisen, J. A. (2004). Phylogenomics of the reproductive parasite *Wolbachia pipientis w*Mel: a streamlined genome overrun by mobile genetic elements. *PLoS Biology* **2**, e69.

Yen, J. H. and Barr, A. R. (1971). New hypothesis of the cause of cytoplasmic incompatibility in *Culex pipiens L. Nature* **232**, 657–658.

Zeh, J. A. and Zeh, D. W. (1996). The evolution of polyan dry I: intragenomic conflict and genetic incompatibility. *Proceedings of the Royal Society of London, Series B* **263**, 1711–1717.

Zeh, J. A. and Zeh, D. W. (1997). The evolution of polyan-dry II: post-copulatory defences against genetic incompatibility. *Proceedings of the Royal Society of London, Series B* **263**, 1711–1717.

Afterword

Lee Ehrman

"Host Manipulation by Parasites," the title of this much anticipated anthology, could accurately be altered, for the chapter to which this afterword is appended, to "Experimental Manipulation…". This is reasonable because I'm the scientist-grandparent to the six endlessly-astonishing *D. paulistorum* semispecies, as well as to their accompanying passengers. Indeed, I have spent my whole adult life studying, and being diverted, by these wee beasties, possessors of but six chromosomes plus omnipresent symbionts, members of the genus *Wolbachia*.

Call it tunnel vision, but how frequently, if ever, are ardent students of evolution gifted with the sight of speciation in progress? This, in laboratory-manipulate-able, short-generation interval-led subjects, was initially presented to me in my teens as a good thesis limited to one sex, the genetics of sterile hybrid males. It was Theodosius Dobzhansky, my PhD supervisor and academic mentor, who first discovered this unique model system depicting incipient speciation in the neotropical *Drosophila paulistorum* superspecies in the late 1950s in Central and South America. At first, mapping each of the three pairs of chromosomes, fortunately low in number, and possessed by all six semispecies, after "mutagenizing" them, I thought I was only dealing with nuclear-cytoplasmic incompatibilities manifest in hybrids, where cytoplasm could not carry out the instructions of all nuclear genes. Then I suspected a chlamydia-like intracellular entity; after that, I envisioned a eubacterial one. My errors were prolific, 'til an Austrian colleague (WM) told me about *Wolbachia* in arthropods. I informed my mentor, before he died, about what I thought about all this, that is both a hereditary and contagious disease. He replied, "Lee, this is too much. Are such extensive evolutionary excursions necessary? To accomplish what?"

What would Dobzhansky have made of our later documented proof that these omnipresent, powerful *Wolbachia* control pheromonal production in both sexes and, in this way, influence mate selection, normally barring any generation of hybrids in nature. And, since grandparents are never young, what a pleasure it is to formally transfer ongoing *D. paulistorum*-based queries to two later generational collaborators, Wolfgang J. Miller and his graduate student, Daniela Schneider, both better trained to cope with mandatory follow-up laboratory procedures in molecular analyses.

To place this multi-organismal, populational story in wider settings, this superspecies manifests an array of extrinsic and intrinsic reproductive isolating mechanisms—none perfect—but, in concert, accomplishing incipient speciation on more than one front, a truly bravura enterprise.

The table below is presented to itemize these chronologically-differentiated isolating mechanisms, all acting in the *D. paulistorum* species complex:

In this list, where does one place symbiosis, of the intimate sort described in our Miller–Schneider chapter? Is it *mutualism*, since this host insect complex cannot survive without semispecific *Wolbachia*? (And what does the symbiont do that is vital for its host?) Is it mutualism because these microorganisms supply fundamental, detailed support for sex pheromone production meticulously adjudicated by choosing females? Is it *parasitic* because tissue-specific gonadal *Wolbachia* overgrowth engenders hybrid male sterility wherein testes develop into *Wolbachia* factories, eschewing spermatogenesis, and even ejaculating *Wolbachia*? Is it *parasitic* because *Wolbachia* interfere and diminish both

EXTRINSIC

Overlapping–partial

Geographic isolation (a good definition of a semispecies)

Habitat isolation due to some allopatricity plus some partial sympatricity

INTRINSIC PRE-MATING

Temporal isolation due to geography, e.g. altitudes and to accompanying slightly different niches

Sexual isolation manifest in programmed, pheromone-supported courtships unique to

each semispecies

INTRINSIC POST-MATING

Hybrid male sterility (premier isolating mechanism)

Hybrid inviability (non-hatching hybrid eggs)

Hybrid breakdown (in later generations, produced by back-crossing relatively fertile hybrid inviability

inviability females to parental males)

---

non-hybrid (semispecies-pure) and hybrid (inter-semispecific) egg hatchability? Is the *Drosophila paulistorum* phenomenon both *mutualism* and *parasitism*, controlled *mutualism* in non-hybrids but rampant *parasitism* in laboratory-produced hybrids (that never occur in nature), such that females consistently produce sterile sons, generation after generation?

I assign these collaborators—Miller and his graduate student, Schneider—tasks involving elucidating the perhaps nutritive roles of *Wolbachia* in their hosts's holometabolous life cycle, for example facilitating iron metabolism and/or other compounds; in elucidating *Wolbachia*'s roles regarding ether, cold, and heat sensitivities, drastically altered when *Wolbachia* are minimized; and, crucially, in the current, active, and predictive, future roles of *Wolbachia* in the unique sort of speciation manifest in this awesome material, currently *in flagrant statu nascendi*!

Permit me to conclude with a prediction: we shall soon see the birth of *Wolbachia*-mediated *species nova*, perhaps only a temporary one (or two), perhaps only laboratory-based, but who knows…I gleefully look forward to this!

CHAPTER 8

Parasites and the superorganism

David P. Hughes

8.1 Introduction

We know that parasites affect the behavior of their hosts in weird and wonderful ways. Contained within this book are surveys of parasites affecting all manner of hosts and reviews of the mechanisms by which these behaviors occur and the ecological and evolutionary significance of such strategies. The field of parasite manipulation of host behavior is maturing quickly (Chapter 1). To date studies have mainly focused on the effects of parasites on the behavior of the individual organism. This is both sensible and logical given the importance of the organism in the long history of evolutionary biology and the on-going discussion of where selection acts (Mayr 1997). But of course behavior is not expressed only by unitary organisms. Behavior is also expressed by superorganisms.

The superorganism is a term used exclusively in the context of eusocial organisms like ants, wasps, bees, and termites (Hölldobler and Wilson 1990, 2009). The eusocial insects account for a low diversity of animal life (<2% of all insects) but by virtue of having multiple individuals in a single colony their biomass is disproportionately large and in some habitats, such as the tropical rainforests, eusocial insects may account for over half of all free-living biomass, including vertebrates (Tobin 1991; Hölldobler and Wilson 2009). As such these groups are interesting to parasitologists because all these bodies moving through the environment present either a large number of hosts to be infected or avoided if social insects are not the target host (Hughes 2005). The word "superorganism" describes how the cooperative group living that we observe in eusocial insects leads to phenotypes that

are a product of multiple individuals that have become specialized to perform separate tasks such that their action can be viewed as distinct parts of the collective. The most fundamental separation of tasks is the division of labor where only a small fraction of the colony reproduces (queen and males) with the majority of individuals performing work and not reproducing directly. Beyond that the non-reproducing majority can be further specialized to perform distinct tasks, which I discuss in more detail below.

In this chapter I want to explore parasites that manipulate social insect behavior. I will ask whether such behavioral changes fit into the framework developed for parasites affecting the behavior of individual hosts. I will then examine a range of collective behaviors expressed by diverse taxa to ask if lessons from studies on superorganisms and their behavior can inform collective behavior more generally. Finally, I will look towards future work that empirically addresses the difference in environments between solitary and eusocial hosts.

8.2 The extended phenotype and the unitary organism

In a landmark book, *The Extended Phenotype*, Richard Dawkins (1982) advocated that the phenotype need not be attached directly to the organism but could be physically distant to the organisms whose genes are encoding it. There are three categories of extended phenotypes (EPs). The first is animal architecture which the Nobel Prize sharing ethologist, Karl Von Frisch called "frozen behavior" (1974). The work of Michael Hansell gives an excellent insight into this little studied, but fascinating component of animal behavior (Hansell 2004, 1996). By the far the most well-known is the beaver dam, which is a physical representation of beaver behavior in wood and mud that increases the fitness of the genes encoding the building behavior. The second EP is parasite manipulation of host behavior. This topic is the focus of this edited volume and has also been extensively reviewed by others (Poulin 2011; Moore 2002; Barnard and Behnke 1990; Poulin 1994). An exemplar of this field is the suicidal behavior of crickets infected by hairworms whereby they jump into water so the adult worm can impressively exit from the thrashing body of its drowning host (Thomas et al. 2002a; see also Chapters 2, 3, and 9). This behavior is controlled by parasite, and not host, genes (Biron et al. 2006). The third and final EP is action at a distance and here a parasite example was used: the manipulation of host behavior by cuckoo chicks (see Chapter 6). In this case the chick is not physically associated with the host, as in the case of hairworms, but influences the expression of its behavioral phenotype nonetheless. Dawkins further discussed how action at a distance need not be confined to parasite–host relationships but can occur elsewhere, such as between conspecifics, as in pheromone based social communication or territorial disputes (Sergio et al. 2011).

The extended phenotype view of behavior is intimately related to the view of the gene as the unit of selection. This paradigm emerged during a period of much debate between advocates of individual and group level selection and through the work of Hamilton (Hamilton 1963, 1964). Dawkins then subsequently developed it as a transparent concept with his selfish gene approach (1976) and it became the foundation for sociobiological theory (Wilson 1975). As an historical aside it was recently emphasized that Wilson's sociobiology stance leant more towards group rather than individual selection and it is Dawkins who deserves the major credit in the current association between sociobiology and gene level/individual selection (Segerstråle 2007). In the last six years Wilson has spoken out against the current sociobiological view that relies heavily on the indirect fitness framework (Hughes 2009, 2011; Wilson and Wilson 2007; Wilson 2005) culminating in a model with mathematical biologists (Nowak et al. 2010) which was formally critiqued in the *Brief Communications* section of *Nature* (24 March 2011 issue).

The debate on individual versus group selection is both important and valuable (Hughes 2011) but it should not obscure the fact that the gene is still the unit of selection. What this paradigm states is that genes alone are transferred between generations; the organisms in which genes reside and their phenotypes are the means by which transmission is secured. Organisms are vehicles and genes are replicators. Natural selection chooses among variation in phenotypes but the information encoding these phenotypes and, ultimately, the unit which is selected is the gene (see discussion by Mayr 1997).

Having discussed the mechanism of genetic material transfer between generations let us return to phenotype. The phenotype has principally been considered a trait of the individual organism. Examples include flower color, head size, butterfly wing spots, behavior, and chemical signals released into the air, to name just a few. But such foci reflect the convenience with which we could study those easily visible attributes of organisms (Dawkins 1990). It also reflects historical effects as modern approaches follow on from the natural history tendencies of previous generations (Burkhardt 2005). Now of course behavioral ecologists are taking advantages of advances in cellular and chemical biology to measure less obvious phenotypes of the organism such as the surface of cells, tissues, and organs. This chain of phenotypes extends down to the transcriptome that affordable next generation sequencing allows to be used for a broad array of non-model taxa (Bonasio et al. 2010).

8.3 The behavior of social insects

When we think of the social insects it is the ants, termites, wasps, and bees that come to mind. The technical term is eusocial, which is defined as having overlapping generations, cooperative care of brood, and division of labor that typically means a reproductive division with the majority of individuals being sterile (Wilson 1971). There have been other definitions of eusociality (e.g., Crespi and Yanega 1995) and there are many other taxa besides ants, termites, wasps, and bees in which we find sociality. Examples of these are: mites, spiders, shrimp, thrips, aphids, beetles, and naked mole rats (Costa 2006; Wilson 1971; Bennett and Faulkes 2000; Crespi and Cho 1997). Even humans and pilot whales have been called social (McAuliffe and Whitehead 2005; Foster and Ratnieks 2005). In this chapter I will restrict myself to the traditionally defined social insects (ants, termites, wasps, and bees). Also, for convenience, and in line with most authors, I will use the term social insects, rather than eusocial insects.

Social insects live in family based groups where a minority of individuals reproduce (queens and kings/males) and the majority (the workers) are functionally sterile and collect resources to provision the offspring of the reproductives. In hymenopterans the male is represented as stored sperm in the female and in termites the male (termed king) is a whole individual that continually mates with the queen. The non-reproducing state of the workers, that is their functional sterility, is an example of altruism. It is considered adaptive for workers since the offspring are usually the full siblings of the workers and by helping to raise future queens and males that begin new colonies they gain indirect fitness benefits (Hamilton 1963, 1964). Social insects live in colonies that vary in size from 10 individuals in hover wasp societies (Turillazzi 1991) to more that 10 million in army ant societies (Hölldobler and Wilson 1990). They can occupy living spaces ranging in size from an acorn (*Temnothorax*) to 5 m high mounds (termites).

Living in societies requires effective communication strategies and studies of social insects have been instrumental in the development of communication theory (Hölldobler and Wilson 1990; Abe et al. 2001; Seeley 1995; Ross and Matthews 1991). Obvious examples are status communication in the linear dominance hierarchy first discovered in paper wasps (Pardi 1948; Turillazzi and West-Eberhard 1996), pheromone communication developed extensively in ants (Wilson 1959; Hölldobler

1995), language among insects in the honeybee waggle dance (Von Frisch 1968; Seeley 1995), and teaching (Richardson et al. 2007; Franks and Richardson 2006). Social insects communicate with other members of the society: signaling identity (which colony they belong to), soliciting of food by larvae, and adults soliciting nutritious regurgitations from larvae; individuals signaling their reproductive status and their position in a hierarchy (submissive posture, badge of status) or describing the location and quality of food (waggle dance). Individuals also communicate with other societies: signaling identity (nest of origin), aggressive displays signaling fighting ability and resource ownership. Collective actions involving many individuals also have communicative roles and are usually used towards potential threats: Asian honey bees (*Apis dorsata*), which form a bee-curtain across their comb, ripple *en masse* to confuse predatory birds (Kastberger and Sharma 2000), paper wasps (*Polistes*) dance *en masse* to threaten parasitoids (West-Eberhard 1969) and, most impressive of all to me, is the production of sound up to 5 m away via cooperative wing beating (*Syanoeca surinama*, a wasp) against the inside of a corrugated carton nest to deter mammalian predators (Rau 1933). The latter report, which is anecdotal, has a parallel in African bees whose sound was shown to deter herds of elephants (King et al. 2007).

Societies also require the evolution of elaborate architecture and social insects are rivaled only by humans in their ability to construct living spaces. No bird nest, spider web, or caddis shell can compare with the multifunctional cathedral mounds built by fungus growing termites; these 5-m high, rock like structures, standing in sun-baked desert brush, contain within them sophisticated natural air-conditioning units, crop fungus growing combs, brood nurseries, refuse piles, networks of passageways, and, at the center, a rock hard protective chamber in which the king and the 3,000 egg per day egg-laying machine that is the queen, reside (Abe et al. 2001). A termite mound is all the more impressive when we recognize that the architectural feat exists as a greenhouse to grow a rainforest adapted fungus in such places as the dry Australian outback (Aanen and Eggleton 2005).

8.4 Behavior of the superorganism

Having provided some brief background to social insects and how they live I now want to discuss the valid use of the metaphor of the colony as a superorganism. The large sizes of social insect societies, the multiple examples of collective action, and the way society members are often behaviorally or morphologically specialized for certain tasks, together with the localization of the colony in a bounded structure that is built by multiple individuals, have lead to the view that the whole colony is a superorganism (Wheeler 1911). This view, though intuitively appealing, lost favor for two reasons. The first was the supposed conflict it had with individual or gene-level selection. However, no such conflict exists so long as the superorganism is viewed within the levels of selection framework (Reeve and Keller 1999; Bourke and Franks 1995, p. 64–66). It is important to be clear and precise when adopting a metaphor and it should be stressed that the superorganism, just like the organism itself, is not a replicator (Dawkins 1990). Colonies can split in two, giving rise to two colonies (e.g., honeybees and army ants) but this is not replication in the strict sense. I would stress that much confusion arises from an inability to parse studies of behavior into mechanistic or functional approaches (Duckworth 2009).

The second reason the superorganism concept declined in popularity was the limitation of a primarily analogical approach (Hölldobler and Wilson 1990, p. 358). The concept was good, but not particularly useful when investigators proceeded to examine the fine details of colony life, such as reproductive decision-making in the light of kin selection. That is because different individuals within a colony may have different goals. Colony members do not come into conflict over resource acquisition but can, and do, conflict over resource allocation (Boomsma and Franks 2006). A clear example is the conflict between workers and queens in hymenopteran societies over the sex ratio of the reproducing offspring; the former favor a 0.75 bias towards females and the latter an equal sex ratio (Bourke and Franks 1995; Bourke 2011). There is also conflict between workers if one decides to reproduce, and here we see the evolution of policing behavior (Ratnieks 1988) where workers "police" the egg laying of other workers because it is in their genetic interests that only the queen reproduces. When examining such conflicts, the individual level view is more useful than a superorganism view. A particularly nice discussion of the superorganism view was recently given by Hamilton et al. (2009).

But in many activities individuals do cooperate and appear to be maximizing something that is usually colony survival or colony propagule production (Queller and Strassmann 2002). So, for example, in seasonally flooded Argentinean habitats, fire ant colonies make a raft of interlinked workers and float to safety; in choosing a new home, swarming bees migrate *en masse* as a single unit; and in rearing its crop fungus leaf-cutting ants have distinct morphological and behavioral castes that transport leaves from the forest to the food fungus in a "Henry Ford-factory like" manner and then process the waste in an extraordinarily efficient division of labor (Anderson et al. 2002). In these cases multiple individuals cooperate because of shared interests and produce phenotypes that cannot be achieved individually. That is, the colony level phenotype. Since the organism is neither the object of selection, nor the replicator, but rather is comprised of cooperating genes that have resolved potential conflicts because of shared interests in gamete production (Dawkins, 1990), then the apparent unity of the superorganism can be explained because it helps genes lever themselves into the next generation (see also Queller and Strassmann 2002).

In a review of this topic Anderson et al. (2002) identified 18 such self-assemblages. There is undoubtedly a genetic basis for this, and no doubt natural selection acted upon variations in rafting ability, for example, to produce an optimal response to seasonally flooded habitats. This phenotype is not an extended one like the physical, abiotic nest walls but rather it is a cumulative effect of the coordinated actions of individuals. The colony level behavior we see is "more than the sum of its parts" (Oster and Wilson 1978, p. 10). In order to produce effective responses to collective goals (e.g., colony survival) the multiple individuals must cooperate irrespective of any gene level conflicts they may have. They may be in conflict later on in the colony cycle (at the timing of reproduction) but when necessary for collective survival the cooperation is necessary and observed. The desiderata, or interests, of the distinct members are aligned for a period of time (Dawkins 1990).

8.5 Parasites divide the interests of superorganism

I have gone to some lengths to stress the biology of social insects and the real unity that exists among social insect society members, because the introduction of parasites into the system leads to cryptic competition within an apparently unified group of colony members. Parasites can live either inside the host body or external to it (and inside the colony). This is shown in Fig. 8.1 for parasites of ants. In the former case which individual is infected is not obvious. Given that many of these colony level activities (house hunting, foraging, defending against predators) are risky pursuits (e.g., Schmid-Hempel and Schmid-Hempel 1984) then conflict is predicted. A parasite infecting a worker will not want its host to exit the colony on foraging trips when such activity entails an appreciable risk of mortality before the parasite is ready to transmit to a new host or complete its development. Note that timing is important since many social insect parasites are trophically transmitted so require the host to be predated upon (e.g., cestodes, trematodes, and nematodes). While we do not generally see conflict in insect societies over resource acquisition (collecting food), but rather over resource allocation (to male vs. female larvae, to own vs. queen reproduction) (Boomsma and Franks 2006) the presence of parasites establishes a conflict scenario over resource acquisition since it entails an appreciable risk.

The superorganism concept is therefore good because it forces us to remember the alignment of interests among non-infected colony members while at the same time erecting a category of aligned members into which the infected individuals may not always fit because of the diverse desiderata of parasites within them. Parasitized individuals in the colony are the ultimate "cheaters" of the cooperative hive but of course, unlike the more well known selfish individuals that want to pursue their own interest (e.g., laying their own eggs), the infected individuals are vehicles for parasite genes. In the next section, I review what behaviorally modifying parasites these chimeric individuals contain.

8.6 Behaviorally modifying parasites of social insects

The keystone concept of social insect biology is the reproductive division of labor. Understanding this is central to all studies on the evolutionary biology of social insects and this is equally true for parasites that infect social insects. The reproductive division of labor means that parasites that infect and kill workers need not necessarily affect the fitness of the worker. This paradoxical statement is resolved when we realize that worker fitness is realized via indirect fitness via helping behavior towards relatives (Hamilton 1964). This is fundamentally different to infection in solitary organisms (Hughes 2005; Hughes et al. 2008). Natural selection might not act on individual defense (e.g., innate immunity for workers) if the cost of that worker's loss from the colony via behavioral manipulation is less than the cost of defense. But if sufficient numbers of workers are lost to infection then we could expect that the colony defends itself against infection. Colonies are well known to be highly adaptable units that rapidly respond to changes: producing more or fewer workers of a certain size for example (Wilson 1983a, 1983b).

Because of the potential for a colony-level response it is correct to view a parasite of a social insect as having two hosts: the individuals in whose body it lives and the colony that the individual belongs to. This means that parasites infecting social insects always infect two hosts at once (Sherman et al. 1988, p. 263; Schmid-Hempel 1998; Hughes et al. 2008). The two-host view of social insects is valid and fully accepted and with this in mind let us progress to examine which parasites infect social insects, and importantly, which manipulate them.

Parasites of social insects have provided prominent and compelling examples of parasite EPs where host behavior is manipulated. The best known example is the "brain-worm" which is a trematode inducing its intermediate ant host to leave the colony and climb blades of grass and bite hard (Carney 1969). The final host is a grazing animal such as a sheep that is presumed to ingest ants along with the grass it is eating (Manga-Gonzalez et al. 2001). So emblematic is this example that it "made the cover" of Janice Moore's excellent review entitled *Parasites and the Behavior of Animals* (2002). Another manipulating parasite that I work on, the fungus *Cordyceps*, which also causes ants to bite onto vegetation, similarly adorned the cover of Paul Schmid-Hempel's book, *Parasites in Social Insects* (1998). There are many parasites in social insect societies (Kistner 1979, 1982; Schmid-Hempel 1998). A sense of this diversity can be had by examining Fig. 8.1 which only shows those infecting ants (called myrmecophiles).

In reviewing here the range of parasites causing behavioral changes among the social insects it will be useful to introduce categories (Table 8.1). There are five categories of behavioral modification in social insects. (1) The first is adaptive manipulation of individual host behavior that favors parasite genes. The above mentioned brain worm is an

example. For many horizontally transmitted or trophically transmitted parasites (i.e. where predation of the host is a necessary requirement for transmission) it is obligatory for the individual host to leave the colony and in these cases nest desertion is the EP of the parasite: Conopids, Strepsiptera, Trematodes, Cestodes, mermithid and rhabtid Nematodes, Entomopthoralean and Clavicipitalean fungi (parasite associations with social insects was extensively reviewed in Schmid-Hempel, 1998 so a full list of references is not presented here due to space constraints). In all cases the manipulation is a multi-step process. Once outside the colony the host is often directed to a particular location where it performs a stereotypical activity: biting vegetation (fungi, trematodes), suicide in water (mermithid nematodes), digging to provide a diapause site for the parasite pupa (conopids), walking in an exposed location so as to be eaten by vertebrates (cestodes, trematodes, nematodes), inactivity in a prominent place to facilitate parasite mating (Strepsiptera), and moving around the environment to disperse parasite propagules from the parent parasite in the social insect. In each of these cases the biology of the

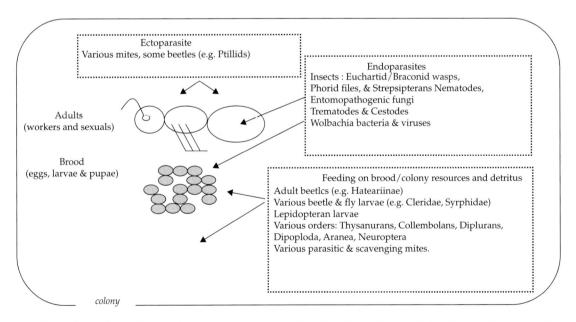

Figure 8.1 A schematic diagram of an ant colony showing a worker and some brood. The relative positions occupied by various parasites are shown. The list of myrmecophiles is not exhaustive.

parasite and its mode of reproduction is such that nest desertion is adaptive to the parasite: remaining in the nest would not lead to infection of other colony members because the parasite is not infective without that necessary departure outside the colony where it either mates or develops in a manner not possible in the colony (e.g., fungi growing through the cuticle or trematodes transferred to a final host).

(2) The second category is adaptive manipulation of more than one individual (i.e., the colony) that favors parasite genes. The entry of social parasites into the colony can be accompanied by the release of chemicals that induce confusion among workers and prevent parasite exclusion (discussed below). Because the parasite is not internal to the individual host (i.e., worker) then we may view this as the action at a distance extended phenotype like the familiar example of cuckoo chicks manipulating their hosts to feed them (Chapter 6). Indeed, the social parasites are often called cuckoo wasps and ants. Recall the justification in considering the colony as a host, in addition to the individual (Sherman et al. 1988, p. 263; Schmid-Hempel 1998).

(3) The third category switches the benefits of the parasite associated behavioral change from the parasite to the host and into defensive behaviors. Individual social insects have a very large repertoire of defensive behavioral reactions against parasites. Most mundane, but evidently important for colony level defense, is self-grooming, allo-grooming and

in the case of crop rearing ants, grooming, or "weeding," their mutualistic fungus (Cremer et al. 2007). More dramatic is cold seeking behavior by conopid-fly infected bees to retard parasite development; every night they move outside of the high temperature nest (Müller and Schmid-Hempel 1993). Conopid larvae live inside bees and cannot infect the siblings of the bee they are infecting, so this nightly self-exclusion is defensive in that it retards the parasite's growth and facilitates a longer working life for the infected individual.

(4) The fourth category is also a defensive behavior against parasites but here it requires the coordinated action of multiple individuals to succeed. The "dancing behavior" of paper wasps in response to the presence of an ovipositing Ichneumonidae wasp is a good example (West-Eberhard 1969). Another example is construction behavior where individuals cooperate to build satellite nests (Jeanne 1979) or walls to quarantine infected areas of the colony (Schultz et al. 2005) and even infected siblings (Epsky and Capinera 1988). The last, that of construction, is an EP of ants and wasps as a defense against parasites and has nice parallels with avian construction such as oven and weaver bird nests as a defense against predators like snakes (Hansell 2004).

(5) The fifth category does not interpret the behavioral change as an adaptive trait of either the parasite or the host but rather as a "boring by-product" of infection (coined by Dawkins 1990; Poulin 1994,

Table 8.1 The five categories of altered behavior in social insects and examples (discussed and referenced in the text).

	Parasite adapted	Host adapted	Byproduct
1	Manipulating the individual e.g., *cordyceps*, *strepsipterans*	-	-
2	Manipulating the colony (Multiple individuals) e.g., ant warfare by parasitoid	-	-
3	-	Individual defense e.g., *Bombus* workers seeking cold temperature against Conopid infection	-
4	-	Multiple individuals defending e.g., Wasp dance against oviposting parasitoids	-
5	-	-	At individual and colony level e.g., Generalist fungi like *Metarizhium*

1998, 2000). This category, though important when considering the EPs of parasites, is often the least satisfactory: it is commonly the one advanced in objection to the adaptationist explanations (the infamous Spandrels of San Marco, by Gould and Lewontin 1979). We can think of lethargy or reduced flying ability when infected as possible examples (Kathirithamby and Hughes 2005).

These are the five categories of behavioral changes among social insects due to the presence of parasites. I will now consider three ways parasites manipulate superorganism behavior.

8.7 Manipulating the whole colony

8.7.1 Parasitizing social resources

When colonies of the Asian army ant *Leptogenys distinguenda* move home (a regular occurrence for such nomadic ants) they pick up a molluscan parasite of their colony in preference to their brood stage siblings (Witte et al. 2002). It appears the mollusc produces an irresistible foam mass that the ants find very attractive (V. Witte, pers. comm.). This sort of super-normal signal (Dawkins and Krebs 1979) appears to be a common strategy among parasites that manipulate the care giving behavior of social insects (Hölldobler et al. 1981; Hölldobler 1971; Als et al. 2001). These are called social parasites and they are exclusively large organisms similar in size to the hosts (e.g., beetles, caterpillars). Social parasitism is also the term applied to brood parasites such as cuckoo chicks where "cuckoos should be selfish because their greed is unconstrained by kinship" (Kilner and Davies 1999). A colony member responding to such signals by a social parasite is misdirecting altruism, and deceptive communication (= behavioral manipulation) is the explanation.

There are many examples of social parasites. A common feature is that the parasite is not internal inside the body of an individual ant, wasp, bee, or termite but only internal in the host colony. In societies of ants, for example, it is possible to find beetles, flies, caterpillars, mites, molluscs, and other macroparasites parasitizing the nest. In some cases these parasites are very well camouflaged to avoid detection or use appeasement substances from specialized glands to avoid being evicted (Fiedler et al.

1996; Pierce 1995). In yet other systems the parasites rely on physical protection to avoid the aggressive overtures of ants, such as protective plates in the fly *Microdon* or a protective parasite-constructed case as in Chrysomelid beetles (Kistner 1982). Social parasites also include other wasps, bees, and ants that ancestrally were social but have lost the ability to build their own nests (Hölldobler and Wilson 1990; Wilson 1971).

Traditionally non-hymenopteran social parasites have been investigated because their biology is unique with respect to close relatives that have not adopted parasitism as a life history trait (Pierce 1995). For hymenopteran social parasites a large focus has been on the phylogenetic relationship between host and parasite (Als et al. 2002). The perspective of the parasite and the effect of behavioral manipulation on the whole colony has not been explicitly considered. There is a great deal of evidence (mainly from studies on beetles, lepidopterans, and hymenopterans in ant societies) of control of individual host behavior (Hölldobler and Wilson 1990). These parasites do affect the behavior of multiple individuals at once. What I feel is interesting here is that such a superorganism effect would not simply be a summation of multiple smaller effects but may, possibly, be some qualitative change we have not yet appreciated. We do not of course know but the point made by Oster and Wilson (1978, p. 9–11) that social insects do not have novel behavior but rather a novel collective phenotype resulting from parallel operating, remains true despite being largely ignored by researchers. How behaviorally manipulating parasites affects the efficiency of an evolved parallel operating modular unit like the superorganism remains to be seen.

8.7.2 Cheating the mutualism

For many the leaf-cutting ants surely represent one of the most powerful and dramatic examples of social insect society (Hölldobler and Wilson 2010). Their attraction holds for social insect biologists also and these marvelous insects have received an enormous amount of attention over the years (Hölldobler and Wilson 1990, Ch 17; Schultz et al. 2005). One of the most impressive features of leafcutters is their coevolved mutualism with a fungus that they eat. They collect leaves, the fungus eats the leaves, and the ants eat the fungus. It is an ancient association extending back 50 million years (Currie et al. 2006; Schultz et al. 2005) and one which is paralleled in the Old World by termites who similarly raise a fungus which they consume (Aanen and Boomsma 2005; Aanen and Eggleton 2005).

In both cases a colony only raises one strain of fungus (Aanen et al. 2009; Poulsen and Boomsma 2005). Since the fungus is liable to infection from specialized or generalist myco-parasites (Currie et al. 1999) then ant societies would be better served by having diverse strains that would reduce the chance of failure if one crop fails. The extent to which greater diversity is desired is a balancing act as too much diversity would select for competitive traits among domesticated fungi which would reduce the food yield (Aanen 2010). As it stands the reliance on one monoculture has selected for expensive chemical (Bot and Boomsma 1996; Bot et al. 2001; Poulsen et al. 2002) and behavioral (Little et al. 2005) defense as well as architectural behavior by the hosts (Aanen et al. 2009; Schultz et al. 2005). It has also selected for the incorporation of other mutualists into the association, and some ants have evolved a relationship with an antibiotic producing bacteria which is then itself infected by a black yeast fungus (Little and Currie 2008). It is reasonable to ask if more diversity would have changed this. Simply having more strains of fungi to feed from would be a better solution. So why don't colonies of fungal growing ants have more strains?

The answer may be because the fungus is manipulating the physiology of the ants, preventing them taking on board new strains. We tend to think that ants (or termites) as having domesticated fungi, but it could also have been the other way around. In the ant system the rounded tips of the hyphae are called Bromatia and in the termites the "bulbous structure developed by fungi cultivated by termites" are Gongylidia (although gongylidia is incorrectly used by ant people (Kirk et al. 2008)). In the ant mutualism the fungus produces a suite of enzymes that pass through the ant's body to act as plant degrading compounds (Schiøtt et al. 2010).

In this system ants defecate on freshly collected leaves and the enzymes from the bromatia have many enzymes like those found in phytopathogenic fungi that degrade plant tissue. The implication that the fungus is manipulating the ant is from a particularly nice study where ants were removed from their symbiont and forced to eat another (Poulsen and Boomsma 2005). They could not, and it required nine days of force feeding before ants could switch cultivar strains; a barrier unlikely to be overcome in the wild. Definitive proof is lacking but given the ability of these fungi to evolve enzymes that can pass through ants to affect plant tissue it is not a stretch to also imagine they affect ant digestion preventing the easy acquisition of other strains, which after all is in the colony's interest. If this is true then fungi could manipulate whole colony behavior just as *Ophipcordyceps* (= *Cordyceps*) manipulates individual behavior (Andersen et al. 2009).

8.7.3 Panicking the crowd

Social parasites need to get into and remain inside the colony. E.O. Wilson, with his usual metaphoric flair, described the colony as a "factory within a fortress" (Wilson 1968) and Schmid-Hempel picked this up leading to the evocative language of parasites "breaking into the fortress" (1998). Many of the entry routes involve chemical signals that affect host behavior. In one example a parasitoid wasp *Ichneumon eumerus* has evolved the ability to induce ant warfare (Thomas et al. 2002b). The parasitoid does not attack ants but rather lays eggs inside a socially parasitic caterpillar *Maculinea rebeli* that infects ant nests (genus Myrmica). The caterpillar uses chemicals to trick ants into bringing it home and continues to use chemicals to obtain food inside the nest (Als et al. 2004). As a consequence, caterpillars are always surrounded by ants. The challenge for the parasitoid wasps is laying an egg in a caterpillar surrounded by formidable bodyguards. What the parasitoid wasp female does, and this is very special, is produce semiochemicals that induce ferocious fighting among ants leaving the caterpillar unattended (Thomas et al. 2002b).

There is a similar example whereby parasitoid flies (Phoridae) cause excessive panicking behavior among trails of ants. Here flies hover above trails and dart down to lay eggs in the heads of workers (Feener et al. 1996). The infection progresses and the head eventually falls off and the fly pupates inside. The loss of the worker for the colony is slight and given that infection levels are very low one cannot imagine the loss at the colony level is high. Yet, the presence of a single fly is enough to cause hundreds of ants to hide under leaves or simply stop foraging (Orr et al. 1995; Feener and Brown 1992). In one study the daily foraging rate decreased by 50% (Mehdiabadi and Gilbert 2002). It is possible that this is just a maladapted trait: the ants just overreact. But what is also possible is that panicking the crowd increases the egg laying ability of parasitoid flies in the same way it does for parasitoid wasps of *Maculinea* discussed above. We don't know, but considering the behavior of the superorganism helps us design research questions to test this.

8.7.4 Shifting foraging ecology

Dipteran and hymenopteran parasitoids are large insects that generally fly and so are visible to foraging social insects. As such, behavioral defenses against ovipositing females, such as ant workers reducing foraging (above) or the coordinated dance behavior among paper wasps in the presence of an ovipositing Ichneumonidae wasp (West-Eberhard 1969) are beneficial. But most parasites of social insect are not surgically delivered into new hosts by ovipositing mothers. Rather, propagules are dispersed in the environment where foraging workers encounter them and bring them back into the colony (e.g., nematodes, cestodes, trematodes, strepsipterans, fungi (including microsporidians), gregarines, coleopterans, and lepidopterans). Therefore, the presence of parasites in the environment could lead to shifts in where social insects forage if avoiding parasites is adaptive to the colony (avoidance may not be adaptive, see Hughes et al. 2008).

There are three ways that parasite-induced shifts in social insect foraging can happen and all three are, not surprisingly, examples of where social

insect behavior is changed to reduce infection (category 4 from Table 8.1). When paper wasps nests (*Polistes canadensis*) are infected by a predaceous moth caterpillar (Tineidae) the workers construct satellite nests to reduce the loss to parasites (Jeanne 1979). In this case the moth burrows though the lower walls of the nest, eating brood as it goes. Once the moth is in it is impossible to stop so satellite nest construction evolved to limit damage. Wasps will have to alter flying patterns in response to two nests. This may not appear to be a big deal but since we know wasps can, and do, accidently shift between colonies (Sumner et al. 2007) the presence of a parasite may increase the incidence of such shifting. In wood ants, *Formica*, from temperate forests workers collect a resin from pine trees, which has been shown to have antibiotic properties and reduce the microbial load inside the colony (Christe et al. 2003). In this case defending against microbial pathogens (i.e., not specialized parasites) will lead to changes in foraging routes as resin is collected. Finally, a group of microbial parasites that are highly specialized are *Cordyceps* fungi (now called *Ophiocordyceps*) (Sung et al. 2007). Here social insects (wasps and ants) are infected when foraging as a spore attaches to and penetrates through the cuticle. Infection takes hours and so the worker has returned to the nest by the time the spore has entered the cuticle. Once through, the parasite proliferates inside the host over days and then instructs the worker to leave the nest since the fungus needs to produce a large stalk from the host's body after it kills it. Since killing the host inside the nest would just result in the dead worker being dumped on the trash pile (midden), selection has led to a range of manipulative strategies where workers desert the nest to die either on the ground outside the nest or attached to leaves and branches of plants. From these platforms spores are released to infect new workers. Anecdotal evidence from one of our field sites in Thailand showed that the principal host, *Camponotus leonardi*, was rarely encountered on the forest floor where spore were released (Pontoppidan et al. 2009). At one site where the fungus was completely absent, the target host was very common on the forest floor. While anecdotal this does suggest parasites can structure where foraging trails can go. Besides these examples there are nematodes (Poinar 2003), strepsipterans (Hughes et al. 2004), and microsporidians (Schmid-Hempel and Loosli 1998) that all distribute propagules in places where social insects forage for food thereby placing selection on colonies to avoid patches of high parasite density. In conclusion, shifting trails is likely not an extended phenotype of the parasite on the collective phenotype of the superorganism. Rather we are likely to discover that changes in the collective phenotype we observe in groups foraging occur when parasite pressure is highly selecting for behavioral defense.

8.8 Future directions and tests

Thinking in terms of superorganisms is not helpful unless by doing so we can be pushed into experimental approaches that a unitary organism view cannot provide. I hope that I have made two essential points obvious is in this chapter. The first is that social insects, through group behavior, are sufficiently distinct from solitary organisms in ways that require a different mindset when examining adaptation by natural selection. This is well established and social insect biology rests upon 400 years of natural history and experimental approaches (Hölldobler and Wilson 2009). The second essential point is obviously less well known but is, I feel, relevant nonetheless. Parasites of social insects experience environments wholly different from those parasites infecting solitary organisms. The great evolutionary transition from solitary living to advanced sociality was also a transition for the parasites that hitchhiked from the solitary ancestor to highly integrated group member (Hughes et al. 2008). Just as we discuss sociobiology so we might also discuss socioparasitology. By socioparasitology I do not mean the ability of parasites to recognize kin and evolve group behaviors as we know parasites can do (Hechinger et al. 2010; Reece et al. 2008). Rather, the name should encourage us to think of what happens to parasites whose hosts are social. How should socioparasitology proceed?

Historically biology has progressed rapidly from its natural history beginnings by using the compara-

tive method. The tools twenty-first century biology offer unparalleled scope to compare manipulative parasites between the social and non-social. First among these tools is comparative genomics, where whole genomes of closely related species manipulating social insect and non-social insect behavior are arrayed and compared. This has proved very popular in identifying gene regions involved in infection. The recent move to sequence hundreds of insect genomes in the next five years (Robinson et al. 2011) would be a great opportunity to test this, as many insect parasites of social insects also infect non-social ones (e.g., strepsiptera, diptera). The comparative genomic approach could be combined with comparative transcriptomics using RNA-seq experimental approaches. These two approaches will identify in gene presence our behavior differences between hosts. To this one can add standard physiological assays such as metabolomics.

The above approaches are of course extensions of the natural phase and are not directed towards an *a priori* experimental approach. For this a very useful approach to adopt would be testing the cumulative effect of manipulator parasites on colony performance. At which stage does the loss of individuals due to manipulation lead to a feedback to the colony resulting in defense? Also, which defense comes first: behavioral, structural or immunological (see also Hughes and Cremer 2007)? Exploring empirically or mathematically the feedback between fitness loss and defense and how this differs between solitary and social insects would be productive. Likewise, since sociality is a gradient from the primitively eusocial with small societies to highly advanced ones with millions of workers, comparisons can also be made within social insects. The long history of social insect research has provided an enormous wealth of knowledge on the "beauty and elegance" of superorganisms (Hölldobler and Wilson 2009). This provides a fabulous opportunity for biologists interested in parasites that change host behavior.

References

Aanen, D. K. (2010) As you weed, so shall you reap: on the origin of algaculture in damselfish. *BMC Biology* 2010, 8:81.

Aanen, D. K. and Boomsma, J. J. (2005) Evolutionary dynamics of the mutualistic symbiosis between fungus-growing termites and Termitomyces fungi. In: *Insect-Fungal Associations: Ecology and Evolution* (eds Vega, F. E., and Blackwell, M.) Oxford University Press, New York.

Aanen, D. K., De Fine Licht, H. H., Debets, A. J. M., Kerestes, N. A. G., Hoekstra, R. F., and Boomsma, J. J. (2009) High symbiont relatedness stabilizes mutualistic cooperation in fungus-growing termites. *Science* **326**, 1103–1106.

Aanen, D. K. and Eggleton, P. (2005) Fungus-growing termites originated in African rain forest. *Current Biology* **15**, 851–855.

Abe, T., Bignell, D. E., and Higashi, M. (eds). (2001) *Termites: Evolution, Sociality, Symbiosis, Ecology.* Kluwer Academic Publishers, Dordrecht.

Als, T. D., Nash, D. R., and Boomsma, J. J. (2001) Adoption of parasitic *Maculinea alcon* caterpillars (Lepidoptera: Lycaenidae) by three *Myrmica* ant species. *Animal Behaviour* **62**, 99–106.

Als, T. D., Nash, D. R., and Boomsma, J. J. (2002) Geographical variation in host-ant specificity of the parasitic butterfly *Maculinea alcon* in Denmark. *Ecological Entomology* **27**, 403–414.

Als, T. D., Vila, R., Kandul, N. P., Nash, D. R., Yen, S. H., Hsu, Y. F., Mignault, A. A., Boomsma, J. J., and Pierce, N. E. (2004) The evolution of alternative parasitic life histories in large blue butterflies. *Nature* **432**, 386–390.

Andersen, S. B., Gerritsma, S., Yusah, K. M., Mayntz, D., Hywel-Jones, N. L., Billen, J., Boomsma, J. J., and Hughes, D. P. (2009) The life of a dead ant: The expression of an adaptive extended phenotype. *American Naturalist* **174**, 424–433.

Anderson, C., Boomsma, J. J., and Bartholdi, J. J. (2002) Task partitioning in insect societies: bucket brigades. *Insectes Sociaux* **49**, 171–180.

Barnard, C. J. and Behnke, J. M. (eds). (1990) *Parasites and Host Behavior.* New York Taylor and Francis.

Bennett, N. C. and Faulkes, C. G. (2000) *African Mole-Rats: Ecology and Eusociality* Cambridge University Press, Cambridge.

Biron, D. G., Ponton, F., Marche, L., Galeotti, N., Renault, L., Demey-Thomas, E., Poncet, J., Brown, S. P., Jouin, P., and Thomas, F. (2006) "Suicide" of crickets harbouring hairworms: a proteomics investigation. *Insect Molecular Biology* **15**, 731–742.

Bonasio, R., Zhang, G. J., Ye, C. Y., Mutti, N. S., Fang, X. D., Qin, N., Donahue, G., Yang, P. C., Li, Q. Y., Li, C., Zhang, P., Huang, Z. Y., Berger, S. L., Reinberg, D., Wang, J., and Liebig, J. (2010) Genomic comparison of the ants *Camponotus floridanus* and *Harpegnathos saltator*. *Science* **329**, 1068–1071.

Boomsma, J. J. and Franks, N. R. (2006) Social insects: from selfish genes to self organisation and beyond. *Trends in Ecology and Evolution* **21**, 303–308.

Bot, A. N. M. and Boomsma, J. J. (1996) Variable metapleural gland size-allometries in Acromyrmex leafcutter ants (Hymenoptera: Formicidae). *Journal of the Kansas Entomological Society* **69**, 375–383.

Bot, A. N. M., Obermayer, M. L., Holldobler, B., and Boomsma, J. J. (2001) Functional morphology of the metapleural gland in the leaf-cutting ant Acromyrmex octospinosus. *Insectes Sociaux* **48**, 63–66.

Bourke, A. F. G. (2011) *Principles of Social Evolution.* Oxford University Press, Oxford.

Bourke, A. F. G. and Franks, N. R. (1995) *Social Evolution in Ants.* Princeton University Press, Princeton.

Burkhardt, R. W. J. (2005) *Patterns of Behavior: Konrad Lorenz, Niko Tinbergen, and the Founding of Ethology* University of Chicago Press, London.

Carney, W. P. (1969) Behavioral and morphological changes in carpenter ants harbouring dicrocoeliid metacercariae. *American Midland Naturalist* **82**, 605–611.

Christe, P., Oppliger, A., Bancala, F., Castella, G., and Chapuisat, M. (2003) Evidence for collective medication in ants. *Ecology Letters* **6**, 19–22.

Costa, J. T. (2006) *The Other Insect Societies.* Harvard University Press, Cambridge.

Cremer, S., Armitage, S. A. O., and Schmid-Hempel, P. (2007) Social immunity. *Current Biology* **17**, R693–R702.

Crespi, B. J. and Cho, J. C. (eds). (1997) *The Evolution of Social Behaviour in Insects and Arachnids.* Cambridge University Press, Cambridge.

Crespi, B. J. and Yanega, D. (1995) The definition of eusociality. *Behavioral Ecology* **6**, 109–115.

Currie, C. R., Mueller, U. G., and Malloch, D. (1999) The agricultural pathology of ant fungus gardens. *Proceedings of the National Academy of Sciences of the United States of America* **96**, 7998–8002.

Currie, C. R., Poulsen, M., Mendenhall, J., Boomsma, J. J., and Billen, J. (2006) Coevolved crypts and exocrine glands support mutualistic bacteria in fungus-growing ants. *Science* **311**, 81–83.

Dawkins, R. (1976) *The Selfish Gene.* Oxford University Press, Oxford.

Dawkins, R. (1982) *The Extended Phenotype.* W. H. Freeman, Oxford.

Dawkins, R. (1990) Parasites, desiderata lists and the paradox of the organism. *Parasitology* **100**, S63–S73.

Dawkins, R. and Krebs, J. R. (1979) Arms races between and within species. *Proceedings of the Royal Society of London Series B-Biological Sciences* **205**, 489–511.

Duckworth, R. A. (2009) The role of behavior in evolution: a search for mechanism. *Evolutionary Ecology* **23**, 513–531.

Epsky, N. D. and Capinera, J. L. (1988) Efficacy of the entomogenous nematode *Steinernema feltiae* against a subterranean termite, *Reticulitermes tibialis* (Isoptera, Rhinotermitidae). *Journal of Economic Entomology* **81**, 1313–1317.

Feener, D. H., Jr. and Brown, B. V. (1992) Reduced foraging of *Solenopsis geminata* (Hymenoptera: Formicidae) in the presence of parasitic *Pseudacteon spp.* (Diptera: Phoridae). *Annals of the Entomological Society of America* **85**, 80–84.

Feener, D. H., Jr., Jacobs, L. F., and Schmidt, J. O. (1996) Specialized parasitoid attracted to a pheromone of ants. *Animal Behavior* **51**, 61–66.

Fiedler, K., Hölldobler, B., and Seufert, P. (1996) Butterflies and ants: the communicative domain. *Experientia* **52**, 14–24.

Foster, K. R. and Ratnieks, F. L. W. (2005) A new eusocial vertebrate? *Trends in Ecology & Evolution* **20**, 363–364.

Franks, N. R. and Richardson, T. (2006) Teaching in tandem-running ants. *Nature* **439**, 153–153.

Gould, S. J. and Lewontin, R. C. (1979) Spandrels of San-Marco and the panglossian paradigm—a critique of the adaptationist program. *Proceedings of the Royal Society of London Series B-Biological Sciences* **205**, 581–598.

Hamilton, A., Smith, N. R., and Haber, M. H. (2009) Social insects and the individuality thesis: Cohesion and the colony as a selectable individual. In: *Organization of Insect Societies: From Genome to Sociocomplexity* (eds Gadau, J., and Fewell, J.) Harvard University Press, Cambridge, MA.

Hamilton, W. D. (1963) Evolution of Altruistic Behaviour. *American Naturalist* **97**, 354–356.

Hamilton, W. D. (1964) The genetical evolution of social behaviour. I and II. *Journal of Theoretical Biology* **7**, 1–52.

Hansell, M. (2004) *Animal Architecture.* Oxford University Press, Oxford.

Hansell, M. H. (1996) Wasp make nests: nests make conditions. In: *Natural History and Evolution of Paper-Wasps* (eds Turillazzi, S., and West-Eberhard, M. J.) Oxford University Press, Oxford.

Hechinger, R. F., Wood, A. C., and Kuris, A. M. (2010) Social organization in a flatworm: trematode parasites form soldier and reproductive castes. *Proceedings of the Royal Society B-Biological Sciences* **278**, 656–665.

Hölldobler, B. (1971) Communication between ants and their guests. *Scientific American* **224**(3), 86–93, 124.

Hölldobler, B. (1995) The chemistry of social regulation: multicomponent signals in ant societies. *Proceedings of*
the National Academy of Sciences of the United States of America **92**, 19–22.

Hölldobler, B., Möglich, M., and Maschwitz, U. (1981) Myrmecophilic relationship of *Pella* (Coleoptera: Staphylinidae) to *Lasius fuliginosus* (Hymenoptera: Formicidae). *Psyche* **88**, 347–374.

Hölldobler, B. and Wilson, E. O. (1990) *The Ants.* Harvard University Press, Cambridge, Mass.

Hölldobler, B. and Wilson, E. O. (2009) *The Superorganism: The Beauty, Elegance, and Strangeness of Insect Societies,* W. W. Norton, London.

Hölldobler, B. and Wilson, E. O. (2010) *The Leafcutter Ants: Civilization by Instinct,* W. W. Norton and Company, London.

Hughes, D. P. (2005) Parasitic manipulation: a social context. *Behavioural Processes* **68**, 263–266.

Hughes, D. P. (2009) Altruists since life began: the superorganism view of life. *Trends in Ecology & Evolution* **24**, 417–418.

Hughes, D. P. (2011) Recent developments in sociobiology and the scientific method. *Trends in Ecology & Evolution,* on line December 28th 2010. vol 26, p 57–58.

Hughes, D. P. and Cremer, S. (2007) Plasticity in antiparasite behaviours and its suggested role in invasion biology. *Animal Behaviour* **74**, 1593–1599.

Hughes, D. P., Kathirithamby, J., and Beani, L. (2004) Prevalence of the parasite Strepsiptera in adult *Polistes* wasps: field collections and literature overview. *Ethology, Ecology and Evolution* **16**, 363–375.

Hughes, D. P., Pierce, N. E., and Boomsma, J. J. (2008) Social insect symbionts: evolution in homeostatic fortresses. *Trends in Ecology & Evolution* **23**, 672–677.

Jeanne, R. L. (1979) Construction and utilization of multiple combs in *Polistes canadensis* in relation to the biology of a predaceous moth. *Behavioral Ecology and Sociobiology* **4**, 293–310.

Kastberger, G. and Sharma, D. K. (2000) The predator-prey interaction between blue-bearded bee eaters (*Nyctyornis athertoni* Jardine and Selby 1830) and giant honeybees (*Apis dorsata* Fabricius 1798). *Apidologie* **31**, 727–736.

Kathirithamby, J. and Hughes, D. P. (2005) Description and biological notes of the first species of *Xenos* (Strepsiptera: Stylopidae) parasitic in *Polistes carnifex* F. (Hymenoptera: Vespidae) in Mexico. *Zootaxa* **1104**, 35–45.

Kilner, R. M. and Davies, N. B. (1999) How selfish is a cuckoo chick? *Animal Behaviour* **58**, 797–808.

King, L. E., Douglas-Hamilton, I., and Vollrath, F. (2007) African elephants run from the sound of disturbed bees. *Current Biology* **17**, R832–R833.

Kirk, P. M., Canon, P. F., Minter, D. W., and Staplers, J. A. (2008) *Dictionary of the Fungi 10th Edition* CABI, Wallingford.

Kistner, D. H. (1979) Social and evolutionary significance of social insect symbionts. In: *Social Insects* (ed. Hermann, H. R.) Academic Press, New York.

Kistner, D. H. (1982) The social insects' bestiary. In: *Social Insects* (ed. Hermann, H. R.) Academic Press, New York.

Little, A. E. F. and Currie, C. R. (2008) Black yeast symbionts compromise the efficiency of antibiotic defenses in fungus-growing ants. *Ecology* **89**, 1216–1222.

Little, A. E. F., Murakami, T., Mueller, U. G., and Currie, C. R. (2005) Defending against parasites: fungus-growing ants combine specialized behaviours and microbial symbionts to protect their fungus gardens. *Biology Letters*, 10.1098/rsbl.2005.0371, 12–16

Manga-Gonzalez, M. Y., Gonzalez-Lanza, C., Cabanas, E., and Campo, R. (2001) Contributions to and review of dicrocoeliosis, with special reference to the intermediate hosts of *Dicrocoelium dendriticum*. *Parasitology* **123**, S91–S114.

Mayr, E. (1997) The objects of selection. *Proceedings of the National Academy of Sciences of the United States of America* **94**, 2091–2094.

Mcauliffe, K. and Whitehead, H. (2005) Eusociality, menopause and information in matrilineal whales. *Trends in Ecology & Evolution* **20**, 650–650.

Mehdiabadi, N. J. and Gilbert, L. E. (2002) Colony-level impacts of parasitoid flies on fire ants. *Proceedings of the Royal Society of London, Series B* **269**, 1695–1699.

Moore, J. (2002) *Parasites and the Behavior of Animals.* Oxford University Press, Oxford.

Muller, C. B. and Schmid-Hempel, P. (1993) Exploitation of cold temperature as defense against parasitoids in bumblebees. *Nature* **363**, 65–67.

Nowak, M. A., Tarnita, C. E., and Wilson, E. O. (2010) The evolution of eusociality. *Nature* **466**, 1057–1062.

Orr, M. R., Seike, S. H., Benson, W. W., and Gilbert, L. E. (1995) Flies suppress fire ants. *Nature* **373**, 292–293.

Oster, G. F. and Wilson, E. O. (1978) *Caste and Ecology in the Social Insects,* Princeton, Princeton University Press.

Pardi, L. (1948) Dominance order in *Polistes* wasps. *Physiological Zoology* **21**, 1–13.

Pierce, N. E. (1995) Predatory and parasitic Lepidoptera: carnivores living on plants. *Journal of the Lepidopterists Society* **49** (4), 412–453.

Poinar, G. (2003) *Formicitylenchus oregonensis* n g, n sp (Allantonematidae: Nematoda), the first tylenchid parasite of ants, with a review of nematodes described from ants. *Systematic Parasitology* **56**, 69–76.

Pontoppidan, M.-B., Himaman, W., Hywel-Jones, N. L., Boomsma, J. J. and Hughes, D. P. (2009) Graveyards on the move: The spatio-temporal distribution of dead *Ophiocordyceps*-infected ants. *PLoS ONE* **4**, e4835.

Poulin, R. (1994) The evolution of parasite manipulation of host behavior: a theoretical analysis. *Parasitology* **109**, S109–S118.

Poulin, R. (1998) *The Evolutionary Ecology of Parasites.* London, Chapman & Hall.

Poulin, R. (2000) Manipulation of host behaviour by parasites: a weakening paradigm? *Proceedings of the Royal Society of London Series B* **267**, 787–792.

Poulin, R. (2011) Parasite manipulation of host behavior: an update and frequently asked questions. In: *Advances in the Study of Behavior* (ed. Brockmann, H. J.) Elsevier, Burlington.

Poulsen, M. and Boomsma, J. J. (2005) Mutualistic Fungi control crop diversity in Fungus growing ants. *Science* **307**, 741–744.

Poulsen, M., Bot, A. N. M., Nielsen, M. G., and Boomsma, J. J. (2002) Experimental evidence for the costs and hygienic significance of the antibiotic metapleural gland secretion in leaf-cutting ants. *Behavioral Ecology and Sociobiology* **52**, 151–157.

Queller, D. C. and Strassmann, J. E. (2002) The many selves of social insects. *Science* **296**, 311–313.

Ratnieks, F. L. W. (1988) Reproductive harmony via mutual policing by workers in eusocial Hymenoptera. *American Naturalist* **132**, 217–236.

Rau, P. (1933). *The jungle bees and wasps of Barro Colorado Island (with notes on other insects).* Missouri: Kirkwood 1–324 pp. [34–36] (*nom. nud.*)

Reece, S. E., Drew, D. R., and Gardner, A. (2008) Sex ratio adjustment and kin discrimination in malaria parasites. *Nature* **453**, 608–615.

Reeve, H. K. and Keller, L. (1999) Levels of selection: Burying the units-of-selection debate and unearthing the crucial new issues. In: *Levels of Selection in Evolution* (ed. Keller, L.) Princeton, Princeton University Press.

Richardson, T. O., Houston, A. I., and Franks, N. R. (2007) Teaching with evaluation in ants. *Current Biology* **17**, 1520–1526.

Robinson, G. E., Hackett, K. J., Purcell-Miramontes, M., Brown, S. J., Evans, J. D., Goldsmith, M. R., Lawson, D., Okamuro, J., Robertson, H. M., and Schneider, D. J. (2011). Creating a buzz About insect genomes. *Science* **331**, 1386.

Ross, K. G. and Matthews, R. W. (1991) *The Social Biology of Wasps.* Cornell University Press, Ithaca.

Schiøtt, M., Rogowska-Wrzesinska, A., Roepstorff, P., and Boomsma, J. J. (2010) Leaf-cutting ant fungi produce cell wall degrading pectinase complexes reminiscent of phytopathogenic fungi. *BMC Biology* **8**, 156.

Schmid-Hempel, P. (1998) *Parasites in Social Insects*. Princeton University Press, Princeton.

Schmid-Hempel, P. and Loosli, R. (1998) A contribution to the knowledge of *Nosema* infections in bumble bees, *Bombus* spp. *Apidologie* **29**, 525–535.

Schmid-Hempel, P. and Schmid-Hempel, R. (1984) Life duration and turnover of foragers in the ant *Cataglyphis bicolor* (Hymenoptera, Formicidae). *Insectes Sociaux* **31**, 345–360.

Schultz, T. R., Mueller, U. G., Currie, C. R., and Rehner, S. A. (2005) Reciprocal illumination: A comparison of agriculture in humans and in fungus-growing ants. In: *Insect-Fungal Associations* (eds Vega, F. E., and Blackwell, M.) Oxford University Press, New York.

Seeley, T. D. (1995) *The Wisdom of the Hive*, Harvard University Press, London.

Segerstråle, U. (2007) An eye on the core: Dawkins and sociobiology. In: *Richard Dawkins: How a Scientist Changed the Way we Think* (eds Grafen, A., and Ridley, M.) Oxford University Press, Oxford.

Sergio, F., Blas, J., Blanco, G., Tanferna, A., LaPez, L., Lemus, J. A., and Hiraldo, F. (2011) Raptor nest decorations are a reliable threat against conspecifics. *Science* **331**, 327–330.

Sherman, P. W., Seeley, T. D., and Reeve, H. K. (1988) Parasites, pathogens, and polyandry in social Hymenoptera. *American Naturalist* **131**, 602–610.

Sumner, S., Lucas, E., Barker, J., and Isaac, N. J. B. (2007) Radio-tagging technology reveals extreme nest drifting in a eusocial insect. *Current Biology* **17**, 140–145.

Sung, G.-H., Hywel-Jones, N. L., Sung, J.-M., Luangsa-Ard, J. J., Shrestha, B., and Spatafora, J. W. (2007) *Phylogenetic Classification of Cordyceps and the Clavicipitaceous Fungi*. CBS, Utrecht.

Thomas, F., Schmidt-Rhaesa, A., Martin, G., Manu, C., Durand, P., and Renaud, F. (2002a) Do hairworms (Nematomorpha) manipulate the water seeking behaviour of their terrestrial hosts? *Journal of Evolutionary Biology* **15**, 356–361.

Thomas, J. A., Knapp, J. J., Akino, T., Gerty, S., Wakamura, S., Simcox, D. J., Wardlaw, J. C., and Elmes, G. W. (2002b) Insect communication: Parasitoid secretions provoke ant warfare—Subterfuge used by a rare wasp may be the key to an alternative type of pest control. *Nature* **417**, 505–506.

Tobin, J. E. (1991) A neotropical rain forest canopy ant community: some ecological considerations. In:. *Ant-Plant Interactions* (eds Huxley, C. R., and Cutler, D. F.) Oxford University Press, Oxford.

Turillazzi, S. (1991) The Stenogastrinae. In: *The Social Biology of the Wasps* (eds Ross, K. G. and Matthews, R. W.) Cornell University Press, Ithaca.

Turillazzi, S. and West-Eberhard, M. J. (1996) *Natural History and Evolution of Paper-Wasps*. Oxford University Press, Oxford.

Von Frisch, K. (1968) Role of dances in recruiting bees to familiar sites. *Animal Behaviour* **16**, 531.

Von Frisch, K. (1974) *Animal Architecture*. Harcourt Brace Jovanovich, New York.

West-Eberhard, M. J. (1969) The social biology of Polistine wasps. *Miscellaneous Publications of the Museum of Zoology, University of Michigan* **140**, 1–101.

Wheeler, W. M. (1911) The ant-colony as an organism. *Journal of Morphology* **22**, 307–325.

Wilson, D. S. and Wilson, E. O. (2007) Rethinking the theoretical foundation of sociobiology. *Quarterly Review of Biology* **82**, 327–348.

Wilson, E. O. (1959) Source and possible nature of the odor trail of Fire Ants. *Science* **129**, 643–644.

Wilson, E. O. (1968) The ergonomics of caste in the social insects. *American Naturalist* **102**, 41–66.

Wilson, E. O. (1971) *The Insect Societies*. Harvard University Press: Cambridge.

Wilson, E. O. (1975) *Sociobiology. The New Synthesis*. Belknap Press Cambridge, Massachusetts.

Wilson, E. O. (1983a) Caste and Division of Labor in Leaf-Cutter Ants (Hymenoptera, Formicidae, Atta) III Ergonomic resiliency in foraging by *Atta cephalotes*. *Behavioral Ecology and Sociobiology* **14**, 47–54.

Wilson, E. O. (1983b) Caste and division of labor in leaf-cutter ants (Hymenoptera: Formicidae: Atta) IV. Colony ontogeny of *A. cephalotes*. *Behavioral Ecology and Sociobiology* **14**, 55–60.

Wilson, E. O. (2005) Kin selection as the key to altruism: Its rise and fall. *Social Research* **72**, 159–166.

Witte, V., Janssen, R., Eppenstein, A., and Maschwitz, U. (2002) *Allopeas myrmekophilos* (Gastropoda, Pulmonata), the first myrmecophilous mollusc living in colonies of the ponerine army ant *Leptogenys distinguenda* (Formicidae, Ponerinae). *Insectes Sociaux* **49**, 301–305.

Afterword

Bert Hölldobler

A superorganism is a colony of individuals self-organized by division of labor and united by a closed system of communication. The eusocial insect society possesses features of organization analogous to the properties of a single organism. The colony is divided into reproductive castes (analogous to gonads) and sterile worker castes (analogous to somatic tissues). Its members may exchange nutrients and pheromones by trophallaxis (analogous to the circulatory systems and signaling with hormones in organisms). Nevertheless, among the thousands of known social insect species, we can find almost every conceivable grade in the division of labor, from hierarchical organizations with competition among nest mates for reproductive status and poorly developed division of labor, to highly complex cooperative networks with specialized worker subcastes. The level of this gradient at which the colony can be called superorganism is perhaps subjective. It may be at the origin of eusociality, or at a higher level, in which within colony competition for reproductive status is greatly reduced or absent (Hölldobler and Wilson 2009).

In my view, insect societies with considerable reproductive competition among nest mates, and as a consequence, poorly developed division of labor among workers, may have some incipient superorganismic traits, but do not deserve to be called fully functional superorganisms. Thus, from this perspective many of the poneromorph ant societies should not be considered true superorganisms, because there is little or no morphological skew between reproductive and non-reproductive individuals, and intra-colony reproductive competition is indeed conspicuously common (see Hölldober and Wilson 2009). In contrast, in true superorgan-isms the size dimorphism (morphological skew) between reproductive individuals (queens) and sterile individuals (workers) is large and reproductive division of labor is deep and not plastic. Although workers receive all their genes from the queen and her mates, they exhibit very different phenotypes, because during their larval development due to social environmental influences, different genes are turned on and expressed in workers than in queens and males. This phenotypic plasticity continues during adult ontogeny. From the behavioral interactions of hundreds, thousands, or even millions of workers colony specific traits emerge which are part of a collective colony phenotype, the phenotype of the superorganism.

The chapter on "Parasitism and Superorganism" in this book makes a strong and convincing case for the concept of "superorganism" when we consider the evolution of social parasitism in ants. Although no ant species appears to be totally free of parasites, social parasites in poneromorph ant societies are rare or absent. This contrasts sharply with the rich fauna of social parasites in ant species, the colony of which can be considered true superorganisms. I name as examples colonies of species belonging to the genera *Formica*, *Lasius*, *Camponotus*, *Oecophylla*, *Myrmica*, or fungus growing attine species, or species of the army and driver ants (Ecitoninae and Dorylinae) (Hölldobler and Wilson 1990). These superorganisms, like any normal organism, are subdivided into functional units or sites that provide special niches for parasites. In a normal organism such niches might be the stomach, intestinal tract, liver, or any other organ or tissue. In the ant super-organism we may identify the nest chambers where eggs, larvae, or pupae are housed, or the queen chamber, the peripheral nest chambers, the kitchen-middens, or the foraging routes as special sites for social parasites specifically adapted to make their living inside these niches.

Most insect superorganisms, like normal organisms, are characterized by precise recognition of "self" and rejection of "non-self" or "foreign." And as is the case for any parasite of normal organisms that have to overcome the organism's immune barriers, social parasites have to conquer social barriers of the superorganism. In other words, they have to break the chemical code by which each colony member is identified as nest mate. In addition social parasites have to evolve behavioral key stimuli that enable them to manipulate the innate behavioral releasing mechanisms that underlie the social life within each niche of the superorganism.

As we have learned from the chapters in this book, parasites often manipulate their host's behavior to their advantage. One of the most striking and first examples was discovered by W. Hohorst in the 1960s. He was a biologist working in the department for pest control of the gigantic chemical company HOECHST in Frankfurt a.M. (Germany), where he investigated the life cycle of the liver fluke (*Dicrocoelium dendriticum*, an important parasite of grazing mammals). Hohorst and his collaborators discovered that *Formica* ants serve as intermediate hosts. The ants inadvertently take up cercaria of this trematode and subsequently some of the cercaria penetrate the pharynx and the gut walls, and develop into metacercaria inside the ants. One of them invades the ant's brain where it settles in the suboesophagial ganglion. Apparently this "brain worm" induces its *Formica* host to leave the nest and climb onto grass stalks where it attaches itself with a firm grip of its mandibles. This exposes the infected ant to be eaten by the grazing animals, the main host of the liver fluke (Hohorst and Graefe 1961; Schneider and Hohorst 1971).

The manipulations by social parasites that exploit the social life of ants are different. These parasites had to acquire the capacity to provide the correct signals to their hosts. During their evolution they have "broken the code" and are thereby able to take advantage of the benefits of social life of their hosts. Among the most simple, but nevertheless striking examples is

that of the phorid fly *Metopina formicomendicula*. This fly is riding on its host ant, the tiny thief ants (*Diplorhoptrum fugax*) and rapidly strokes the ant's mouth parts with its forelegs to elicit regurgitation of food. The tactile stimulation is a crude imitation of the host ant's food exchange behavior, but the imitation is good enough to work (Hölldobler 1948).

However, in order to be able to live inside the ant colony's brood chambers and to prey unimpeded on the ant brood, and entice the nurse ants to groom and feed the parasites and raise the parasites' brood, the parasitic species have to decode a rich repertoire of chemical and behavioral communication signals employed by the host ants inside the brood chambers. At first glance it appears almost impossible to imagine how such complex parasitic adaptations could have evolved by gradual natural selection.

About 35–40 years ago I devoted much of my research efforts to trace the evolution of such social parasitic adaptations. The focus of my research was myrmecophilous beetles of the staphylinid subfamily Aleocharinae. I undertook a comparative experimental analysis of aleocharine species adapted to different niches in their host superorganism, and I hoped to discover different evolutionary grades of myrmecophilic adaptations which would allow me to at least reconstruct an evolutionary pathway from relatively simple to highly complex social parasitism. Indeed, different species of these aleocharine myrmecophiles occupy different sites within an ant colony. Some live along the trails of the ants, some at the garbage dumps outside the nest, others within the outermost nest chambers, while still others are found within the brood chambers (Hölldobler 1967, 1970, 1971, 1977; Hölldobler et al 1981). In each case the requirements of interspecific communication are different, and with each evolutionary advance towards the center of the superorganism, the brood chambers, the parasite added new features to its "tool box," such as new exocrine glands that produced either appeasement secretion or adoption signals (most likely imitations of ant brood pheromones) and behavioral patterns, such as tactile signals that elicit regurgitation behavior in host ants. In fact, in some of the most accomplished social parasites among the aleocharine staphylinids, the beetles and their larvae produce "super-normal releasers" with which they elicit much stronger response in the host ants than the ants' nest mates do. Let me illustrate this with a few examples.

The aleocharine beetles that live in the kitchen middens and along the ants' trails are scavengers and predators. They evolved the chemical tools to repel or appease the ants. For example, *Pella laticollis* lives near trails of *Lasius fuliginosus* and hunts ants. When attacked by the ants, it quickly provides the appeasement secretions from a gland at the tip of its abdomen. However, it uses the moment's pause to jump on the back of the ant and kill her by biting between the head and the thorax. The beetle then drags the ant away from the trail and devours it.

The aleocharine beetle of the genus *Dinarda* is usually found in more peripheral nest chambers of *Formica* species, where food exchange occurs between foragers and nest workers. It is here that *Dinarda* is able to participate in the social food flow. Occasionally they insert themselves between two workers exchanging food and literally snatch the food droplet from the donor's mouth, or they use a simple begging behavior in order to obtain food from returning food-laden foragers. The ant, however, after having regurgitated liquid from its social stomach often recognizes the beetle as alien and commences to attack it. At the first sign of hostility the beetle raises its abdomen and offers the ant a tiny droplet of appeasement secretion, a proteinaceous substance, which is quickly licked up by the ant, and almost immediately the attack ceases. During this brief interval the beetle makes its escape.

Some of the most advanced myrmecophilic relationships are found in the aleocharine beetles genera *Lomechusa* and *Atemeles* which live inside the brood chambers of their *Formica* and *Myrmica* hosts. They, too, have the appeasement and repellent glands, but in addition they are equipped with dorsolateral adoption glands, the secretion of which entices the host ants to carry the beetles into the brood chambers of the nests. These myrmecophiles also have a rich behavioral repertoire which enables them to elicit regurgitation from their host ants and quantitative measurements of the social food flow inside the nest reveal that these beetles and their larvae employ supernormal releasers that entice the nurse ants to pay more attention to them than to the ants' sister brood.

All these remarkable social parasitic adaptations only exist in true ant superorganisms, and the chapter by David Hughes on "Parasites and the Superorganism" presents an excellently reasoned argument supporting the superorganism concept in the context of parasitism in social insects.

References

Hohorst, W. and Graefe, G. (1961). Ameisen—obligate Zwischenwirte des Lanzettegels (*Dicrocoelium dendriticum*). *Naturwissenschaften* **58**, 327–328.

Hölldobler, B. (1967). Zur Physiologie der Gast-Wirt-Beziehungen (Myrmecophylie) bei Ameisen. I: Das Gastverhältnis der *Atemeles*- und *Lomechusa*-Larven (Col. Staphylinidae) zu *Formica* (Hym. Formicidae). *Zeitschrift für Vergleichende Physiologie* **56**, 1–21.

Hölldobler, B. (1970). Zur Physiologie der Gast-Wirt-Beziehungen (Myrmecophylie) bei Ameisen. II: Das Gastverhältnis der imaginalen *Atemeles pubicollis* Bris. (Col. Staphylinidae) zu *Myrmica* und *Formica* (Hym. Formicidae). *Zeitschrift für Vergleichend Physiologie* **66**, 215–250.

Hölldobler, B. (1971). Communication between ants and their guests. *Scientific American* **224**, 86–93.

Hölldobler, B. (1977). Communication in social Hymenoptera. In: *How Animals Communicate* (ed. Sebeok, T. A.) pp. 418–471. Indiana University Press, Bloomington.

Hölldobler, B. and Wilson, E. O. (1990). *The Ants.* Harvard University Press, Cambridge, Mass.

Hölldobler, B. and Wilson, E. O. (2009). *The Superorganism.* W. W. Norton, New York, London.

Hölldobler, B., Möglich, M., and Maschwitz, U. (1981). Myrmecophilic relationship of *Pella* (Coleoptera, Staphylinidae) to *Lasius fuliginosus* (Hymenoptera: Formicidae). *Psyche* **88**, 347–374.

Hölldobler, K. (1948). Über ein parasitologisches Problem: Die Gastpflege der Ameisen und die Symphilieinstinkte. *Zeitschrift für Parasitologie* **14**, 3–26.

Schneider, G. and Hohorst, W. (1971). Wanderungen der Metacercarien des Lanzett-Egels in den Ameisen. *Naturwissenschaften* **58**, 327–328.

CHAPTER 9

Ecological consequences of manipulative parasites

Kevin D. Lafferty and Armand M. Kuris

9. 1 Introduction

Parasitic "puppet masters", with their twisted, self-serving life history strategies and impressive evolutionary takeovers of host minds, capture the imagination of listeners—even those that might not normally find the topic of parasitism appealing (which includes most everyone). A favorite anecdote concerns the trematode *Leucochloridium paradoxum* migrating to the eyestalks of its intermediate host snail and pulsating its colored body, presumably to attract the predatory birds that are the final hosts for the worm. Identifying a parasite as "manipulative" infers that a change in host behavior or appearance is a direct consequence of the parasite's adaptive actions that, on average, will increase the fitness of the parasite. The list of parasites that manipulate their hosts is long and growing. Holmes and Bethel (1972) presented the earliest comprehensive review and brought the subject to mainstream ecologists. Over two decades ago, Andy Dobson (1988) listed seven cestodes, seven trematodes, ten acanthocephalans, and three nematodes that manipulated host behavior. Fifteen years later, Janice Moore (2002) filled a book with examples. The five infectious trophic strategies, typical parasites (macroparasites), pathogens, trophically transmitted parasites, parasitic castrators, and parasitoids (Kuris and Lafferty 2000; Lafferty and Kuris 2002, 2009) can modify host behavior, but the likelihood that a parasite manipulates behavior differs among strategies. The most studied infectious agents, non-trophically transmitted pathogens and macroparasites, have enormous public health, veterinary, and wildlife disease importance, yet few manipulate host behavior. The best-

studied manipulative infectious agents are trophically transmitted parasites in their prey intermediate hosts. Parasitoids and parasitic castrators can also manipulate host behavior, but for different purposes and with different implications. Several studies of manipulative parasites conclude with phrases such as "may ultimately influence community structure" (Kiesecker and Blaustein 1999), yet few demonstrate ecological effects. Here, we consider the conditions under which manipulative parasites might have a substantial ecological effect in nature and highlight those for which evidence exists (see also Chapter 10).

Some changes in host behavior can result from pathological side effects that do not increase or can even decrease parasite fitness, or can result from an adaptive response (e.g., a defensive response) by the host to minimize the fitness cost of the infectious agent (Poulin 1995). For instance, the females of some species of phorid flies oviposit an egg behind the head of an ant. The fly larva penetrates the host cuticle and develops inside the head, soon causing the ant to decapitate itself. The phorid continues to feed on the tissues in the detached head, completes its development, pupates, and emerges as an adult through an opening at the base of the head. The presence of phorids disrupts worker ant activities, and a colony under attack engages in a frenzied, disoriented suite of behaviors, so much so that colonies of fire ants (*Solenopsis* spp.) fail to thrive in the presence of phorids (Feener and Brown 1995). Perhaps the best documented example of the ecological effect of defensive behaviors against infectious agents comes from research on cleaner wrasses such as *Labroides dimidiatus* on coral reefs (Grutter et al. 2003). Infested fish visit cleaners several times a day, and the diversity of fishes decreases on patch reefs after removing cleaner wrasses. Many large predators and herbivores choose not to visit patch reefs lacking cleaners. Hence, fishes try to reduce ectoparasite abundance and this alters the structure of fish communities among reefs. Hosts can alter spatial patterns of foraging to avoid infection (e.g., leading to ungrazed "roughs" in pastures) and increase the time spent grooming, as opposed to other activities, such as watching for predators (Hart 1990). Such examples, though interesting in their own right, are not the subject of this review.

9.2 What makes a manipulator important ecologically?

Several factors can determine if a manipulative parasite will have ecological effects. (1) Changes to host individuals should scale with the strength of manipulation. (2) A high incidence of infection will have a greater effect on the host population. (3) Parasites that infect common or otherwise important hosts are more able to leave a mark on ecological communities. Unfortunately, we cannot yet evaluate how many parasites are manipulative or have strong effects, use common hosts, or have high incidence. Further, assessments of the "importance" of the ecological role of potential hosts are available for only a few communities.

Strong manipulations should benefit the parasite. In mathematical models, the probability that a parasite can invade a host population increases with the strength of manipulation (Dobson 1988). Strong manipulation also has consequences for host populations. For models of a trophically transmitted parasite, manipulation decreases the equilibrium abundance of prey, whereas the abundance of predators increases with manipulation (Lafferty 1992). This suggests that the ecological effects of manipulation will increase with the magnitude of the behavioral changes associated with parasitism. Limited access to physiological systems that influence host behavior, energetic costs of manipulation, and host counter-adaptations will constrain the strength of manipulation.

Several studies have measured the magnitude of manipulation by contrasting the behaviors or the susceptibility to predation of infected and uninfected hosts. A review of eight studies found that parasites of prey increased predation rates by a factor of 1.62 to 7.5 (Dobson 1988). These estimates are sensitive to sampling without replacement, suggesting such values might be underestimates (Lafferty and Morris 1996). A trematode metacercaria that encysts on the brain of killifish has perhaps the strongest effect documented in the literature. At average parasite intensities, infected killifish are 30 times more likely to be eaten by birds, a value much higher than the four-fold increase in the frequency of conspicuous behaviors observed in the aquarium (Lafferty and Morris 1996). In other words, a parasite-induced change in behavior can lead to an even greater change in transmission.

If manipulation is intensity-dependent (as it can be for typical parasites and trophically transmitted parasites), the behavior of the host depends on both the per-parasite manipulation and parasite intensity. In their mathematical models of macroparasite manipulators, Dobson and Keymer (1985) defined manipulation as α, a per-parasite multiplier of predation risk to final hosts such that, for no manipulation $\alpha = 1$, and, if a single parasite doubles the risk of predation, $\alpha = 2$. Intensity can make up for strength if manipulation is intensity dependent. The per-parasite effect translates to changes in behavior as a power function of intensity ($\alpha^{\text{Intensity}}$), meaning that parasites with small individual effects can, as a group, alter behavior if they reach high intensities. For instance, a weak manipulation of $\alpha = 1.0025$ by a single trematode metacercaria can lead to a thirty-fold increase in predation risk for killifish because infected hosts have a mean intensity of 1,400 metacercariae in the brain case (Lafferty and Morris 1996). Intensity-dependent manipulation can also lead to a high prevalence of the parasite in the intermediate host population because predators are not as likely to remove hosts with low intensity infections.

An indirect way to evaluate the strength of manipulation from field data is to examine the shape of the frequency distribution of parasites in the host population. Although most typical parasites have an "aggregated" or negative binomial distribution (Crofton 1971), the distribution of manipulative parasites in prey hosts appears less aggregated because the tail end of the distribution is truncated (e.g., Adjei et al. 1986; Shaw et al. 2010; Lafferty and Morris 1996; Crofton 1971; Joly and Messier 2004), suggesting that manipulation delivers the most infected prey to predators. For aggregated distributions fitted to a negative binomial distribution, the lower the k-value, the greater the degree of aggregation, with $k < 1$ considered highly aggregated (Shaw et al. 1998). Compiling data from Shaw et al. (1998), for 40 typical parasites and trophically transmitted parasites in their predator hosts, mean $k = 0.41$, whereas for 10 trophically transmitted parasites in their prey intermediate hosts known or prone to behavior modification, mean $k = 1.30$. Therefore, reduced aggregation appears to be a general feature of manipulative parasites.

The ecological role played by manipulative parasites should increase with the frequency that hosts become infected. For parasitoids, castrators, and predator hosts, parasite prevalence is a good measure of abundant manipulation. However, for manipulated prey hosts infected with trophically transmitted parasites, intense manipulation decreases prevalence because predators remove hosts soon after infection (Lafferty 1992). For example, larval acanthocephalans that cause terrestrial isopods to leave shelter are rare in the field because starlings eat infected isopods soon after the behavioral change (Moore 1984). Therefore, in many cases, it can be challenging to determine how abundant manipulation is just by looking at parasite prevalence. Incidence is a far better metric, but one that is harder to measure.

Indirect effects can result from a manipulative parasite if the host plays an important role in the ecosystem. Some hosts are too rare or do not have a disproportionate effect on the ecosystem. Other hosts do not interact much with other potential predators or competitors. However, abundant or interactive hosts can play important roles in ecosystems. For example, the effect of the strong manipulator, *Euhaplorchis californiensis*, is magnified in importance because its killifish host is often the most common fish in the estuarine systems of

California and Baja California (Kuris et al. 2008). In three case studies below, we explore what happens when common parasites strongly manipulate hosts with important roles in ecosystems.

9.3 Parasitic castrators and parasitoids as host behavior manipulators

Parasitic castrators and parasitoids can take over the identity of their hosts. Once the parasite prevents reproduction of the host (and this often happens soon after infection), the genotype of the host no longer matters. If manipulation is weak or absent, parasites can compete with the uninfected host population (Lafferty 1993). The intensity of this competition depends on how similarly infected and uninfected hosts use resources. Manipulation can, therefore, reduce competition for castrators and parasitoids.

If parasitoid or castrator life histories differ from those of their hosts, host behaviors can change after infection. Such changes can cause the host to occupy a different niche where it might interact differently with its potential consumers, prey, and competitors. For instance, immigration of castrated snails adds a new phenotype to a community. As an example, when infected with a heterophyid trematode, the snail, *Batillaria cumingi*, moves lower in the intertidal habitat (Miura et al. 2006). Submergence seems adaptive for the parasite's life history because heterophyid cercariae shed from these snails seek to penetrate fishes. In another example, freshwater snails (*Physa acuta*) castrated by certain trematodes abandon their refuge under leaf litter and rise to the surface, perhaps improving the success of cercariae shed by these snails but also increasing their risk of predation by birds (Bernot 2003). In both cases, abundant infected snails occupy a new ecological niche.

Parasitic castrators can alter host-feeding rates with potential indirect effects on lower trophic levels. On rocky shores along the east coast of North America, the most common intertidal snail, *Littorina littorea*, has important impacts on algal communities and can be frequently parasitized by larval trematodes. Infected snails graze less, increasing colonization by ephemeral algae in areas where the parasite is more prevalent (Wood et al. 2007). In contrast, *Physa acuta* infected with *Posthodiplostomum minimum* graze more than do uninfected snails, reducing periphyton biomass by 20% when the prevalence of infected snails exceeds 50% (Bernot and Lamberti 2008). This intensified grazing also alters the species composition of the periphyton community. These opposite effects of castrators on grazing rates illustrate how the ecological consequences of manipulation can vary among parasites.

9.3.1 Nematomorphs, endangered charr, and crickets in Japanese streams

Japanese fisheries biologists are making a dedicated effort to protect the Kirikuchi charr (a trout), *Salvelinus leucomaenis japonicas* (Fig. 9.1). Threatened by over-fishing and habitat destruction, the charr occurs in a few remaining watersheds (Sato 2007). To understand how to save the charr, biologists

track tagged fish, monitoring their diet and growth rate. Charr, like most stream-dwelling salmonids, feed on aquatic invertebrates as well as terrestrial insects that fall into the stream from the surrounding forest (Kawaguchi et al. 2003). This diet changes in the late summer and fall when fish start consuming terrestrial camel crickets (*Tachycines* spp.). Takuya Sato noticed that several of the crickets eaten by trout had once hosted nematomorph worms, *Gordionus chinensis*, in their abdomens (Sato et al. 2008).

Nematomorphs are parasitoids with a complex life cycle. Non-feeding nematomorph adults live in streams and produce larval worms that infect aquatic insects such as mayflies. Mayflies leave the stream and metamorphose into short-lived adults. However, the larval nematomorphs stay alive inside the mayfly carcasses that fall to the forest floor, infecting crickets that scavenge the infected carcasses. As it matures in the cricket, a growing nematomorph consumes almost all non-essential

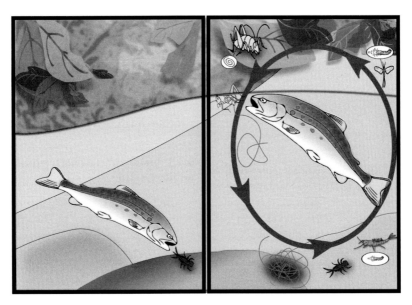

Figure 9.1 Hypothesized effects of a nematomorph worm on a Japanese stream ecosystem. In the left panel, the parasite is absent and charr forage on a sparse prey base of benthic arthropods. In the right panel, adult nematomorph worms mate in the stream. Their larvae infect larval insects and leave the stream as the insects mature and disperse into the forest. Crickets scavenging on dead insects ingest the larvae, and become the second host. As the worm matures in the cricket, it drives its host from the forest into the stream, attracting predation by charr. Such crickets provide 60% of the charr's energetic intake. Satiated by crickets, the charr eat fewer benthic insects, resulting in a higher production of adult insects that leave the stream for the forest. Artwork K. D. Lafferty. See also Plate 14.

fats and reproductive organs. Then comes a considerable challenge. Nematomorphs are aquatic as adults, yet crickets are terrestrial. A dramatic behavioral manipulation solves this problem; the worm causes the cricket to seek water (Thomas et al. 2002) (see also Chapter 2). Infected crickets jump into the stream, and the mature nematomorph worm explodes through the anus of the cricket. The parasitoid leaves its dying host twitching on the surface and swims away to find a mate. Charr attack insects falling into the stream and many consume the worm along with the cricket (Sato et al. 2008). This must have been a long-standing obstacle in the evolution of nematomorphs because the worms escape from the cricket's predators by squirming out through their gills, mouth, or anus (Ponton et al. 2006).

Realizing that nematomorphs were driving crickets into streams, Sato et al. (2011a) measured the contribution of manipulated crickets to the annual energy budget of the charr population. They pumped charr stomachs, divided the prey into several categories, and estimated the caloric content of the prey every month. The researchers set out baited traps in the forest for crickets and also measured the rate that insects fell into the stream. Crickets were common year round in the forest; however, infections with nematomorphs occurred in the summer and fall. Infected crickets were 20 times more likely to fall into streams than were uninfected crickets, confirming that the worm altered cricket behavior, and explaining the seasonal pulse of food for charr. Sato et al. (2011a) found that crickets delivered to the stream by nematomorphs contributed an amazing 60% of the charr's annual calories. This endangered fish might even depend on the nematomorph for its long-term persistence in Japan. Unfortunately for charr, the nematomorph is less prevalent in the conifer plantations that are increasingly replacing native forest (Sato et al. 2011b). There are other potential indirect effects of the nematomorph (Sato et al. 2011a). The nematomorph moves a substantial amount of energy from the forest to the stream. In return, satiated by crickets, charr consume fewer aquatic insects. This shifts energy from the stream back to the forest because many surviving aquatic insects, such as mayflies and damselflies, metamorphose into flying adults that move back to the forest

and become predators and prey for terrestrial animals (Sato et al. 2011a).

9.4 Trophically transmitted parasites as host behavior manipulators

Trophic transmission is a common life history strategy for parasites and seems to select for manipulation. For most parasites, a common cause of death is probably predation of the host (Johnson et al. 2010). However, some parasites have adaptations to survive predation and use the predator as the next host in the life cycle. At worst, parasitizing a predator makes the best of a bad situation. However, if the predator is large and long-lived, the parasite can extend its lifespan and have increased access to food (Lafferty 1999). Alternatively, parasites of predatory hosts might find it advantageous to add prey hosts to their life cycle (Choisy et al. 2003). Prey are more abundant than their predators and are therefore more likely to contact a free-living infectious stage of the parasite. The predator host then unintentionally contacts the parasite as it hunts for prey. For either evolutionary pathway, a trophically transmitted parasite will benefit from increasing predation rates on infected prey so long as the parasite delays manipulation until its maturation and can focus increased predation on suitable predatory hosts. Next, we provide two examples of how manipulation by trophically transmitted parasites might have ecological effects.

9.4.1 Tapeworms, wolves, moose, and forests on Isle Royale

Moose (*Alces alces*) probably first crossed the 30 kilometers to Isle Royale, in Lake Superior from the mainland around 1900, increasing in number due to an abundance of food and absence of wolves (*Canis lupus*) (Fig. 9.2). Population explosions led to periods of over-grazing and mass starvation. In 1949, wolves colonized the island over a rare ice bridge and began to reduce the number of moose. Since 1958, ecologists such as David Mech and Rolf Peterson have studied this 540 square-kilometer outdoor laboratory (see the comprehensive book by Peterson 1995). They have found that many outside

Figure 9.2 Hypothesized effects of a tapeworm on the Isle of Royale ecosystem. In the left panel, the parasite is absent, and wolves have a hard time persisting on moose. Moose, unchecked by predators, over-graze the forest. Right panel, the adult worms of *Echinococcus granulosus* live in the intestines of wolves where they cause little pathology. Tapeworm eggs contaminate the soil and are incidentally ingested by moose. Larval tapeworms form hydatid cysts in the lungs of moose. Wolves can more easily prey on infected moose, reducing the abundance of the moose population and allowing re-vegetation of the island. Artwork K. D. Lafferty. See also Plate 15.

forces affect the success of wolves and moose. For instance, an increase in moose density followed an epidemic of canine parvovirus that almost extirpated wolves in the 1990s. The moose over-grazed the forests and again began to starve. In poor condition, the moose suffered from an outbreak of ticks, and extreme cold weakened their health even more. At the same time, the wolf population began to rebound. This back and forth pattern results in cycles of wolves tracking the abundance of moose, offset by about a decade.

On Isle Royale, the tapeworm, *Echinococcus granulosus*, uses moose as the intermediate host and wolves as the final host. Moose ingest tapeworm eggs from soil and water contaminated with wolf feces. Larval tapeworms form large, debilitating hydatid cysts in the lungs, braincase, liver, and other organs of the intermediate host. Wolves acquire the tapeworms when they eat an infected moose. In the wolf, the small adult tapeworms live in the intestine and cause no measurable harm. Throughout its range, prevalence of

E. granulosus in wolves ranges from 14–72% (Rausch 1995).

The debilitating effects of hydatid cysts likely make it easier for wolves to kill moose. Evidence for this is indirect. Hunters shoot more infected moose earlier in the hunting season (Rau and Caron 1979), and heavily infected moose are less common than expected (Joly and Messier 2004), suggesting hunters or predators remove infected individuals from herds.

How might manipulation by this tapeworm alter the Isle Royale ecosystem? Simple mathematical models hypothesize that there could be situations where wolves could not persist on moose as prey without the assistance of the debilitating parasite (Hadeler and Freedman 1989). More recent models have suggested that manipulative parasites do not affect the invasion criteria for a predator population. This is because at the moment of first contact between a predator and prey population, the prey population is uninfected and unmanipulated (Fenton and Rands 2006). In other words, for wolves to benefit from infected moose, the wolves must first bring the parasite with them and establish the life cycle. On Isle Royale, the initial pack of wolves encountered an uninfected moose population and wolves established without the initial help of the parasite. However, it is probable that the presence of the tapeworm enables wolves to drive the moose population to lower levels than would be possible without the tapeworm in the system. Consequently, the tapeworm might indirectly favor forests on Isle Royale. A further theoretical effect of a parasite that increases susceptibility to predation is an increase in oscillations between predator and prey (Fenton and Rands 2006), suggesting that the tapeworm could influence the period and amplitude of the moose–wolf cycle seen on Isle Royale.

9.4.2 Trematodes, cockles, limpets, and anemones in New Zealand mudflats

On the mudflats of New Zealand, the most noticeable inhabitant is the "cockle", *Austrovenus stutchburyi* (Stewart and Creese 2002) (Fig. 9.3). This little neck clam reaches 6 cm in width and can attain densities of thousands per square meter, supporting a recreational and commercial harvest (Hartill et al. 2005). As the most abundant component of the biomass on these flats, cockles are also a common resource for birds, fishes, and crabs (Thompson et al. 2005).

Cockle shells can protrude from the sediment, creating a habitat for several epibionts, including an anemone (*Anthopleura aureoradiata*), chitons, the estuarine barnacle *Elminius modestus*, tubicolous amphipods, and serpulid worms (Thomas et al. 1998). Exposed shells also are substrates for algae that support a small limpet, *Notoacmea helmsi*. In the bays of New Zealand, there are few alternative natural substrates for this rich and distinctive epibiont community, and the provision of novel habitat makes *A. stutchburyi* an "ecosystem engineer" (Thomas et al. 1998).

Pied oystercatchers (*Haematopus ostralegus finschi*) foraging on the mudflats carry adult trematode worms in their intestines. For many trematode species in New Zealand mudflats, the first intermediate host snails are either

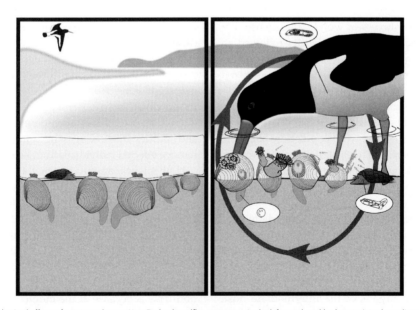

Figure 9.3 Hypothesized effects of a trematode on a New Zealand mudflat ecosystem. In the left panel, cockles burrow into the sediment with only their siphons protruding. In the right panel, adult trematodes live in the intestine of shorebirds. Birds defecate trematode eggs onto the mudflat where they infect an estuarine snail as the first intermediate host. Trematode cercariae emerge from the snail and then seek out a cockle as a second intermediate host, forming a cyst in the foot. Infected cockles have impaired digging abilities, making them easier prey for birds. Raised above the mud surface, the cockles provide hard substrate for a community of invertebrates and algae. Artwork K. D. Lafferty. See also Plate 16.

Zeacumantus subcarinatus or *Cominella glandi-formis*. Trematode cercariae emerge from the infected snail and then encyst in a second inter-mediate host (which varies among trematode species), and the life cycle is completed when an oystercatcher eats an infected second intermedi-ate host. Two trematode genera (*Curtuteria* and *Acanthoparyphium*) with six cryptic species use the cockle as a second intermediate host, encyst-ing in the tip of the foot (Babirat et al. 2004).

Key to the understanding of this system was the discovery that the trematodes, by encysting in the foot, reduce the burrowing ability of the cockles, thereby stranding them on the surface where they become easy prey for oystercatchers (Thomas and Poulin 1998). The trematodes, therefore, are the mechanism by which the cockles increase the avail-able substrate for epibionts to colonize. In addition, by digging less, infected cockles modify properties of the sediment, which alters infaunal communities (Mouritsen and Poulin 2005). Two types of evidence show a clear cause-effect relationship between the trematode and changes to the ecosystem. Experimentally increasing or decreasing the number of stranded cockles alters the mudflat community (Mouritsen and Poulin 2005). Also, by comparing 17 sheltered bays around Otago Harbour, Mouritsen and Poulin (2010) showed that spatial variation in trematode infections was associated with corre-sponding variation in the intertidal community. It appears that New Zealand mudflats would have less biodiversity without these manipulative parasites.

9.5 The ecological reach of host behavior manipulators

Where might we find other cases where manipula-tive parasites have ecological effects? Some wide-spread infectious wildlife diseases with major human public health concerns can manipulate the behavior of their prey hosts. The most notable of these are *Toxoplasma gondii* and other two-host coc-cidians, taeniid tapeworms such as *Taenia solium*, *T. multiceps*, and *Echinococcus* spp., the raccoon round-worm, *Baylisascaris procyonis*, and *Trichinella* species. These parasites often have low host specificity for

the prey hosts. In prey hosts, all are either neuro-tropic (e.g., *T. gondii*, *B. procyonis*), or infect the lungs, diaphragm, or other key organs needed for stamina (e.g., *Echinococcus* spp. *T. solium*, *Trichinella* spp.). Their ability to modify host behavior, either through involvement with brain chemistry or mus-cle physiology, probably makes hosts less wary or more risk tolerant, or impairs escape responses (see Chapter 3). A combination of negligible host specifi-city and increased susceptibility to predators ena-bles these parasites to be widespread. For instance, *T. gondii* infects any warm-blooded vertebrate (ter-restrial or aquatic) as an intermediate host on every continent (Tenter et al. 2000). Infections with *T. gon-dii* can be prevalent, sometimes with substantial pathology (Dubey and Beattie 1988). Such a com-mon and sizeable manipulation suggests *T. gondii* has the potential for large-scale ecological effects. Similarly, *Baylisascaris procyonis* infects more than 100 species of potential prey for raccoons and is common in North America and Europe (Kazacos 2001). *Echinococcus multilocularis* occurs throughout arctic ecosystems (Rausch 1995) and *Trichinella* spp. infections build through the food chain via car-nivory and carrion feeding (Pozio et al. 1992). These parasites meet the criteria for having strong ecologi-cal effects.

9.6 Testing for the ecological effects of manipulative parasites

A research program on the ecological effects of manipulative parasites will require collaboration between ecologists and parasitologists. Ecologists identify free-living species that play important roles in ecosystems. Parasitologists can then search these important hosts for parasites that are abundant and have the ability to manipulate host behavior. Together, ecologists and parasitologists could con-duct studies that link behavioral changes to ecologi-cal effects.

The hypothesized ecological role of *Echinococcus granulosus* indicates the importance of working in systems like Isle Royale where indirect effects cas-cade through the food web. The prediction that the tapeworm could indirectly affect forest growth comes from substantial observational work on the prevalence of the tapeworm in moose and wolves, the potential effect of the parasite on moose, the effect of moose on vegetation, and the effect of wolves on moose. This system is simple enough to explore with mathematical models, which can reveal additional predictions (e.g., about increased cycling). Carnivore reintroduction programs manipulate the presence of wolves and parasites and might lend insight into the effect of *E. ganulosus* at other locations. Wolves and other large carnivores have been extirpated from large areas of their former ranges and are sometimes reintroduced (such as into Yellowstone National Park in 1995). Before initiating such programs, it should be possible to determine if parasites such as *E. granulosus* are circulating in the ecosystem. Because hydatid disease has severe human health consequences, veterinarians involved in relocations treat wolves at least twice with antihelminth drugs to eliminate tapeworms. Reintroductions, therefore, are an opportunity to observe wolf–prey dynamics without *E. granulosus*. Specifically, comparing dynamics in worm-free areas with source areas where the worm occurs could indicate how this tapeworm affects forest ecosystems. Other species might be more tractable for studying the ecological effects of manipulative tapeworms. *E. multilocularis*, for instance, uses smaller carnivores like foxes and coyotes as final hosts and rodents as intermediate hosts. One could explore how the presence or absence of this tapeworm across island habitats affects predator–prey dynamics.

Research on the New Zealand cockle required a variety of approaches. Observations of stranded, fouled cockles combined with parasitological investigations led to the hypothesis that the parasite manipulated the cockle in a way that altered the ecosystem. Researchers also studied across many sites, permitting an understanding of how variation in parasitism drove changes to the ecosystem. This system was amenable to experimentation because the behavioral manipulation could be mimicked in the laboratory and field, as could parasitization rates (by manipulating snail densities). Finally, by creating a food web for the system (Thompson et al. 2005), researchers had a model for various direct and indirect relationships among

species potentially connected to the stranding of parasitized cockles on the mudflat surfaces. These efforts have resulted in this system being the most cited example of how parasites can affect ecosystems.

The Japanese nematomorph study indicates how non-parasitologists can reveal effects of parasites. Through careful and repeated quantification of charr stomach contents, anomalies were discovered that pointed to the role of the nematomorphs. These biologists were focused on the flow of energy among ecosystems, and were thus able to discover the dramatic contribution of parasitized crickets to the diet of endangered charr.

9.7 Conclusions

Numerous studies have shown that the direct effects of parasitism significantly affect populations, community structure, and ecosystem energetics (e.g., Hudson et al. 1998; Waldie et al. 2011; Lafferty et al. 2006; Kuris et al. 2008; Hechinger et al. 2011). In addition, manipulative parasites are more than just entertaining cocktail party anecdotes. They can exert effects across hierarchical ecological levels. Those that have strong manipulative effects on their hosts can alter aspects of the distribution and abundance of their host populations. If parasitism is common, effects on the host population can be strong. If the host is common or interacts with other species in the system, indirect effects on the food web could occur through the alteration of trophic cascades, creation of new habitats, or new niches, or by altering the flow of energy among habitats. It will be a while before we have a systematic understanding of the importance of manipulative parasites at the ecosystem level. Further insight into the effects of manipulative parasites will require challenging experiments and observations, ideally with strong collaboration among parasitologists and ecologists.

Acknowledgements

We thank R. Hechinger, K. Miles, R. Poulin, T. Sato, J. Shaw, S. Sokolow, K. Weinersmith, and S. Weinstein for valuable discussion of these ideas.

References

Adjei, E. L., Barnes, A., and Lester, R. J. G. (1986). A method for estimating possible parasite-related host mortality, illustrated using data from *Callitetrarhynchus gracilis* (Cestoda, Trypanorhyncha) in lizardfish (*Saurida* spp). *Parasitology* **92**, 227–243.

Babirat, C., Mouritsen, K. N., and Poulin, R. (2004). Equal partnership: two trematode species, not one, manipulate the burrowing behaviour of the New Zealand cockle, *Austrovenus stutchburyi. Journal of Helminthology* **78**, 195–199.

Bernot, R. J. (2003). Trematode infection alters the anti-predator behavior of a pulmonate snail. *Journal of the North American Benthological Society* **22**, 241–248.

Bernot, R. J. and Lamberti, G. A. (2008). Indirect effects of a parasite on a benthic community: an experiment with trematodes, snails and periphyton. *Freshwater Biology* **53**, 322–329.

Choisy, M., Brown, S., Lafferty, K. D., and Thomas, F. (2003). Evolution of trophic transmission in parasites: why add intermediate hosts? *American Naturalist* **162**, 172–181.

Crofton, H. D. (1971). A quantitative approach to parasitism. *Parasitology* **62**, 179–193.

Dobson, A. P. (1988). The population biology of parasite-induced changes in host behavior. *Quarterly Review of Biology* **63**, 139–165.

Dobson, A. P. and Keymer, A. E. (1985) Life history models. In: *Acanthocephalan Biology* (eds Crompton, D. and Nickol, B.) Cambridge University Press, Cambridge.

Dubey, J. P. and Beattie, C. P. (1988) *Toxoplasmosis of Animals and Man.* CRC Press, Boca Raton, Florida.

Feener, D. H. and Brown, B. V. (1995). Reduced foraging of *Solenopsis geminata* (Hymenoptera: Formicidae) in the presence of parasitic *Pseudacteon* spp. (Diptera: Phoridae). *Annals of the Entomological Society of America* **85**, 80–84.

Fenton, A. and Rands, S. A. (2006). The impact of parasite manipulation and predator foraging behavior on predator-prey communities. *Ecology* **87**, 2832–2841.

Grutter, A. S., Murphy, J., and Choat, H. (2003). Cleaner fish drives local fish diversity on coral reefs. *Current Biology* **13**, 64–67.

Hadeler, K. P. and Freedman, H. I. (1989). Predator-prey populations with parasitic infection. *Journal of Mathematical Biology* **27**, 609–631.

Hart, B. L. (1990). Behavioral adaptations to pathogens and parasites: five strategies. *Neuroscience and Biobehavioral Review* **14**, 273–294.

Hartill, B. W., Cryer, M., and Morrison, M. A. (2005). Estimates of biomass, sustainable yield, and harvest: neither necessary nor sufficient for the management of non-commercial urban intertidal shellfish fisheries. *Fisheries Research* **71**, 209–222.

Hechinger, R. F., Lafferty, K. D., Dobson, A. P., Brown, J. H., and Kuris, A. M. (2011). A common scaling rule for abundance, energetics and production of parasitic and free-living species. *Science*, **333**, 445–448.

Holmes, J. C. and Bethel, W. M. (1972) Modification of intermediate host behavior by parasites. In: *Behavioural Aspects of Parasite Transmission* (eds Canning, E. U. and Right, C. A.) Zoological Journal of the Linnean Society, London.

Hudson, P. J., Dobson, A. P., and Newborn, D. (1998). Prevention of population cycles by parasite removal. *Science* **282**, 2256–2258.

Johnson, P. T. J., Dobson, A., Lafferty, K. D., Marcogliese, D. J., Memmott, J., Orlofske, S. A., Poulin, R., and Thieltges, D. W. (2010). When parasites become prey: ecological and epidemiological significance of eating parasites. *Trends in Ecology and Evolution* **25**, 362–371.

Joly, D. O., and Messier, F. (2004). The distribution of *Echinococcus granulosus* in moose: evidence for parasite-induced vulnerability to predation by wolves? *Oecologia* **140**, 586–590.

Kawaguchi, Y., Nakano, S., and Taniguch, I. Y. (2003). Terrestrial invertebrate inputs determine the local abundance of stream fishes in a forested stream. *Ecology* **84**, 701–708.

Kazacos, K. R. (2001) *Baylisascaris procyonis* and related species. In: *Parasitic Diseases of Wild Mammals*, 2nd edn. (eds Samuel, W. M., Pybus, M. J., and Kocan, A. A) Iowa State University Press, Ames.

Kiesecker, J. M. and Blaustein, A. R. (1999). Pathogen reverses competition between larval anurans. *Ecology* **80**, 2142–2148.

Kuris, A. M., Hechinger, R. F., Shaw, J. C., Whitney, K. L., Aguirre-Macedo, L., Boch, C. A., Dobson, A. P., Dunham, E. J., Fredensborg, B. L., Huspeni, T. C., Lorda, J., Mababa, L., Mancini, F., Mora, A. B., Pickering, M., Talhouk, N. L., Torchin, M. E., and Lafferty, K. D. (2008). Ecosystem energetic implications of parasite and free-living biomass in three estuaries. *Nature* **454**, 515–518.

Kuris, A. M. and Lafferty, K. D. (2000) Parasite-host modeling meets reality: adaptive peaks and their ecological attributes. In: *Evolutionary Biology of Host-Parasite Relationships: Theory Meets Reality* (eds Poulin, R., Morand, S., and Skorping, A.) Elsevier, Amsterdam.

Lafferty, K. D. (1992). Foraging on prey that are modified by parasites. *American Naturalist* **140**, 854–867.

Lafferty, K. D. (1993). Effects of parasitic castration on growth, reproduction and population dynamics of the marine snail *Cerithidea californica. Marine Ecology Progress Series* **96**, 229–237.

Lafferty, K. D. (1999). The evolution of trophic transmission. *Parasitology Today* **15**, 111–115.

Lafferty, K. D. and Kuris, A. M. (2002). Trophic strategies, animal diversity and body size. *Trends in Ecology & Evolution* **17**, 507–513.

Lafferty, K. D. and Kuris, A. M. (2009). Parasitic castration: the evolution and ecology of body snatchers. *Trends in Parasitology* **25**, 564–572.

Lafferty, K. D. and Morris, A. K. (1996). Altered behavior of parasitized killifish increases susceptibility to predation by bird final hosts. *Ecology* **77**, 1390–1397.

Miura, O., Kuris, A. M., Torchin, M. E., Hechinger, R. F., and Chiba, S. (2006). Parasites alter host phenotype and may create a new ecological niche for snail hosts. *Proceedings of the Royal Society B-Biological Sciences* **273**, 1323–1328.

Moore, J. (1984). Altered behavioral responses in intermediate hosts: an acanthocephalan parasite strategy. *American Naturalist* **123**, 572–577.

Moore, J. (2002) *Parasites and the Behavior of Animals*, Oxford University Press, Oxford.

Mouritsen, K. N. and Poulin, R. (2005). Parasites boost biodiversity and change animal community structure by trait-mediated indirect effects. *Oikos* **108**, 344–350.

Mouritsen, K. N. and Poulin, R. (2010). Parasitism as a determinant of community structure on intertidal flats. *Marine Biology* **157** 201–207.

Peterson, R. O. (1995) *The Wolves of Isle Royale: A Broken Balance*. Willow Creek Press, Minocqua WI.

Ponton, F., Lebarbenchon, C., Lefèvre, T., Biron, D. G., Duneau, D., Hughes, D. P., and Thomas, F. (2006). Parasitology: parasite survives predation on its host. *Nature* **440**, 756.

Poulin, R. (1995). "Adaptive" changes in the behaviour of parasitized animals: A critical review. *International Journal for Parasitology* **25**, 1371–1383.

Pozio, E., La Rosa, G., Rossi, P., and Murrell, K. D. (1992). Biological characterization of *Trichinella* isolates from various host species and geographical regions. *Journal of Parasitology* **78**, 647–653.

Rau, M. E., and Caron, F. R. (1979). Parasite-induced susceptibility of moose to hunting. *Canadian Journal of Zoology-Revue Canadienne De Zoologie* **57**, 2466–2468.

Rausch, R. L., (1995) Life cycle patterns and geographic distribution of *Echinococcus* species. In: *Echinococcus and Hydatid Disease*. (eds Thompson, R. C. C., and Lymbery, A. J.) CAB International, Wallingford, Oxon.

Sato, T. (2007). Threatened fishes of the world: Kirikuchi charr, *Salvelinus leucomaenis japonicus* Oshima, 1961 (Salmonidae). *Environmental Biology of Fishes* **78**, 217–218.

Sato, T., Arizono, M., Sone, R., and Harada, Y. (2008). Parasite-mediated allochthonous input: do hairworms enhance subsidised predation of stream salmonids on crickets? *Canadian Journal of Zoology-Revue Canadienne De Zoologie* **86**, 231–235.

Sato, T., Watanabe, K., Kanaiwa, M., Niizuma, Y., Harada, Y., and Lafferty, K. D. (2011a). Nematomorph parasites drive energy flow through a riparian ecosystem. *Ecology* **91**, 201–207.

Sato, T., Watanabe, K., Tokuchi, N., Kamauchi, H., Harada, Y., and Lafferty, K. D. (2011b). A nematomorph parasite explains variation in terrestrial subsidies to trout streams in Japan. *Oikos* 120, 1596–1599.

Shaw, D. J., Grenfell, B. T., and Dobson, A. P. (1998). Patterns of macroparasite aggregation in wildlife host populations. *Parasitology* **117**, 597–610.

Shaw, J. C., Lafferty, K. D., Hechinger, R. F., and Kuris, A. M. (2010). Ecology of the brain trematode *Euhaplorchis californiensis* and its host, the California killifish (*Fundulus parvipinnis*). *Journal of Parasitology* **96**, 482–490.

Stewart, M. J. and Creese, R. G. (2002). Transplants of intertidal shellfish for enhancement of depleted populations: preliminary trials with the New Zealand little neck clam. *Journal of Shellfish Research* **21**, 21–27.

Tenter, A. M., Heckeroth, A. R., and Weiss, L. M. (2000). *Toxoplasma gondii*: from animals to humans. *International Journal for Parasitology* **30**, 1217–1258.

Thomas, F. and Poulin, R. (1998). Manipulation of a mollusc by a trophically transmitted parasite: convergent evolution or phylogenetic inheritance? *Parasitology* **116**, 431–436.

Thomas, F., Renaud, F., De Meeüs, T., and Poulin, R. B. (1998). Manipulation of host behaviour by parasites: ecosystem engineering in the intertidal zone? *Proceedings of the Royal Society of London Series B-Biological Sciences* **265**, 1091–1096.

Thomas, F., Schmidt-Rhaesa, A., Martin, G., Manu, C., Durand, P., and Renaud, F. (2002). Do hairworms (Nematomorpha) manipulate the water seeking behaviour of their terrestrial hosts? *Journal of Evolutionary Biology* **15**, 356–361.

Thompson, R. M., Mouritsen, K. N., and Poulin, R. (2005). Importance of parasites and their life cycle characteristics in determining the structure of a large marine food web. *Journal of Animal Ecology* **74**, 77–85.

Waldie, P. A., Blomberg, S. P., Cheney, K. L., Goldizen, A. W., and Grutter, A. S. (2011). Long- term effects of the cleaner fish *Labroides dimidiatus* on coral reef fish communities. *PLoS ONE*, **6**, e21201.

Wood, C. L., Byers, J. E., Cottingham, K. L., Altman, I., Donahue, M. J., and Blakeslee, A. M. H. (2007). Parasites alter community structure. *Proceedings of the National Academy of Sciences of the United States* **104**, 9335–9339.

Afterword

Michel Loreau

Do manipulative parasites matter?

"Are manipulative parasites more than just interesting cocktail party anecdotes?" The question asked by Lafferty and Kuris in their chapter prompts two comments. The first is that parasitologists seem to attend rather special cocktail parties. Second, and more seriously, we obviously still know much too little about the ecological consequences of the bulk of biodiversity on our planet. Although we have made tremendous progress, both empirically and theoretically, in our understanding of the ecological and societal consequences of changes in biodiversity in recent years (Naeem et al. 2009; Loreau 2010b), the mere fact that Lafferty and Kuris ask this question shows that the ecological significance of manipulative parasites is poorly known and understood. And in this respect, manipulative parasites are the rule rather than the exception. While the taxonomy and ecology of large-sized plants and vertebrates are comparatively well known, the huge diversity of small-sized organisms on Earth still remains invisible to us despite considerable recent technological advances. Yet there is mounting evidence that microorganisms largely shape both the phylogenetic tree of life and the Earth's global biogeochemical cycles. Admittedly, free-living bacteria account for the bulk of this newly discovered phylogenetic and functional diversity. So, are parasites doomed to ecological insignificance and to receiving attention only in parasitologists' cocktail parties?

The laws of physics and ecology do limit the biomass of parasites and pathogens. First, like any other consumer trophic level, parasites and pathogens must have a lower production than their hosts.

Second, their life style forces them to have a small body mass, and hence a high metabolism per unit mass, that is a high turnover rate of materials and energy. Since at steady state biomass is equal to the ratio between production and turnover rate, parasites and pathogens are constrained to have a much smaller biomass than their hosts. Kuris et al. (Kuris et al. 2008) recently reported exceptionally high parasite biomass in three estuaries; it is important to keep in mind, however, that even their exceptionally high figures represented only 0.2-1.2% of total animal biomass in these ecosystems. Using biomass as an indicator of a species' importance in an ecosystem will rarely show parasites in a favorable light.

But David defeated Goliath despite a much smaller biomass, and lions are portrayed as king of the jungle, also despite their having a smaller biomass than their prey. A few years ago, we concluded a review of the role of parasites in ecosystems thus:

"The role of parasites in ecosystem functioning has usually been underestimated and poorly investigated because of their low biomass, low visibility, and small direct contribution to energy and material flows in natural ecosystems. (…) They may nevertheless have significant indirect impacts on ecosystem properties, by controlling numerically dominant host species, by exerting top–down control and maintaining the diversity of lower trophic levels, by shifting from parasitic to mutualistic interactions with their hosts, and by channelling limiting nutrients to more or less efficient recycling pathways" (Loreau et al. 2005).

While our review underestimated the contribution of parasites to energy and material flows, which now appears to be of the same magnitude as that of free-living animals with the same trophic position (Hechinger et al. 2011), there can be no doubt that parasites, like top predators, have considerable indirect impacts on ecosystem properties. A single virus with negligible biomass, the rinderpest virus, was responsible for some of the most dramatic ecosystem changes observed in the Serengeti over the past century, decimating huge herds of large herbivores, thereby modifying the behavior of large predators, causing migrations of local human populations, and altering vegetation patterns and fire regimes (Sinclair and Norton-Griffiths 1979). Lafferty and Kuris's chapter provides several fascinating examples in the same vein, showing significant indirect effects of manipulative parasites on ecosystem processes through behavioral alterations of their hosts.

It is highly likely that, the more we study parasites and pathogens, the more we will find examples of this type of indirect effect in natural ecosystems. Increasing evidence suggests that parasites strongly affect the structure and dynamics of food webs (Lafferty et al. 2008). Recent studies now suggest that soil pathogens play a significant part in plant community composition, diversity, and ecosystem processes. Interactions between plants and their soil pathogens predict tree-species relative abundances in some tropical forests (Comita et al. 2010; Mangan et al. 2010), and fungal pathogens appear to drive the positive diversity-productivity relationship in some temperate grasslands (Maron et al. 2011).

Manipulative parasites are not only likely to alter the dynamics and functioning of natural ecosystems through their effects on their hosts' behavior, they might also contribute indirectly to changing our own perception of ecosystems, and hence ultimately our behavior toward them. Parasitic "puppet masters" that take control of other organisms and make them lose their identity challenge the traditional organism-centered view that permeates ecology and evolutionary biology. This view hinges on the basic assumption that individual organisms are self-contained discrete entities governed by a simple genotype. There are two non-exclusive potential ways to resolve the theoretical difficulty generated by the parasitic takeover of host organisms. One is the "selfish gene" theory, which views organisms as mere vehicles of genes. On this view, the manipulated host organism would simply be a vehicle that has changed driver. Although a powerful theory that has brought new insights in evolutionary biology, this theory is also, in my opinion, extremely reductionist, not to say simplistic.

Another way to resolve this issue would be to view the organism itself as a complex ecological system in which the organism's genotype plays a key but non-exclusive structuring role. Although this view might seem iconoclastic at first sight, it is not inconsistent with recent trends in systems biology and human microbial ecology. Systems biology is beginning to unveil the enormously complex networks of interactions that make cells and organisms work as organized systems. At the same time, human microbial ecology is revealing an astonishing diversity of microbes in the human body. Current estimates suggest that bacteria living within the body of an average healthy adult human outnumber human cells ten to one and that the number of microbial species is about 180 on the skin, 700 in the mouth, and 1,000 or more in the gastrointestinal tract. Alterations in the ecology of the gastrointestinal tract may be related to incidence and severity of food allergy, asthma, eczema, inflammatory bowel disease, and obesity (Bernstein and Ludwig 2008). Thus, the human body increasingly resembles an ecosystem with its myriad of species and species interactions. Microbial diversity may play an important role in human disease regulation just as biodiversity does in the stability and regulation of ecosystem processes. On this view, host manipulation by parasites would be an extreme instance of ecological change in the host driven by an invasive species.

As is often the case, a single perspective is likely not to be enough to fully account for the complexity of reality. But a shift to a more ecological approach in the biological, medical, and social sciences would be particularly useful today. Humans are changing the world to such an extent and at such a rate that a major global ecological crisis is now looming. To overcome this crisis, we should stop thinking of ourselves as godlike creatures designed to rule the world and accept our limits within the biosphere (Loreau 2010a). Recognizing that we are ecological systems embedded in larger ecological systems might help us see our interdependency with the rest of nature and hasten the emergence of innovative ways out of this crisis. If manipulative parasites can foster this awareness, their indirect ecological impact will be enormous.

References

Bernstein, A. S. and Ludwig, D. S. (2008). The importance of biodiversity to medicine. *JAMA* **300**, 2297–2299.

Comita, L. S., Muller-Landau, H. C., Aguilar, S., and Hubbell, S. P. (2010). Asymmetric density dependence shapes species abundances in a tropical tree community. *Science* **329**, 330–332.

Hechinger, R. F., Lafferty, K. D., Dobson, A. P., Brown, J. H., and Kuris, A. M. (2011). A common scaling rule for abundance, energetics, and production of parasitic and free-Living species. *Science* **333**, 445–448.

Kuris, A. M., Hechinger, R. F., Shaw, J. C., Whitney, K. L., Aguirre-Macedo, L., Boch, C. A., Dobson, A. P., Dunham, E. J., Fredensborg, B. L., Huspeni, T. C., Lorda, J., Mababa, L., Mancini, F. T., Mora, A. B., Pickering, M., Talhouk, N. L., Torchin, M. E., and Lafferty, K. D. (2008). Ecosystem energetic implications of parasite and free-living biomass in three estuaries. *Nature* **454**, 515–518.

Lafferty, K. D., Allesina, S., Arim, M., Briggs, C. J., Leo, G. D., Dobson, A. P., Dunne, J. A., Johnson, P. T. J., Kuris, A. M., Marcogliese, D. J., Martinez, N. D., Memmott, J., Marquet, P. A., McLaughlin, J. P., Mordecai, E. A., Pascual, M., Poulin, R., and Thieltges, D. W. (2008). Parasites in food webs: the ultimate missing links. *Ecology Letters* **11**, 533–546.

Loreau, M. (2010a). *The Challenges of Biodiversity Science.* International Ecology Institute, Oldendorf/Luhe.

Loreau, M. (2010b). *From Populations to Ecosystems: theoretical foundations for a new ecological synthesis.* Princeton University Press, Princeton, New Jersey.

Loreau, M., Roy, J., and Tilman, D. (2005). Linking ecosystem and parasite ecology. In: *Parasitism and ecosystems* (eds Thomas, F., Guégan, J-F. and Renaud, F.) Oxford University Press Oxford, pp. 13–21.

Mangan, S. A., Schnitzer, S. A., Herre, E. A., Mack, K. M. L., Valencia, M. C., Sanchez, E. I., and Bever, J. D. (2010). Negative plant-soil feedback predicts tree-species relative abundance in a tropical forest. *Nature* **466**, 752–755.

Maron, J. L., Marler, M., Klironomos, J. N., and Cleveland, C. C. (2011). Soil fungal pathogens and the relationship between plant diversity and productivity. *Ecology Letters* **14**, 36–41.

Naeem, S., Bunker, D. E., Hector, A., Loreau, M., and Perrings, C. (2009). *Biodiversity, Ecosystem Functioning, and Human Wellbeing: An Ecological and Economic Perspective.* Oxford University Press, Oxford.

Sinclair, A. R. E. and Norton-Griffiths, M. (eds). (1979). *Serengeti: Dynamics of an Ecosystem.* University of Chicago Press, Chicago.

CHAPTER 10

Applied aspects of host manipulation by parasites

Robert Poulin and Edward P. Levri

10.1 Introduction

As other chapters in this book make very clear, the manipulation of host phenotype by parasites is a ubiquitous phenomenon with ramifications into most branches of biology. There are several reasons why manipulative parasites should be of interest to applied scientists, especially in areas of conservation, economic, or medical relevance. First, manipulative parasites belong to all major parasite taxa, from viruses to worms, and have phenotypic effects on a broad range of host taxa, including both invertebrates and vertebrates (Moore 2002). We are therefore not dealing with a phenomenon limited to a few unusual species of purely academic interest. Second, manipulation is not merely a component of parasite virulence; it represents effects above and beyond the direct negative impact of parasitic infection on host fitness. From the perspective of host health, therefore, manipulation is not subsumed within virulence, but is instead a parallel strategy of host exploitation with additive effects of its own (Poulin 2007). The genes responsible for the manipulative abilities of parasites are no doubt distinct from those leading more directly to host debilitation or pathology; the two sets of genes are not necessarily under identical selection pressures and probably evolve independently. Third, the impact of manipulative parasites on the phenotype of their hosts range from subtle to drastic; in some cases, manipulated hosts are clearly morphologically distinct or ecologically segregated from normal hosts (Poulin and Thomas 1999). Manipulated hosts can occupy different microhabitats, feed on different prey, and be eaten by different predators compared

to uninfected hosts. In extreme cases and many less pronounced ones, having manipulated hosts in a population is not unlike having two related host species present in the community, sharing many traits but differing sharply with respect to others, and occurring at relative abundances corresponding to the proportions of infected and uninfected individuals.

It is therefore inevitable that manipulative parasites affect some species of interest to society, inducing in them changes that are not merely direct consequences of the pathology usually expected from infection, and altering their ecology in ways that affect their productivity, their status as pest species, or their threat to human health or wellbeing. This chapter will explore the potential impact of manipulative parasites on conservation efforts, on animal productivity in agriculture and aquaculture, and on disease transmission to humans. We will also discuss cases where manipulative parasites accidentally infect people and alter their personality or mental health. We focus on a small set of case studies to illustrate general concepts. Our goal is to bring host manipulation out from the esoteric closet and into the scientific mainstream by highlighting its relevance to key social concerns.

10.2 Manipulative parasites, biological invasions, and conservation

As they induce pronounced changes in the phenotype of their hosts, manipulative parasites can have measurable effects on host population dynamics and on community-level phenomena (Dobson 1988;

Lefèvre et al. 2009, see also Chapter 9). Thus, in many situations, manipulative parasites can affect, either directly or indirectly, species of relevance to conservation biology and wildlife management.

For instance, parasitism in general can influence the outcome of biological invasions, and the establishment of unwanted exotic species. Biological invasions represent the second most important proximate cause of biodiversity loss worldwide, after habitat destruction (Molnar et al. 2008), and they impose huge economic costs on society (Pimentel et al. 2005). The recent increase in the rate at which species are introduced to new areas is most likely a consequence of the globalization of the world's transport systems (Cohen and Carlton 1998; Ruiz et al. 2000; Ricciardi 2007). However, the successful establishment of an exotic species in its new area depends on its ecological properties and those of the local ecosystem. In particular, the success of exotic species is often attributed to the fact that they leave behind their natural enemies, including parasites, during translocation to a new area (Torchin et al. 2003; Colautti et al. 2004). There are several other ways in which parasites can affect the invasive process: exotic parasites can be transferred from introduced hosts to native hosts, native parasites can infect exotic hosts following their introduction to a new area, and parasites can indirectly mediate competitive or predatory interactions determining whether an exotic species will become established and subsequently spread to adjacent areas (Prenter et al. 2004; Taraschewski 2006; Dunn 2009). Manipulative parasites may also intervene in the invasion process. In fact, given how they impact other ecosystem properties ranging from community structure to food web dynamics (Lefèvre et al. 2009), there is every reason to believe that manipulative parasites can modulate the success or failure of biological invasions.

A series of recent studies on invasive freshwater amphipods illustrate well how manipulative parasites may influence the success of exotic species. In these case studies, the parasites involved were acanthocephalans. These worms have a two-host life cycle, in which transmission to the definitive vertebrate host occurs by predation on the arthropod intermediate host. Practically all acanthocephalan species investigated in detail show an ability to alter the behavior of their intermediate host in ways that enhance predation by the definitive host, making them master manipulators (Moore 1984; Kennedy 2006).

In Irish freshwaters, the native amphipod *Gammarus duebeni celticus* is gradually being replaced by the exotic *G. pulex*, a phenomenon apparently mediated by mutual predation between the two amphipod species. Although large individuals of either species can prey on small individuals of the other, the process is biased in favor of the exotic invader under most field conditions (Dick et al. 1993). However, both amphipod species are parasitized by the acanthocephalan *Echinorhynchus truttae*, which must reach a fish definitive host via predation to complete its life cycle. In the field, infections by this parasite are more common in the exotic *G. pulex* than in the native *G. d. celticus* (MacNeil et al. 2003a). In addition, in both field and laboratory experiments, as a side-effect of behavioral manipulation by the parasite, acanthocephalan infections in the exotic amphipods significantly lowered their predatory impact on the native species (MacNeil et al. 2003a). In this system, the net effect of a manipulative parasite may be to slow down the rate at which a native species is being replaced by an introduced one. In contrast, another invasive amphipod, *G. tigrinus*, also competing with the native *G. d. celticus* for the same ecological niche, appears to benefit from the actions of a manipulative parasite. In rivers where both amphipods co-occur, the acanthocephalan *Polymorphus minutus* alters phototropism in *G. d. celticus* and greatly increases its probability of drifting, which exposes amphipods to greater predation by fish and waterfowl (MacNeil et al. 2003b). Somehow, the exotic *G. tigrinus* is unaffected, and may thus gain a slight edge over its native rival.

In French rivers, the tables are turned on the amphipod *G. pulex*: there, it is in its native range, and facing competition from an Eastern European invader, *G. roeseli*. Although both amphipods are equally susceptible to infection by the acanthocephalan *Pomphorhynchus laevis*, the parasite can only successfully manipulate the native species: it induces a switch from strong photophobia to pronounced photophilia in *G. pulex*, but has no behavioral effect on the exotic species (Bauer et al. 2000). In the field, this translates into greatly increased predation rates by fish on infected native amphipods, but not on exotic ones (Lagrue et al. 2007). In this case, too, manipulative parasites can facilitate the invasion of exotic species.

The above scenarios, summarized in Fig. 10.1, are probably played out in numerous other situations, since the number of exotic species in aquatic and terrestrial ecosystems is increasing all the time while our knowledge of manipulative parasites is still somewhat limited to a few well-studied model species. The amphipod studies summarized above illustrate clearly some of the various ecological pathways through which the effect of manipulative parasites can modulate the spread of exotic pests. Beyond the mere pathological effects generally

associated with parasitic infections, manipulative parasites can alter interspecific interactions and directly or indirectly affect the dynamics of species invasion. Their potential effects should be added to the list of factors that must be taken into account for the management of biological invasions (Hulme 2006).

Outside the context of biological invasions, manipulative parasites can still create concerns for applied wildlife ecologists. Parasitic diseases are increasingly recognized as serious threats to the conservation of endangered species (Daszak et al. 2000; Lafferty and Gerber 2002; Lebarbenchon et al. 2009). Given their effect on host behavior, manipulative parasites can pose special challenges, since changes in host social behavior or activity levels following infection can drastically alter parasite transmission dynamics. For instance, rabies is seen as a

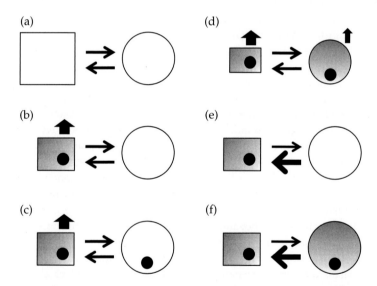

Figure 10.1 Possible outcomes of biological invasions mediated by a manipulative parasite. In all scenarios, a native species (square) faces competition from an invading species (circle); invasion success is determined by the relative population sizes of the two species, indicated by the relative sizes of squares and circles. Interspecific competition (arrows between squares and circles) and predation (upward arrows) are the main determinants of population sizes in these scenarios, with their strength proportional to the thickness of the arrows. (a) The parasite-free scenario, in which the native and the invader are equally-matched competitors, providing a benchmark for comparisons. (b) A manipulative parasite (black dot) induces an alteration (shading) in the phenotype of the native species, causing it to incur higher predation from the parasite's definitive host, while the invader is not infected. (c) Same as in the previous scenario, except that the invader is also susceptible to infection, but the parasite is incapable of altering its phenotype. (d) Both the native and invading species are infected and manipulated by the parasite, but the increase in predation rates is much higher for the native than the invader. (e) Assuming insignificant predation, manipulation of the native species by the parasite may reduce its competitive abilities, to the benefit of the uninfected invader. (f) When both the native and invader are parasitized and manipulated, the effect on their competitive abilities may be asymmetrical, with one benefiting at the expense of the other. In all these hypothetical scenarios, the invader is portrayed as benefiting from the action of a manipulating parasite; in reality, either host species may benefit.

major threat to free-ranging populations of carnivores (Murray et al. 1999; Haydon et al. 2002). The virus that causes this disease also induces a marked change in host behavior. Rabid animals typically, though not always, become more aggressive and more likely to wander out of their territory, leading to a greater likelihood of fights with other individuals of the same or other species. Since the viruses move to the host's salivary glands following the induction of this behavioral change, and get transmitted when the host bites a susceptible animal, this may represent a case of host manipulation to enhance transmission success (see Rupprecht et al. 2002, see also Chapter 3). The altered behavior of rabid animals increases their contact rates with susceptible animals beyond those expected from pre-existing social networks or movement patterns, and this consequence of host manipulation must be integrated into both epidemiological models and conservation practices to protect declining populations of wild carnivores, especially canids (Murray et al. 1999), from the risks posed by this disease.

Other manipulative parasites are also of relevance to conservation biology. Consider the apicomplexan protozoan *Toxoplasma gondii*, which is transmitted by predation from a rat intermediate host to a feline definitive host (see Section 10.5). *Toxoplasma gondii* is famous for its ability to manipulate rat behavior, changing a rat's normal aversion to cat odor into an attraction toward that predator's smell (Berdoy et al. 2000, see also Chapter 3). This not only increases transmission success of the parasite under normal conditions, but also affects predator–prey dynamics. In addition, *T. gondii* can affect a wide range of non-host species. The parasite shows remarkably little specificity toward its intermediate host, and infections have been reported in numerous mammal species from different orders, including humans (Dubey 1998, see Section 10.5). Although these alternative hosts are transmission dead-ends, they can nonetheless suffer from the manipulative efforts of the parasite. In its intermediate host, whether a rat or some other species unsuitable for transmission, *T. gondii* encysts within the central nervous system, from where the parasites can exert their influence to induce what would normally be a behavioral change in the rat host. In humans, this may result in changes

in a person's psychological profile (see Section 10.5); the effects on wildlife remain poorly known but can include encephalitis and other neurological diseases (Dubey 1998). There are several recent reports of *T. gondii* infections in wildlife, often implicating domestic cats and dogs, and thus human presence, as the main source (Fiorello et al. 2004; Suzán and Ceballos 2005). The parasite also impacts marine mammals, reaching coastal habitats in contaminated runoff water and accumulating in filter-feeding invertebrates before being acquired by sea otters (Gaydos et al. 2007; Miller et al. 2008; Johnson et al. 2009). In large part due to human activities, *T. gondii* now infects several novel host species, some of them endangered, at that stage of its life cycle where it would normally sit within a rat's nervous system wreaking havoc with its host's normal behavior and antipredator defenses. The population-level and/or ecosystem-wide consequences of *T. gondii*'s failed manipulation attempts in these novel hosts are not yet understood, but they highlight why the relevance of host manipulation extends well beyond the walls of academia.

10.3 Manipulative parasites in agriculture and aquaculture

Only a relatively small number of animal species have been domesticated and are currently raised for meat or other animal products like milk or wool. Even after centuries of animal husbandry, domesticated animals are still plagued by many of the same parasites that infected their ancestors. These parasites cause important production losses, and therefore require the implementation of various control strategies. Controlling parasite infections is very costly, however. For instance, New Zealand sheep and cattle farmers spend close to $100 million each year on anthelmintic drugs against nematode parasites, yet these can only mitigate the effects of parasitism, with close to 20% of potential sheep production still not being realized (Vlassoff et al. 2001). With resistance to these drugs spreading slowly but steadily (Kaplan 2004), the fight against livestock parasites is not about to ease.

Some of the parasites of livestock are known manipulators of host behavior (Lagrue and Poulin

2010). For example, the trematode *Dicrocoelium dendriticum* is a common liver fluke parasite of wild and domestic ruminants, with a worldwide distribution (Otranto and Traversa 2002). It is also the poster child of host manipulation by parasites, and its most prominent textbook examples. Larval stages (metacercariae) of *D. dendriticum* encyst within intermediate insect hosts, usually ants, and somehow induce them to remain fixed to the tip of grass blades, awaiting accidental ingestion by the grazing herbivores serving as definitive hosts (Carney 1969; Romig et al. 1980; Moore 2002). Sheep and cattle themselves are not manipulated by the parasite, but their acquisition of the parasite is mediated by host manipulation. In this system, the epidemiology of the disease is critically influenced by the increased ant-to-livestock transmission rate resulting from the parasitic manipulation of ants, and any effort to mitigate transmission by pasture or grazing management must take this into account.

Farmed animals may themselves be manipulated by parasites. Consider the cestodes *Echinococcus* spp., the parasites responsible for hydatid disease (echinococcosis) in sheep and other domestic animals, and even sometimes in humans. These parasites have a life cycle requiring two hosts, both mammals. Definitive hosts are carnivores in which adult tapeworms occupy the small intestine. Eggs released in host feces are accidentally ingested by an intermediate host in which they hatch and develop into hydatid cysts. These metacestodes are found in various internal organs (mainly lungs and liver). The definitive host becomes infected when preying on an infected intermediate host (Jenkins et al. 2005; Thompson 2008). *Echinococcus* spp. have a narrow range of definitive hosts, mainly domestic and wild canids, but a wide geographical distribution, occurring virtually on every continent (Lymbery and Thompson 1996; Jenkins et al. 2005). They are not particularly host-specific for their intermediate host and can develop in a wide range of herbivorous and omnivorous animals. *Echinococcus granulosus* and *E. multilocularis* are the most important species of concern for domestic animal health (Jenkins et al. 2005; Schweiger et al. 2007). These species are responsible for cystic and

alveolar echinococcosis, respectively, both resulting, through asexual reproduction, in increasingly large and abundant cysts that eventually become debilitating, if not lethal (Jenkins et al. 2005). This is where host manipulation may occur. Infected intermediate hosts seem more vulnerable to predators, including the definitive host of the parasite (Moore 2002). The hypothesis that echinococcosis may increase host vulnerability to predation came from an unusual incident reported by Crisler (1956) in which a caribou cow failed to escape and laid down in front of young wolves. Postmortem examination revealed that its lungs were infected with several large cysts of *E. granulosus*. Also, moose harboring numerous *E. granulosus* cysts are more likely to be shot by hunters than uninfected moose (Rau and Caron 1979). Of course, the link between *E. granulosus* infection and aberrant escape response is unclear and difficult to quantify (Messier et al. 1989). However, in moose (*Alces alces*), a large proportion of cysts are found in the lungs and severe infections reduce the endurance of moose trying to escape from grey wolves (*Canis lupus*), one of the definitive hosts of *E. granulosus* (Messier et al. 1989). Moose infected by *E. granulosus* may not be handicapped when there is no need for exertion, but the condition could seriously impair individuals forced to flee or defend themselves. Consequently, moose with cystic hydatid disease should be more vulnerable to predation; wolves and other carnivores are also known to selectively prey on weaker animals (Messier et al. 1989; Joly and Messier 2004). However, no study to date has empirically shown the direct link between hydatid infection, increased predation risk, and increased transmission rates for the parasite. Nevertheless, the strong association between infection by *E. granulosus* and altered behavior is an additional complication, beyond simple pathology, of concern for veterinary medicine.

Domesticated mammals are not the only animals of commercial importance that can be victims of manipulative parasites. The growth of aquaculture has lead to the domestication and husbandry of numerous species of molluscs, crustaceans, and fish. Under the conditions typical of fish farms and other aquaculture facilities, that is high densities, genetically homogeneous stock, and so on, many parasites have prospered and certain particularly virulent strains are often favored (Nowak 2007). The resulting mortalities and losses in production are a constant concern for the industry. The more subtle but potentially deleterious effects of manipulative parasites have received little attention in the context of aquaculture. Some farmed invertebrates harbor species of parasites whose close relatives are known to manipulate the behavior of their hosts; these include microphallid trematodes in the crabs *Scylla* spp. in south-west Asia (Hudson and Lester 1994) and acanthocephalans in several freshwater crayfish species (Edgerton et al. 2002).

A good example of how manipulative parasites can impact aquaculture is provided by the trematode genus *Diplostomum*, which consists of a vast complex of cryptic species (Locke et al. 2010). After leaving their snail first intermediate host, cercarial stages of *Diplostomum* penetrate the skin or gills of fish; in most species, such as the well-studied *D. spathaceum*, the parasites then proceed to migrate toward the lenses of their host's eyes, where they develop into long-lived metacercariae. There is little host specificity at this stage of the life cycle, and numerous fish species are physiologically suitable as second intermediate host. The parasite's life cycle is completed when an infected fish is ingested by any piscivorous bird, in which the parasites develop into adult worms.

Diplostomum spp. are distributed across all temperate regions of the northern hemisphere, and often reach high prevalence and intensities of infection in many fish species (Valtonen and Gibson 1997; Locke et al. 2010). Under normal conditions, this parasite induces no measurable pathology in fish. However, it does so indirectly via its manipulation of fish vision which results in greater predation of fish by birds, that is higher transmission success for the parasite. Indeed, fish infected by *Diplostomum* develop cataracts following mechanical damage to their lens and metabolic products excreted by the parasites lodged in the lens (Karvonen et al. 2004). This leads to several changes in fish behavior related to predator avoidance, such as reduced responsiveness to visual stimuli, movement toward the better-lit surface waters, and impaired crypsis, that is

matching of body coloration with that of the substrate (Crowden and Broom 1980; Seppälä et al. 2004, 2005). As a consequence, infected fish are more susceptible to predation than control fish under both laboratory and field conditions (Seppälä et al. 2004, 2006).

It is very difficult to keep this parasite away from fish farms, as it would require eradication of snails or treatment of water to remove infective cercarial stages. Therefore, *Diplostomum* infections are common in freshwater salmonid aquaculture, where high fish densities facilitate host-finding by the parasites. One obvious implication is that avian predation on farmed fish may be more intense as an outcome of parasite manipulation, thus necessitating nets above culture ponds, or other protective measures. In addition, the impaired visual acuity of infected fish reduces not only their antipredator defenses, but also their feeding efficiency. Infected fish fail to detect prey and have a lower strike rate when capturing food than uninfected conspecifics (Crowden and Broom 1980; Owen et al. 1993). Ironically, rainbow trout, *Oncorhynchus mykiss*, infected by *Diplostomum* show low responsiveness to the artificial lures used by anglers (Moody and Gaten 1982). For fish farmers, though, the consequence may be a lower feeding efficiency, lower fish growth, and greater waste of food.

The above examples are not a complete list, but only some illustrative case studies of the ways in which manipulative parasites can affect animal productivity in agriculture and aquaculture. If domesticated animals can feel the influence of manipulative parasites, so can humans. The next two sections explore how host manipulation can impact our wellbeing more directly, by affecting our health rather than our economy.

10.4 Parasite manipulation of disease vectors

A key area where alterations of hosts have important effects on humans is the effect of parasites on the behavior and life history of disease vectors. In many vector-borne diseases of humans, one of the most common control strategies is to attack the vector. Thus, knowledge of how infected and uninfected vectors differ could be very valuable in combating some of the deadliest diseases known to humanity. It is becoming increasingly evident that vector-transmitted parasites can cause subtle and sometimes substantial changes in the phenotypes of their vector hosts, which appear to increase their transmission rate (reviewed by Moore 1993; Hamilton and Hurd 2002; Moore 2002; Hurd 2003; Lefèvre et al. 2006; Schaub 2006; Lefèvre and Thomas 2008).

The life cycle of vector-borne pathogens presents several opportunities to influence their hosts (both vector and vertebrate) to help facilitate transmission (Fig. 10.2). We are assuming here that the parasite is transmitted by the bite of the vector, as is true in most cases. In order for a vector-borne disease to be transmitted, the vector must first feed from an infected host. Then, for transmission to occur, the parasite must avoid the vector's defenses (Stage 1), develop appropriately within the vector (this often involves reproduction) surviving long enough for the vector to take another meal (Stage 2), and find a vertebrate host (Stage 3). Once a vertebrate is located, the parasite must get from the vector into the new host (Stage 4). The parasite must then avoid

the vertebrate host's defenses (Stage 5) and survive long enough to allow for transmission to the next vector (Stage 6). For the parasite to get in a vector, the latter must locate the infected host (Stage 7), and the parasite must be taken up by the vector (Stage 8). The parasite may influence its transmission at any one of these stages, some of which are summarized in Table 10.1.

The most likely stages where the parasite may alter its host are those where there is a conflict of interest between the host and parasite. For example, changes in host finding behaviors in the vector are not likely in most cases because both parasite and vector benefit from host finding in the same way. Thus, selection on the vector has likely already optimized that behavior for the parasite. However, a behavior such as biting has different costs and benefits for the parasite and the vector (Koella 1999; Schwartz and Koella 2001). From the vector's point of view, the best strategy is to get only the meal size that is required, and then get away from the host to avoid the dangers of vertebrate host defenses. From the parasite's point of view, however, the optimal strategy may be to stay on the host longer to assure transmission, and to

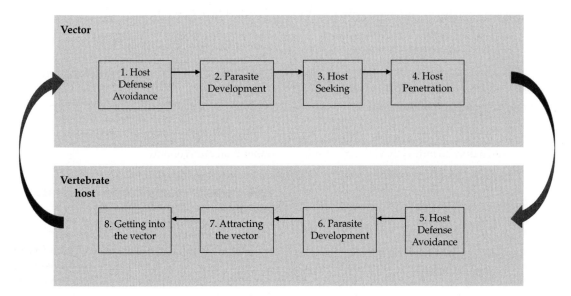

Figure 10.2 Stages in the life cycle of a vector-transmitted parasite where phenotypic alterations of either the vector or the vertebrate host could potentially increase transmission. See Table 8.1 for further details.

visit multiple vertebrate hosts even after a meal has been taken. Various parasites of vectors of human diseases have been shown to influence their vectors in ways that may increase the probability of transmission, including the parasites that cause Bubonic plague, Dengue fever, Babesiosis, and LaCrosse encephalitis. Here, we briefly survey three different vector-transmitted diseases of humans, leishmaniasis, African sleeping sickness, and malaria, to examine how the parasite uses host manipulation to influence its transmission in each case (see summary in Table 10.1).

10.4.1 Leishmaniasis

Leishmaniasis is caused by flagellated protozoans of the genus *Leishmania*. The genus includes about 30 taxa of which the majority infect humans, resulting in various types of disease. Leishmaniasis is a major public health problem in many underdeveloped countries, and the annual incidence of new cases is estimated to be between 1.5 and 2 million worldwide (Dedet and Pratlong 2003). The disease is transmitted by sandfly vectors of the family

Psychodidae. Typically the flies feed on plants, but the females must take a blood meal before they lay eggs, and it is during these feedings that the parasite is transmitted among its vertebrate hosts.

Leishmania appears to concentrate its manipulative effort on the feeding behavior of its vector (Stage 4 in Fig. 10.2) (Rogers and Bates 2007; Ready 2008). Infected sandflies produce promastigote secretory gel (PSG), which blocks the foregut. This blockage results in an accumulation of *Leishmania* promastigotes behind the plug (Rogers et al. 2004). The blockage limits feeding of the fly by inhibiting the intake of blood until the blockage is removed by regurgitation. Infected flies probe the skin more frequently and spend more time feeding as they attempt to regurgitate, resulting in increased transmission rates (Killick-Kendrick et al. 1977; Beach et al. 1985). Transmission rate is further increased because infected flies are more likely to feed on multiple hosts than uninfected individuals (Rogers and Bates 2007). Included in the gel is filamentous proteophosphoglycan (fPPG), which has been demonstrated to positively influence parasite growth in the vertebrate host (Rogers et al. 2004). In addition,

Table 10.1 Potentially beneficial alterations of host phenotype at various stages in the life cycle of a vector-transmitted parasite.

	Stage	Potentially altered trait	Leishmania	Trypanosoma	Plasmodium
Vector	1. Host defense avoidance	Vector immune function			Yes (Lefèvre et al. 2008)
	2. Parasite development and reproduction	Increased life span of vector or decreased activity until infectious		? (Hamilton and Hurd 2002; Hurd 2003)	? (Hamilton and Hurd 2002; Hurd 2003)
	3. Host seeking	Blood meals taken from more than one host			Yes
	4. Vertebrate host penetration	Feeding behaviour	Yes (Hamilton and Hurd 2002; Hurd 2003)	Yes (Hamilton and Hurd 2002; Hurd 2003)	Yes (Hamilton and Hurd 2002; Hurd 2003)
Vertebrate host	5. Host defense avoidance	Vertebrate host immune function	Yes (see Lefèvre et al. 2008)	Yes (Lefèvre et al. 2008)	Yes (Lefèvre et al. 2008)
	6. Parasite development and reproduction	Reduced host activity			Yes
	7. Attracting the vector	Host odor altered to attract vector	Yes		Yes
	8. Getting into the vector	Easier uptake of blood meal by vector		Yes	Yes

the behavioral change in the vector is correlated with parasite development. Flies with infective stages of the parasite feed more persistently and produce more aggressive infections (Rogers and Bates 2007). Salivary peptides and fPPG are potential targets for vaccine development (Valenzuela et al. 2001; Rogers et al. 2006). Thus understanding the physiological mechanisms for manipulation could be of great medical benefit in these cases.

Leishmania also appears to be capable of influencing host attractiveness to the vector (Stage 7). Odor from hamsters infected with *Leishmania* were found to be more attractive to sandflies than odor from uninfected hamsters (Rebollar-Tellez 1999 cited in Hamilton and Hurd 2002; O'Shea et al. 2002). Thus the parasites' manipulative effects extend through its vertebrate host to the vector.

10.4.2 African sleeping sickness

African sleeping sickness (trypanosomiasis) is a blood and lymph infection affecting an estimated 300,000 to 500,000 people, primarily in Africa. It is caused by the flagellated protist *Trypanosoma brucei*, and is transmitted by the bite of tsetse flies of the genus *Glossina* (Burri and Brun 2003).

Like *Leishmania*, *Trypanosoma* also appears to influence the feeding behavior of its vector (Stage 4). Tsetse flies infected with *T. brucei* probe more often and are more likely to take blood than uninfected flies (Jenni et al. 1980). The effect appears to be caused by a concentration of parasites around mechanoreceptors in the labrum resulting in reduced blood flow into the fly, leading to more attempts at feeding. This effect may not be universal, however (e.g., Makumi and Moloo 1991). Lefèvre et al. (2007a) used a proteomics approach to determine what molecular changes were occurring in the central nervous systems of tsetse flies infected with *T. brucei*. They found changes in expression of three glycolytic enzymes suggesting that the parasite influences energy metabolism. Changes in metabolism could directly or indirectly influence feeding rates by increasing the need for energy, and thus affect transmission rates. There were also changes in proteins found associated with signal transduction and neurotransmitter synthesis, both of which could easily be related to behavioral changes. These results are exciting as they provide a starting point in the search for the mechanisms of behavioral changes in this system.

In addition, the parasite may influence its vertebrate host to make the blood meal easier to get for the vector (Stage 8). Several studies have found that cattle infected with *T. congolense* were more easily fed upon by tsetse flies than uninfected cattle, probably due to increased vasodilation in infected cattle (Baylis and Nambiro 1993; Baylis and Mbwabi 1995; Moloo et al. 2000). It is possible that *T. brucei* could have similar effects on humans and other vertebrate hosts. Older studies suggested that *Trypanosoma* increased the longevity of its vector host (Stage 2), although more recent studies cast doubt on this assertion (reviewed in Lefèvre and Thomas 2008).

10.4.3 Malaria

Malaria is one of the world's great parasitic challenges. Approximately 300 million people worldwide are infected, and approximately one million people die from the disease each year. There are several different species of *Plasmodium* that cause human malaria, but the most problematic is *P. falciparum*. The disease is transmitted between humans by mosquitoes of the genus *Anopheles* when they take a blood meal. Of the 400 or so species in the genus, about 45 are considered important vectors of the parasite (White 2003).

The malarial parasite uses a diverse array of strategies to influence its own transmission. Malaria alters the biting behavior of its mosquito vector (Stage 4). Infected mosquitoes bite for longer periods of time and more often than uninfected mosquitoes (Rossignol et al. 1986; Koella et al. 1998), resulting in infected mosquitoes being more likely to take larger blood meals (Koella and Packer 1996). At least some of these effects appear to be due to the malarial parasite's influence on apyrase activity. Apyrase is an enzyme in the saliva of the vector that decreases platelet aggregation and aids in blood vessel location (Ribeiro et al. 1984). Reducing apyrase activity results in the vector having greater difficulty in taking a blood meal, thus increasing bite duration and frequency. The parasite also influences host-seeking behavior in the vector (Stage 3). Engorged uninfected mosquitoes avoid taking a second blood meal, whereas engorged mosquitoes infected with transmissible sporozoites of *P. gallinaceum* are more likely to probe for a second meal. Interestingly, non-transmissible oocyst stage-infected mosquitoes were less likely to probe for a second meal (Koella et al. 2002). In a preliminary study examining the mechanism of behavioral change, Lefèvre et al. (2007b) used a proteomics approach similar to that of Lefèvre et al.'s (2007a) study on tsetse flies. They found evidence from the head region of *Anopheles* indicating different protein expression in infected and uninfected mosquitoes. The protein differences included proteins known to be important in central nervous system function. Some of the biochemical pathways altered in the mosquito are similar to those altered in the tsetse fly, suggesting some sort of evolutionary convergence (Lefèvre et al. 2007a).

Several studies have examined the effect of the malarial parasite on the life span of its vector host (Stage 2). Malaria takes a relatively long time to become transmissible once acquired by the mosquito, causing many to hypothesize that the parasite should manipulate the host to increase its life span. However, most results point toward a shorter life span in infected mosquitoes or no difference compared to uninfected vectors (Ferguson and Read 2002).

For decades, arguments have been made that infected human hosts of malaria are more attractive to the mosquito vector than uninfected individuals (Stages 7 and 8). It has been suggested that the symptoms induced by malaria in humans make transmission to the mosquito easier and safer for the vector (Nacher 2005). The symptoms of increased skin temperature, sweating, and increased CO_2 expiration all have been shown to be attractants for mosquitoes, and the incapacitation caused by the disease makes it more likely that a blood-sucking mosquito will not be swatted. It appears, though, that the parasite has taken this a step further. An experiment by Lacroix et al. (2005) clearly demonstrated that the parasite induces pheromonal cues in its human host that attract the mosquito vector when the parasite is in its transmissible gametocyte stage. In the study, three groups of children (one uninfected, one infected with gametocyte stage

malaria, and one infected with non-infective stage malaria) were placed in tents. Mosquitoes were given a choice of which tent to move toward, based on odor; they preferred the gametocyte-infected children over the other two groups. Interestingly, when the children were treated with anti-malarial drugs to clear the parasite, there was no longer a preference by the mosquitoes for those children.

The exploration of the mechanisms of behavioral changes in these systems could provide valuable information about the important chemicals involved in the behaviors that affect parasite transmission. The result could be new approaches to vector control, perhaps by creating pesticides that influence important behaviors associated with vector host location and feeding. Most epidemiological models assume no difference in behavior between infected and uninfected vectors and no difference in attractiveness between infected and uninfected human hosts. More reliable epidemiological models, based on understanding and quantifying these effects, are urgently needed to incorporate aspects of host manipulation into control strategies.

10.5 Parasite manipulation in humans: the case of *Toxoplasma gondii*

One of the most interesting examples of applied impacts of parasite manipulation of hosts is the case of the intracellular protozoan *Toxoplasma gondii*. Toxoplasmosis is very common in human populations in both the developed and developing world with reported national incidences ranging from about 0% to 100% (Tenter et al. 2000), and a worldwide average commonly estimated at 30–40%. For decades, latent infection with this parasite has been considered asymptomatic in individuals with functioning immune systems. However, a multitude of studies performed in the past two decades have provided considerable evidence that many human behavioral traits are influenced by this parasite, and the parasite has been linked to several neurological abnormalities (for reviews see Holliman 1997; Webster 2001; Yolken and Torrey 2006; Flegr 2007; da Silva and Langoni 2009; Henriquez et al. 2009; Yolken et al. 2009; Zhu 2009; Fekadu et al. 2010; Flegr 2010; Lagrue and Poulin 2010).

The life cycle of *Toxoplasma gondii* can be either simple or complex (Fig. 10.3). The parasite matures in intestinal cells of felids, and oocysts are released in the cat's feces where they contaminate the surrounding environment. Oocysts can persist in the soil for more than a year (Frenkel et al. 1995). Oocysts can complete their life cycle by being consumed by other felids, or they can be consumed by any other warm-blooded vertebrate, usually a rodent, which then serves as an intermediate host. In the intermediate host, the parasite reproduces asexually and forms many thin-walled cysts in host cells, especially in the brain where they persist for the remainder of the host's life. The life cycle is completed when an infected intermediate host is consumed by a felid. Humans can acquire the parasite by accidentally ingesting it from the environment (often by coming in contact with cat litter) or by consuming undercooked, infected meat (Beverley 1976). The parasite can also be passed congenitally from mother to offspring in utero, which may in some cases be enough to maintain the parasite in rodent populations even without the proximity of felids (Webster 1994b). Vertical transmission (from mother to offspring) can occur if the mother herself acquires the infection 4–6 months or less before conception or during pregnancy. Congenital toxoplasmosis can be problematic in humans as it can result in premature birth, abortion, neonatal death, fetal abnormalities, deafness, seizures, retinal damage, and/or abnormal brain and nervous system function (Tenter et al. 2000; Jones et al. 2001; Holliman 2003). Symptoms of acute adult-acquired infection are mostly mild, but can include fever, headache, lymphadenopathy, myalgia, and seizures (Carme et al. 2009).

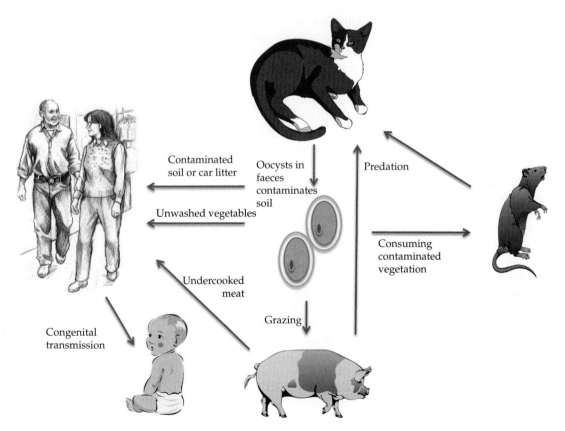

Figure 10.3 The life cycle of *Toxoplasma gondii*, summarizing the various routes of transmission including human infection.

There is substantial evidence detailing the influence of the parasite on the behavior of its rodent intermediate host (reviewed in Berdoy et al. 1995a; Webster 2001; Webster 2007). In most cases, behaviors in infected mice or rats have been found to increase the probability of encounter with cats (see also Chapter 3). Early studies revealed that infected mice and rats had significantly higher activity levels, enhanced exploratory behaviors, or were less neophobic than uninfected counterparts (Hutchinson et al. 1980 a,b; Hay et al. 1983, 1984; Webster 1994a; Webster et al. 1994; Gonzalez et al. 2007), while other parasites infecting the central nervous system caused no such effects (Webster 1994a). Others noticed that only behaviors that would seem to increase the encounter rate with cats were influenced by infection. Berdoy et al. (1995b) found that *T. gondii*-infected rats increased their exploratory behaviors but experienced no change in social status or mating behaviors. Arnott et al. (1990) found that infection resulted in mice being less cautious towards uninfected conspecifics and increased territorial aggression in males. The effects of the parasite were found to be even more specific as *T. gondii* reverses antipredator avoidance behaviors to felids in rodents. While uninfected rodents have an innate aversion to cat odor, infected rats are actually attracted to the smell of cats (Berdoy et al. 2000). Further studies showed that the attraction was cat-specific as mink odor (Lamberton et al. 2008) and dog odor (Kannan et al. 2010) did not induce attraction. In addition, when rats were cured of infection with anti-*Toxoplasma* drugs, the preference for cat odor was lost (Lamberton et al. 2008). Vyas et al. (2007) found that *T. gondii* also influenced learned fear driven by cat odor in rats but not learning associated with other stimuli. Thus, the effect of this parasite is to specifically eliminate one component of the fear response in rats (aversion to cats) and replace it with an attraction response, while apparently leaving all other aspects of fear-related behaviors unchanged (Vyas and Sapolsky 2010). The results above make a strong case for manipulation of the rodent intermediate host (however, at least some of these behavioral changes have been interpreted by some as byproducts of infection (Hrda et al. 2000)).

Because of the considerable effect of the parasite on rodent behavior, its residence in the brain, and its high prevalence in the human population, researchers began questioning whether the parasite also influences human behavior. Individuals exposed prenatally to *Toxoplasma* (and influenza) exhibited impaired reasoning and problem solving abilities in mental tests (Brown et al. 2009). Similar to rodents, infected humans have been found to have longer reaction times than uninfected counterparts, and these differences have been suggested to be part of the manipulative strategy of the parasite (Havlicek et al. 2001). Humans are likely to be a dead-end host for the parasite, but behavioral changes induced by *T. gondii* may have been beneficial to the parasite in ancestral human populations that encountered felid predators. In modern society, the longer reaction times of *T. gondii*-infected individuals appear to lead to higher rates of infected individuals being involved in automobile accidents (Flegr et al. 2002; Klein 2003; Yereli et al. 2006; Kocazeybek et al. 2009). Some behavioral alterations, such as a reduction in novelty seeking, have been suggested to be byproducts of infection in humans because other non-trophically transmitted parasites of the central nervous system (e.g., cytomegalovirus) also induce a similar behavioral change (Novotna et al. 2005).

Additionally, the influence of the parasite on human personality traits appears to be pronounced. Infected men and women have higher levels of apprehension than uninfected individuals (Flegr 2010), though many of the effects seem to be gender dependent. Numerous personality tests comparing *T. gondii*-infected and uninfected individuals have revealed infected men have lower superego strength, higher vigilance, are more likely to disregard rules, are more expedient, suspicious, jealous, and dogmatic while women have higher warmth and higher superego strength resulting in infected women being more warm-hearted, outgoing, conscientious, persistent, and moralistic than uninfected women (summarized in Flegr 2010). In addition, infected men scored significantly lower than uninfected men in "self control" and "clothes tidiness," while no such effect was found in women. Interestingly, the effects of latent *Toxoplasma* infection seem to disproportionately affect human behavioral traits dealing with how individuals interact with others in society (Lindova et al. 2006).

The apparent substantial influence of *T. gondii* on human personality has caused some to examine the potential for the parasite to influence human culture (Lafferty 2005). Lafferty (2006) compared seroprevalence of *T. gondii* in different nations to aggregate personality traits. He found a significant positive correlation between seroposivity to *T. gondii* globally and aggregate neuroticism, suggesting that high prevalence of latent *T. gondii* infection in some countries leads to greater overall anxiousness and obsessiveness. In western nations, he found positive correlations between parasite prevalence and uncertainty avoidance and masculine sex roles. Cultures with high uncertainty avoidance may be more likely to try to avoid uncertain situations by developing more rule-oriented societies. Similarly, cultures rating higher in masculine sex roles are generally more likely to have more gender-specific roles in work and family. These cultures would likely be more ego-driven and materialistic and focus more on individual achievement than on interpersonal relationships and quality of life (Lafferty 2006).

In addition to personality changes in individuals infected with *T. gondii*, there is increasing evidence from many parts of the world that the parasite may be associated with a variety of behavioral and psychiatric disorders. Alvarado-Esquivel et al. (2006) found that psychiatric patients in general are more likely to test positive for *T. gondii* IgG antibodies compared to controls. *Toxoplasma* infection has been implicated in specific cases of dementia (Freidel et al. 2002; Habek et al. 2009), bipolar disorder (Hinze-Selch 2002), obsessive-compulsive disorder (Miman et al. 2010b), autism spectrum disorder, and, possibly, attention-deficit/hyperactivity disorder and cryptogenic epilepsy (Prandota 2010). An examination of patients with intellectual disability (mental retardation) found no association between latent toxoplasmosis and intellectual disability (Sharif et al. 2007), although intellectual disabilities are common in patients with congenital infections. *Toxoplasma* also appears to be positively associated with attempting suicide as psychiatric patients with

clinical depression that had a history of suicide attempts were significantly more likely to be infected with *T. gondii* than similar patients without suicidal tendencies or control individuals without a history of depression (Arling et al. 2009; Yagmur et al. 2010). Recent studies have also suggested a link between the parasite and Parkinson's disease (Celik et al. 2010; Miman et al. 2010a) and also brain cancer (Thomas et al. 2011).

By far the best studied relationship between *Toxoplasma* infection and human psychiatric disorder is the link between infection by the parasite and schizophrenia. Numerous studies (over 40) have demonstrated that schizophrenia patients are more likely than others to be seropositive for anti-*Toxoplasma* antibodies (for reviews see Ledgerwood et al. 2003; Torrey and Yolken 2007; Torrey et al. 2007; da Silva and Langoni 2009; Henriquez et al. 2009; Yolken et al. 2009; Zhu 2009). Individuals with first-episode schizophrenia have been found to have significantly elevated levels of anti-*T. gondii* antibodies in serum and cerebro spinal fluid (Yolken et al. 2001). Increased levels of anti-*T. gondii* antibodies have been observed up to three years prior to the onset of schizophrenia, with a peak of six months suggesting a causative relationship (Niebuhr et al. 2008). Many of these cases appear to be related to congenital toxoplasmosis (Meyer and Feldon 2009) as women with elevated *Toxoplasma* IgG antibody concentrations have been found to be significantly more likely to have children that eventually develop schizophrenia (Brown et al. 2005; Mortensen et al. 2007), with the onset typically occurring sometime after puberty (Kinney et al. 2010). In one case, women seropositive for *T. gondii* were 2.5 times more likely than other women to have children that would eventually develop schizophrenia (Brown et al. 2005). However, in a smaller study, no such relationship was found (Buka et al. 2001), and post-mortem analyses have failed to find greater levels of *T. gondii* in orbital frontal brains of schizophrenia patients (though the parasites may have been present at undetectable levels) (Conejero-Goldberg et al. 2003). Also, in China *T. gondii* was not found to be associated with schizophrenia (Xiao et al. 2010). However, meta-analyses of studies relating the parasite to the disorder revealed positive associations in China (Zhu et al. 2007) and elsewhere (Torrey et al. 2007). Other evidence for the association between *T. gondii* and schizophrenia include the facts that acute *T. gondii* infection often results in symptoms similar to schizophrenia, and exposure to cats in childhood has long been considered a risk factor for the development of schizophrenia (Torrey and Yolken 2003). Interestingly, among schizophrenia patients, seropositivity to *Toxoplasma* was correlated with a higher mortality rate due to natural causes (Dickerson et al. 2007). Several medications used to treat schizophrenia inhibit the in vitro replication of *T. gondii*. Thus, the mechanism of action of these drugs may, in part, be due to their ability to harm the parasite (Jones-Brando et al. 2003). Schizophrenia patients with increased levels of IgG antibody to *T. gondii* that were treated with antipsychotic medications eventually developed antibody levels similar to controls, suggesting that the drugs may have inhibited the parasite (Leweke et al. 2004).

In addition to effects on behavior and personality in humans, *T. gondii* has also been suggested to hasten the onset of puberty (Setian et al. 2002), increase the probability of Down's Syndrome in offspring of infected mothers (Hostomska et al. 1957), influence the human sex ratio (James 2010), and explain the enigmatic Rh factor polymorphism in humans (Flegr et al. 2010). The initial effect of the parasite in mothers is to skew the sex ratio toward males, and this effect has been observed in both rodents and humans. It has been suggested that a male-biased sex ratio in mice may be adaptive for the parasite because males are more migratory than females (Kankova et al. 2007b). In any event, this effect has been evident in humans, as the sex ratio produced by women seems to be correlated with the concentration of anti-*T. gondii* antibodies, with antibody levels being positively correlated with more male-biased sex ratios (James 2010). In fact, women with the highest concentrations of anti-*T. gondii* antibodies during early phase infection, had male offspring about 72% of the time (Kankova et al. 2007a). Interestingly, later phases of infection resulted in a female bias among offspring (Kankova et al. 2007a).

The existence of the polymorphism in Rh factor in human blood has been an evolutionary puzzle.

When RhD-negative women have offspring with an RhD-positive man, the offspring is at a higher risk of fetal or newborn death from hemolytic disease. Thus, in the absence of other factors, the rarer of the two alleles should be selected against in a population. However, it has been found that RhD-positive males have faster reaction times than RhD-negative males, possibly leading toward selection for the positive allele. In addition, when infected by *T. gondii*, men heterozygous for RhD did not experience the prolongation of reaction times typical of individuals with *T. gondii* infection (Novotna et al. 2008). Another study found similar results in females (Flegr et al. 2008b). Thus, balancing selection through heterozygote advantage mediated by *T. gondii* may be an important part of the explanation for the Rh factor polymorphism in human populations.

The mechanism of behavioral alteration in rodents and humans is not entirely clear (for reviews, see Henriquez et al. 2009; Vyas and Sapolsky 2010; Webster and McConkey 2010; Fekadu et al. 2010). In rodents, the behavioral alterations appear to be tied to the localization of infection in the brain, as some areas of the brain appear to accumulate disproportionate numbers of cysts. These include the olfactory bulbs, amygdala, and the nucleus accumbens. The specificity of the behavioral change in rats, as previously outlined with respect to cat odor, suggests fine manipulation of the olfactory center (Lamberton et al. 2008; Webster and McConkey 2010). The concentration of cysts in the amygdala may likely influence the fear response (Webster and McConkey 2010). High concentrations of cysts may affect the nucleus accumbens, which is thought to influence reward, pleasure, and fear (Schwienbacher et al. 2004). Interestingly, there is a link between function of the nucleus accumbens and the aetiology of schizophrenia (Gray et al. 1991). The influence of the parasite in these areas is most likely due to alterations of neurotransmission or neuromodulatory pathways (Berdoy et al. 2000; Webster et al. 2006). At least some of the behavioral effects of *T. gondii* in rats appear to be related to elevated dopamine levels (Skallova et al. 2006). Webster et al. (2006) demonstrated that a treatment of *T. gondii*-infected rats with the dopamine antagonist,

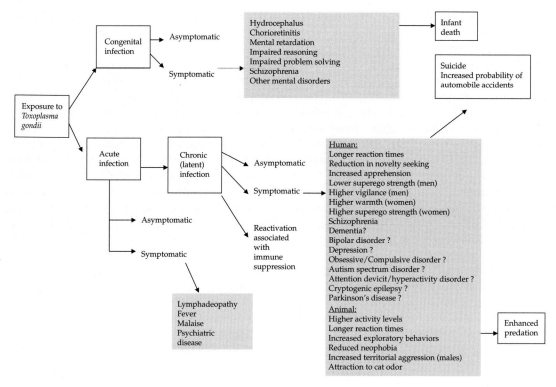

Figure 10.4 The effects of different types of *Toxoplasma gondii* infection on its host, with symptoms summarized in the shaded boxes (modified from Holliman 1997).

haloperidol, resulted in a reduction in *T. gondii*-related behavioral alterations. Further, some evidence suggests that *T. gondii* may be able to synthesize dopamine (Gaskell et al. 2009). Disruptions in dopamine levels have also been found in human *T. gondii* infection (Flegr et al. 2003) and have been associated with changes in personality (Flegr et al. 2003; Skallova et al. 2005). Imbalances in dopamine are also characteristic of schizophrenia patients (Torrey and Yolken 2003). This relationship has caused some to propose that elevated dopamine induced by *T. gondii* may increase the probability of developing schizophrenia (reviewed in Yolken et al. 2009). This has resulted in a line of research to investigate the possibility that anti-*T. gondii* drugs may be effective in the treatment of schizophrenia (Yolken et al. 2009). Additionally, some of the behavioral changes may have to do with altered testosterone levels in humans. *Toxoplasma gondii*-infected

males have greater testosterone levels than uninfected males while infected females possess lower levels (Flegr et al. 2005; Flegr et al. 2008a). The cause and effect in this system is not clear. Elevated testosterone results in impaired immune function, which could result in higher rates of *T. gondii* infection, or the parasite could induce higher testosterone to impair the immune system in order to allow the infection to persist (Flegr 2007). In addition, recent studies have demonstrated that the behavioral and psychiatric effects of the parasite may be strain-dependent in both rodents (Kannan et al. 2010) and humans (Xiao et al. 2009).

It seems clear from this brief review that the impacts of *T. gondii* on human behavior are both extensive and complex. In one of the first reviews on the topic, Holliman (1997) organized the effects of *T. gondii* based on the sequence of the disease. In Fig. 10.4, we attempt to bring his synthesis up to date. Congenital and acute *T. gondii* infections have been well studied for their clinical manifestations. It appears that it is time to stop assuming that latent infections are "asymptomatic," and make some difficult decisions about how to deal with these cases. While most will agree that the effects of latent infections by the parasite on human males are negative, many would consider some of the effects on females to be positive. Should attempts be made to alleviate these effects? In an infected person, where does human personality stop and *Toxoplasma*'s effects begin? To what degree can these effects be alleviated? What proportion of latent infections are symptomatic in some way? These important questions remain without satisfactory answers.

10.6 Conclusion

Far from being merely a natural history curiosity, host manipulation by parasites directly or indirectly affects several areas of conservation, agricultural, or medical importance. To date, there have been very few empirical studies addressing host manipulation specifically in a veterinary or medical context. However, as this chapter makes clear, there is a pressing need to understand the selective pressures under which this trait has evolved and the proximate conditions under which it is expressed if we are to successfully deal with its consequences. For instance, in the case of *T. gondii* and its influence on human personality, we must not think of maladaptation; instead, we need to consider what physiological or biochemical alterations in the usual intermediate host, a rat, have been selected over the course of evolutionary time to understand what mechanisms may operate during accidental human infection and how to mitigate their cascading effects. The manipulation of host behavior by parasites is yet one more reason to endorse a Darwinian approach to medicine, and to seek remedies informed by a solid understanding of host–parasite coevolution (Stearns and Koella 2008).

In addition to their direct impact on people, manipulative parasites can clearly influence the rate at which invasive species expand their range, the productivity of farmed animals, and the trans-mission rates of vector-borne diseases. These are key parameters in models of biological invasions, agricultural economics, and disease epidemiology, respectively. Yet we generally lack quantitative estimates of the precise effect of manipulative parasites on these key parameters, and it is therefore difficult to integrate host manipulation as a factor in the design of strategies to control invasions or diseases, or to improve productivity. Acknowledging that manipulative parasites have some influence is only the first step; the next challenge will be to measure how much influence they do have, and take this into account during strategic planning in conservation biology, veterinary science, and disease epidemiology.

Acknowledgements

We thank Maureen Levri for valuable comments on an earlier draft, and the editors for inviting us to contribute to this book.

References

Alvarado-Esquivel, C., Alanis-Quinones, O., Arreola-Valenzuala, M., Rodriguez-Briones, A., Piedra-Nevarez, L., Duran-Morales, E., Estrada-Martinez, S., Martinez-Garcia, S., and Liesenfeld, O. (2006). Seroepidemiology of *Toxoplasma gondii* infection in psychiatric inpatients in a northern Mexican city. *BMC Infectious Diseases* **6**, 178–184.

Arling, T. A., Yolken, R. H., Lapidus, M., Langenberg, P., Dickerson, F. B., Zimmerman, B. S., Balis, T., Cabassa, J. A., Scrandis, D. A., Tonelli, L. H., and Postolache, T. T. (2009). *Toxoplasma gondii* antibody titers and history of suicide attempts in patients with recurrent mood disorders. *Journal of Nervous and Mental Disease* **197**, 905–908.

Arnott, M. A., Cassella, J. P., Aitken, P. P., and Hay, J. (1990). Social interactions of mice with congenital *Toxoplasma* infection. *Annals of Tropical Medicine and Parasitology* **84**, 149–156.

Bauer, A., Trouvé, S., Grégoire, A., Bollache, L., and Cézilly, F. (2000). Differential influence of *Pomphorhynchus laevis* (Acanthocephala) on the behaviour of native and invader gammarid species. *International Journal for Parasitology* **30**, 1453–1457.

Baylis, M. and Mbwabi, A. L. (1995). Feeding behaviour of tsetse flies (*Glossina pallidipes* Austen) on *Trypanosoma*-infected oxen in Kenya. *Parasitology* **110**, 297–305.

Baylis, M. and Nambiro, C. O. (1993). The effect of cattle infection by *Trypanosoma congolense* on the attraction, and feeding success, of the tsetse fly *Glossina pallidipes*. *Parasitology* **106**, 357–361.

Beach, R., Kiilu, G., and Leeuwenburg, J. (1985). Modification of sand fly biting behavior by *Leishmania* leads to increased parasite transmission. *American Journal of Tropical Medicine and Hygiene* **34**, 278–282.

Berdoy, M., Webster, J. P., and MacDonald, D. W. (1995a). The manipulation of rate behaviour by *Toxoplasma gondii*. *Mammalia* **59**, 605–613.

Berdoy, M., Webster, J. P., and MacDonald, D. W. (1995b). Parasite-altered behaviour: is the effect of *Toxoplasma gondii* on *Rattus norvegicus* specific? *Parasitology* **111**, 403–409.

Berdoy, M., Webster, J. P., and Macdonald, D. W. (2000). Fatal attraction in rats infected with *Toxoplasma gondii*. *Proceedings of the Royal Society of London B* **267**, 1591–1594.

Beverly, J. K. A. (1976). Toxoplasmosis in animals. *Veterinary Record* **99**, 123–127.

Brown, A. S., Schaefer, C. A., Quesenberry, C. P., Liu, L. Y., Babulas, V. P., and Susser, E. S. (2005). Maternal exposure to toxoplasmosis and risk of schizophrenia in adult offspring. *American Journal of Psychiatry* **162**, 767–773.

Brown, A. S., Vinogradov, S., Kremen, W. S., Poole, J. H., Deicken, R. F., Penner, J. D., McKeague, I. W., Kochetkova, A., Kern, D., and Schaefer, C. A. (2009). Prenatal exposure to maternal infection and executive dysfunction in adult schizophrenia. *American Journal of Psychiatry* **166**, 683–690.

Buka, S. L., Tsuang, M. T., Torrey, E. F., Klebanoff, M. A., Bernstein, D., and Yolken R. H. (2001). Maternal infections and subsequent psychosis among offspring. *Archives of General Psychiatry* **58**, 1032–1037.

Burri, C. and Brun, R. (2003). Human African Trypanosomiasis. In *Manson's Tropical Diseases*, 21st edn. (eds G. C. Cook and A. Zumla), pp. 1303–1324. Saunders, London.

Carme, B., Demar, M., Ajzenberg, D., and Darde, M. L. (2009). Severe acquired toxoplasmosis caused by wild cycle of *Toxoplasma gondii*, French Guiana. *Emerging Infectious Diseases* **15**, 656–658.

Carney, W. P. (1969). Behavioral and morphological changes in carpenter ants harboring dicrocoelid metacercariae. *American Midland Naturalist* **82**, 605–611.

Celik, T., Kamisli, O., Babur, C., Cevik, M. O., Oztuna, D., and Altinayar, S. (2010). Is there a relationship between *Toxoplasma gondii* infection and idiopathic Parkinson's disease? *Scandinavian Journal of Infectious Disease* **42**, 604–608.

Cohen, A. N. and Carlton, J. T. (1998). Accelerating invasion rate in a highly invaded estuary. *Science* **279**, 555–558.

Colautti, R. I., Ricciardi, A., Grigorovich, I. A., and MacIsaac, H. J. (2004). Is invasion success explained by the enemy release hypothesis? *Ecology Letters* **7**, 721–733.

Conejero-Goldberg, C., Torrey, E. F., and Yolken, R. H. (2003). Herpesviruses and *Toxoplasma gondii* in orbital frontal cortex of psychiatric patients. *Schizophrenia Research* **60**, 65–69.

Crisler, L. (1956). Observations of wolves hunting caribou. *Journal of Mammalogy* **37**, 337–346.

Crowden, A. E. and Broom, D. M. (1980). Effects of the eye fluke, *Diplostomum spathaceum*, on the behaviour of dace (*Leusiscus leusiscus*). *Animal Behaviour* **28**, 287–294.

da Silva, R. C. and Langoni, H. (2009). *Toxoplasma gondii*: host-parasite interaction and behavior manipulation. *Parasitology Research* **105**, 893–898.

Daszak, P., Cunningham, A. A., and Hyatt, A. D. (2000). Emerging infectious diseases of wildlife: threats to biodiversity and human health. *Science* **287**, 443–449.

Dedet, J. P. and Pratlong, F. (2003). Leishmaniasis. In: *Manson's Tropical Diseases*, 21st edn (eds Cook, G. C., and Zumla, A.) pp. 1339–1364. Saunders, London.

Dick, J. T. A., Montgomery, W. I., and Elwood, R. W. (1993). Replacement of the indigenous amphipod *Gammarus duebeni celticus* by the introduced *G. pulex*: differential cannibalism and mutual predation. *Journal of Animal Ecology* **62**, 79–88.

Dickerson, F., Boronow, J., Stallings, C., Origoni, A., and Yolken R. (2007). *Toxoplasma gondii* in individuals with schizophrenia: Association with clinical and demographic factors and with mortality. *Schizophrenia Bulletin* **33**, 737–740.

Dobson, A. P. (1988). The population biology of parasite-induced changes in host behavior. *Quarterly Review of Biology* **63**, 139–165.

Dubey, J. (1998). Advances in the life cycle of *Toxoplasma gondii*. *International Journal for Parasitology* **28**, 1019–1024.

Dunn, A. M. (2009). Parasites and biological invasions. *Advances in Parasitology* **68**, 161–184.

Edgerton, B. F., Evans, L. H., Stephens, F. J., and Overstreet, R. M. (2002). Synopsis of freshwater crayfish diseases and commensal organisms. *Aquaculture* **206**, 57–135.

Fekadu, A., Shibre, T., and Cleare, A. J. (2010). Toxoplasmosis as a cause for behaviour disorders—overview of evidence and mechanisms? *Folia Parasitologica* **57**, 105–113.

Ferguson, H. M. and Read, A. F. (2002). Why is the effect of malaria parasites on mosquito survival still unresolved? *Trends in Parasitology* **18**, 256–261.

Fiorello, C. V., Deem, S. L., Gompper, M. E., and Dubovi, E. J. (2004). Seroprevalence of pathogens in domestic carnivores on the border of Madidi National Park, Bolivia. *Animal Conservation* **7**, 45–54.

Flegr, J. (2007). Effects of *Toxoplasma* on human behavior. *Schizophrenia Bulletin* **33**, 757–760.

Flegr, J. (2010). Influence of latent toxoplasmosis on the phenotype of intermediate hosts. *Folia Parasitologica* **57**, 81–87.

Flegr, J., Havlicek, J., Kodym, P., Maly, M., and Smahel, Z. (2002). Increased risk of traffic accidents in subjects with latent toxoplasmosis: a retrospective case-control study. *BMC Infectious Diseases* **2**, 11.

Flegr, J., Hruskova, M., Hodny, Z., Novotna, M., and Hanusova J. (2005). Body height, body mass index, waist-hip ratio, fluctuating asymmetry and second to fourth digit ratio in subjects with latent toxoplasmosis. *Parasitology* **130**, 621–628.

Flegr, J., Lindova, J., and Kodym, P. (2008a). Sex-dependent toxoplasmosis-associated differences in testosterone concentration in humans. *Parasitology* **135**, 427–431.

Flegr, J., Novotna, M., Lindova, J., and Havlicek, J. (2008b). Neurophysiological effect of the Rh factor: Protective role of the RhD molecule against *Toxoplasma*-induced impairment of reaction times in women. *Neuroendocrinology Letters* **29**, 475–481.

Flegr, J., Novotna, M., Fialova, A., Kolbekova, P., and Gasova, Z. (2010). The influence of RhD phenotype and toxoplasmosis- and age-associated changes in personality profile of blood disorders. *Folia Parasitologica* **57**, 143–150.

Flegr, J., Preiss, M., Klose, J, Havlicek, J., Vitakova, M., and Kodym, P. (2003). Decreased level of psychobiological factor novelty seeking and lower intelligence in men latently infected with the protozoan parasite *Toxoplasma gondii*: Dopamine, a missing link between schizophrenia and toxoplasmosis? *Biological Psychology* **63**, 253–268.

Freidel, S., Martin-Solch, C., and Schreiter-Gasser, U. (2002). Alzheimer's dementia or cerebral toxoplasmosis? Case study of dementia following toxoplasmosis infection. *Nervenarzt* **73**, 874–878.

Frenkel, J. K., Hassanein, K. M., Hassanein, R. S., Brown, E., Thulliez, P., and Quinteronunez, R. (1995). Transmission of *Toxoplasma gondii* in Panama-City, Panama. *American Journal of Tropical Medicine and Hygiene* **53**, 458–468.

Gaskell, E. A., Smith, J. E., Pinney, J. W., Westhead, D. R., and McConkey, G. A. (2009). A unique dual activity amino acid hydroxylase in *Toxoplasma gondii*. *PLoS One* **4**, e4801.

Gaydos, J. K., Conrad, P. A., Gilardi, K. V. K., Blundell, G. M., and Ben-David, M. (2007). Does human proximity affect antibody prevalence in marine-foraging river otters (*Lontra canadensis*)? *Journal of Wildlife Diseases* **43**, 116–123.

Gonzalez, L. E., Rojnik, B., Urrea, F., Urdaneta, H., Petrosino, P., Colasante, C., Pino, S., and Hernandez, L. (2007). *Toxoplasma gondii* infection lower anxiety as measured in the plus-maze and social interaction tests in rats—A behavioral analysis. *Behavioural Brain Research* **177**, 70–79.

Gray, J. A., Feldon, J., Rawlins, J. N. P., Hemsley, D. R., and Smith, A. D. (1991). The neuropsychology of schizophrenia. *Behavior and Brain Science* **14**, 1–84.

Habek, M., Ozretic, D., Zarkovic, K., Djakovic, V., and Mubrin, Z. (2009). Unusual cause of dementia in an immunocompetent host: toxoplasmic encephalitis. *Neurological Sciences* **30**, 45–49.

Hamilton, J. G. C. and Hurd, H. (2002). Parasite manipulation of vector behaviour. In: *The Behavioural Ecology of Parasites* (eds Lewis, E. E., Campbell, J. F., and Sukhdeo, M. V. K.) pp. 259–281. CABI Publishing, Wallingford, UK.

Havlicek, J., Gasova, Z., Smith, A. P., Zvara, K., and Flegr, J. (2001). Decrease of psychomotor performance in subjects with latent "asymptomatic" toxoplasmosis. *Parasitology* **122**, 515–520.

Hay, J., Aitken, P. P., Hair, D. M., Hutchison, W. M., and Graham, D. I. (1984). The effect of congenital *Toxoplasma* infection on mouse activity and relative preference for exposed areas over a series of trials. *Annals of Tropical Medicine and Parasitology* **78**, 611–618.

Hay, J., Aitken, P. P., Hutchison, W. M., and Graham, D. I. (1983). The effect of congenital and adult-acquired *Toxoplasma* infections on activity and responsiveness to novel stimulation in mice. *Annals of Tropical Medicine and Parasitology* **77**, 483–495.

Haydon, D. T., Laurenson, M. K., and Sillero-Zubiri, C. (2002). Integrating epidemiology into population viability analysis: managing the risk posed by rabies and canine distemper to the Ethiopian wolf. *Conservation Biology* **16**, 1372–1385.

Henriquez, S. A., Brett, R., Alexander, J., Pratt, J., and Roberts, C. W. (2009). Neuropsychiatric Disease and *Toxoplasma gondii* infection. *Neuroimmunomodulation* **16**, 122–133.

Hinze-Selch, D. (2002). Infection, treatment and immune response in patients with bipolar disorder versus patients with major depression, schizophrenia or healthy controls. *Bipolar Disorders* **4**(Suppl. 1), 81–83.

Holliman, R. E. (1997). Toxoplasmosis, behaviour and personality. *Journal of Infection* **35**, 105–110.

Holliman, R. E. (2003). Toxoplasmosis. In: *Manson's Tropical Diseases*, 21st edn (eds Cook, G. C., and Zumla, A.), pp. 1365–1371. Saunders, London.

Hostomska, L., Jirovec, O., Horackova, M., and Hrubcova, M. (1957). The role of toxoplasmosis in the mother in development of mongolism in the child. *Cesko-Slovenska Pediatrie* 12, 713–723.

Hrda, S., Votypka, J., Kodym, P., and Flegr, J. (2000). Transient nature of *Toxoplasma gondii*-induced behavioral changes in mice. *Journal of Parasitology* 86, 657–663.

Hudson, D. A. and Lester, R. J G. (1994). Parasites and symbionts of wild mud crabs *Scylla serrata* (Forskal) of potential significance in aquaculture. *Aquaculture* 120, 183–199.

Hulme, P. E. (2006). Beyond control: wider implications for the management of biological invasions. *Journal of Applied Ecology* 43, 835–847.

Hurd, H. (2003). Manipulation of medically important insect vectors by their parasites. *Annual Review of Entomology* 48, 141–161.

Hutchison, W. M., Aitken, P. P., and Wells, B. W. P. (1980a). Chronic *Toxoplasma* infections and familiarity-novelty discrimination in the mouse. *Annals of Tropical Medicine and Parasitology* 74, 145–150.

Hutchison, W. M., Aitken, P. P., and Wells, B. W. P. (1980b). Chronic *Toxoplasma* infections and motor-performance in the mouse. *Annals of Tropical Medicine and Parasitology* 74, 507–510.

James, W. H. (2010). Potential solutions to problems posed by the offspring sex ratios of people with parasitic and viral infections. *Folia Parasitologica* 57, 114–120.

Jenkins, D. J., Romig, T., and Thompson, R. C. A. (2005). Emergence/re-emergence of *Echinococcus* spp., a global update. *International Journal for Parasitology* 35, 1205–1219.

Jenni, L., Molyneux, D. H., Livesey, J. L., and Galun, R. (1980). Feeding behaviour of tsetse flies infected with salivarian trypanosomes. *Nature* 283, 383–385.

Johnson, C. K., Tinker, M. T., Estes, J. A., Conrad, P. A., Staedler, M., Miller, M. A., Jessup, D. A., and Mazet, J. A. K. (2009). Prey choice and habitat use drive sea otter pathogen exposure in a resource-limited coastal system. *Proceedings of the National Academy of Sciences of the USA* 106, 2242–2247.

Joly, D. O. and Messier, F. (2004). The distribution of *Echinococcus granulosus* in moose: evidence for parasite-induced vulnerability to predation by wolves. *Oecologia* 140, 586–590.

Jones, J. L., Lopez, A., Wilson, M., Schulkin, J., and Gibbs, R. (2001). Congenital toxoplasmosis: a review. *Obstetrical and Gynecological Survey* 56, 296–305.

Jones-Brando, L., Torrey, E. F., and Yolken, R. (2003). Drugs used in the treatment of schizophrenia and bipolar disorder inhibit the replication of *Toxoplasma gondii*. *Schizophrenia Research* 62, 237–244.

Kankova, S., Kodym, P., Frynta, D., Vavrinova, R., Kubena, A., and Flegr, J. (2007a). Influence of latent toxoplasmosis on the secondary sex ratio in mice. *Parasitology* 134, 1709–1717.

Kankova, S., Sule, J., Nouzova, K., Fajfrlik, K., Frynta, D., and Flegr, J. (2007b). Women infected with the parasite *Toxoplasma* have more sons. *Naturwissenschaften* 94, 122–127.

Kannan, G., Moldovan, K., Xiao, J. C., Yolken, R. H., Jones-Brando, L., and Pletnikov, M. V. (2010). *Toxoplasma gondii* strain-dependent effects on mouse behaviour. *Folia Parasitologica* 57, 151–155.

Kaplan, R. M. (2004). Drug resistance in nematodes of veterinary importance: a status report. *Trends in Parasitology* 20, 477–481.

Karvonen, A., Seppälä, O., and Valtonen, E. T. (2004). Eye fluke-induced cataract formation in fish: quantitative analysis using an ophthalmological microscope. *Parasitology* 129, 473–478.

Kennedy, C. R. (2006). *Ecology of the Acanthocephala*. Cambridge University Press, Cambridge.

Killick-Kendrick, R., Leaney, A. J., Ready, P. D., and Molyneux, D. H. (1977). *Leishmania* in phlebotomid sand flies. IV. The transmission of *Leishmania mexicana amazonensis* to hamsters by the bite of experimentally infected *Lutzomyia longipalpis*. *Proceedings of the Royal Society of London B* 196, 105–115.

Kinney, D. K., Hintz, K., Shearer, E. M., Barch, D. H., Riffin, C., Whitley, K., and Butler, R. (2010). A unifying hypothesis of schizophrenia: Abnormal immune system development may help explain roles of prenatal hazards, post-pubertal onset, stress, genes, climate, infections, and brain dysfunction. *Medical Hypotheses* 74, 555–563.

Klein, S. L. (2003). Parasite manipulation of the proximate mechanisms that mediate social behavior in vertebrates. *Physiology & Behavior* 79, 441–449.

Kocazeybek, B., Oner, Y. A., Turksoy, R., Babur, C., Cakan, H., Sahip, N., Unal, A., Ozaslan, A., Kilic, S., Saribas, S., Aslan, M., Taylan, A., Koc, S., Dirican, A., Uner, H. B., Oz, V., Ertekin, C., Kucukbasmaci, O., and Torun, M. M. (2009). Higher prevalence of toxoplasmosis in victims of traffic accidents suggest increased risk of traffic accident in *Toxoplasma*-infected inhabitants of Istanbul and its suburbs. *Forensic Science International* 187, 103–108.

Koella, J. (1999). An evolutionary view of the interactions between anopheline mosquitoes and malarial parasites. *Microbes and Infection* 1, 303–308.

Koella, J. and Packer, M. J. (1996). Malaria parasites enhance blood-feeding of their naturally infected vector *Anopheles punctulatus*. *Parasitology* **113**, 105–109.

Koella, J., Sorensen, F. L., and Anderson, R. A. (1998). The malaria parasite, *Plasmodium falciparum*, increases the frequency of multiple feeding of its mosquito vector, *Anopheles gambiae*. *Proceedings of the Royal Society of London B* **265**, 763–768.

Koella, J. C., Rieu, L., and Paul, R. E. L. (2002). Stage-specific manipulation of a mosquito's host-seeking behavior by the malaria parasite *Plasmosdium gallinaceum*. *Behavioral Ecology* **13**, 816–820.

Lacroix, R., Mukabana, W. R., Gouagna, L. C., and Koella, J. C. (2005). Malaria infection increases attractiveness of humans to mosquitoes. *PLoS Biology* **3**, 1590–1593.

Lafferty, K. D. (2005). Look what the cat dragged in: do parasites contribute to human cultural diversity? *Behavioural Processes* **68**, 279–282.

Lafferty, K. D. (2006). Can the common brain parasite, *Toxoplasma gondii*, influence human culture? *Proceedings of the Royal Society of London B* **273**, 2749–2755.

Lafferty, K. D. and Gerber, L. R. (2002). Good medicine for conservation biology: the intersection of epidemiology and conservation theory. *Conservation Biology* **16**, 593–604.

Lagrue, C., Kaldonski, N., Perrot-Minnot, M. J., Motreuil, S., and Bollache, L. (2007). Modification of hosts' behavior by a parasite: field evidence for adaptive manipulation. *Ecology* **88**, 2839–2847.

Lagrue, C. and Poulin, R. (2010). Manipulative parasites in the world of veterinary science: implications for epidemiology and pathology. *Veterinary Journal* **184**, 9–13.

Lamberton, P. H. L., Donnelly, C. A., and Webster, J. P. (2008). Specificity of the *Toxoplasma gondii*-altered behaviour to definitive versus non-definitive host predation risk. *Parasitology* **135**, 1143–1150.

Lebarbenchon, C., Poulin, R., and Thomas, F. (2009). Parasitism, biodiversity, and conservation biology. In: *Ecology and Evolution of Parasitism* (eds Thomas, F., Renaud, F., and Guégan, J.-F.) pp. 149–160. Oxford University Press, Oxford.

Ledgerwood, L. G., Ewald, P. W., and Cochran, G. M. (2003). Genes, germs, and schizophrenia—an evolutionary perspective. *Perspectives in Biology and Medicine* **46**, 317–348.

Lefèvre, T. and Thomas, F. (2008). Behind the scene, something else is pulling the strings: Emphasizing parasitic manipulation in vector-borne diseases. *Infection, Genetics and Evolution* **8**, 504–519.

Lefèvre, T., Koella, J., Renaud, F., Hurd, H., Biron, D. G., and Thomas, F. (2006). New prospects for research on manipulation of insect vectors by pathogens. *PLoS Pathogens* **2**, 633–635.

Lefèvre, T., Thomas, F., Ravel, S., Patrel, D., Renault, L., Le Bourligu, L., Cuny, G., and Biron, D. G. (2007a). *Trypanosoma brucei brucei* induces alteration in the head proteome of the tsetse fly vector *Glossina palpalis gambiensis*. *Insect Molecular Biology* **16**, 651–660.

Lefèvre, T., Thomas, F., Schwartz, A., Levashina, E., Blandin, S., Brizard, J-P., Le Bourligu, L., Demettre, E., Renaud, F., and Biron, D. G. (2007b). Malaria *Plasmodium* agent induces alteration in the head proteome of their *Anopheles* mosquito host. *Proteomics* **7**, 1908–1915.

Lefèvre, T., Roche, B., Poulin, R., Hurd, H., Renaud, F., and Thomas, F. (2008). Exploiting host compensatory responses: the 'must' of manipulation? *Trends in Parasitology* **24**(10), 435–439.

Lefèvre, T., Lebarbenchon, C., Gauthier-Clerc, M., Missé, D., Poulin, R., and Thomas, F. (2009). The ecological significance of manipulative parasites. *Trends in Ecology and Evolution* **24**, 41–48.

Leweke, F. M., Gerth, C. W., Koethe, D., Klosterkotter, J., Ruslanova, I., Krivogorsky, B., Torrey, E. F., and Yolken, R. H. (2004). Antibodies to infectious agents in individuals with recent onset schizophrenia. *European Archives of Psychiatry and Clinical Neurosciences* **254**, 4–8.

Lindova, J., Novotna, M., Havlicek, J., Jozifkova, E., Skallova, A., Kolbekova, P., Hodny, Z., Kodym, P., and Flegr, J. (2006). Gender differences in behavioural changes induced by latent toxoplasmosis. *International Journal for Parasitology* **36**, 1485–1492.

Locke, S. A., McLaughlin, J. D., Dayanandan, S., and Marcogliese, D. J. (2010). Diversity and specificity in *Diplostomum* spp. metacercariae in freshwater fishes revealed by cytochrome *c* oxidase I and internal transcribed spacer sequences. *International Journal for Parasitology* **40**, 333–343.

Lymbery, A. J. and Thompson, R. C. A. (1996). Species of *Echinococcus*: pattern and process. *Parasitology Today* **12**, 486–491.

MacNeil, C., Fielding, N. J., Dick, J. T. A., Briffa, M., Prenter, J., Hatcher, M. J., and Dunn, A. M. (2003a). An acanthocephalan parasite mediates intraguild predation between invasive and native freshwater amphipods (Crustacea). *Freshwater Biology* **48**, 2085–2093.

MacNeil, C., Dick, J. T. A., Hatcher, M. J., and Dunn, A. M. (2003b). Differential drift and parasitism in invading and native *Gammarus* spp. (Crustacea: Amphipoda). *Ecography* **26**, 467–473.

Makumi, J. N. and Moloo, S. K. (1991). *Trypanosoma vivax* in *Glossina palpalis gambiensis* do not appear to affect feeding behaviour, longevity, or reproductive performance of the vector. *Medical and Veterinary Entomology* **5**, 35–42.

Messier, F., Rau, M. E., and McNeill, M. A. (1989). *Echinococcus granulosus* (Cestoda: Taeniidae) infections and moose-wolf population dynamics in southwestern Quebec. *Canadian Journal of Zoology* **67**, 216–219.

Meyer, U. and Feldon, J. (2009). Prenatal exposure to infection: a primary mechanism for abnormal dopaminergic development in schizophrenia. *Psychopharmacology* **206**, 587–602.

Miller, M. A., Miller, W. A., Conrad, P. A., James, E. R., Melli, A. C., Leutenegger, C. M., Dabritz, H. A., Packham, A. E., Paradies, D., Harris, M., Ames, J., Jessup, D. A., Worcester, K., and Grigg, M. E. (2008). Type X *Toxoplasma gondii* in a wild mussel and terrestrial carnivores from coastal California: new linkages between terrestrial mammals, runoff and toxoplasmosis of sea otters. *International Journal for Parasitology* **38**, 1319–1328.

Miman, O., Kusbeci, O. Y., Aktepe, O. C., and Cetinkaya, Z. (2010a). The probable relation between *Toxoplasma gondii* and Parkinson's disease. *Neuroscience Letters* **475**, 129–131.

Miman, O., Mutlu, E. A., Ozcan, O., Atambay, M., Karlidag, R., and Unal, S. (2010b). Is there any role of *Toxoplasma gondii* in the etiology of obsessive-compulsive disorder? *Psychiatry Research* **177**, 263–265.

Molnar, J. L., Gamboa, R. L., Revenga, C., and Spalding, M. D. (2008). Assessing the global threat of invasive species to marine biodiversity. *Frontiers in Ecology and the Environment* **6**, 485–492.

Moloo, S. K., Sabwa, C. L., and Baylis, M. (2000). Feeding behaviour of *Glossina pallidipes* and *G. morsitans centralis* on Boran cattle infected with *Trypanosoma congolense* or *T. vivax* under laboratory conditions. *Medical and Veterinary Entomology* **14**, 290–299.

Moody, J. and Gaten, E. (1982). The population dynamics of eyeflukes *Diplostomum spathaceum* and *Tylodelphys clavata* (Digenea: Diplostomatidae) in rainbow trout in Rutland Water: 1974–1978. *Hydrobiologia* **88**, 207–209.

Moore, J. (1984). Altered behavioral responses in intermediate hosts: an acanthocephalan parasite strategy. *American Naturalist* **123**, 572–577.

Moore, J. (1993). Parasites and the behaviour of biting flies. *Journal of Parasitology* **79**, 1–16.

Moore, J. (2002). *Parasites and the Behavior of Animals*. Oxford University Press, Oxford.

Mortensen, P. B., Norgaard-Pedersen, B., Waltoft, B. L., Sorensen, T. L., Hougaard, D., Torrey, E. F., and Yolken, R. H. (2007). *Toxoplasma gondii* as a risk factor for early-onset schizophrenia: analysis of filter paper blood samples obtained at birth. *Biological Psychiatry* **61**, 688–693.

Murray, D. L., Kapke, C. A., Evermann, J. F., and Fuller, T. K. (1999). Infectious disease and the conservation of free-ranging large carnivores. *Animal Conservation* **2**, 241–254.

Nacher, M. (2005). Charming the mosquito: Do malaria symptoms increase the attractiveness of the host for the vector? *Medical Hypotheses* **64**, 788–791.

Niebuhr, D. W., Millikan, A. M., Cowan, D. N., Yolken, R., Li, Y. Z., and Weber, N. S. (2008). Selected infectious agents and risk of schizophrenia among U. S. military personnel. *American Journal of Psychiatry* **165**, 99–106.

Novotna, M., Hanusova, J., Klose, J., Preiss, M., Havlicek, J., Roubalova, K., and Flegr, J. (2005). Probable neuroimmunological link between *Toxoplasma* and cytomegalovirus infections and personality changes in human host. *BMC Infectious Diseases* **5**, 54.

Novotna, M., Havlicek, J., Smith, A. P., Kolbekova, P., Skallova, A., Klose, J., Gasova, Z., Pisacka, M., Sechovska, M., and Flegr, J. (2008). *Toxoplasma* and reaction time: role of toxoplasmosis in the origin, preservation and geographical distribution of Rh blood group polymorphism. *Parasitology* **135**, 1253–1261.

Nowak, B. F. (2007). Parasitic diseases in marine cage culture: an example of experimental evolution of parasites? *International Journal for Parasitology* **37**, 581–588.

O'Shea, B., Rebollar-Tellez, E., Ward, R. D., Hamilton, J. G. C., El Naiem, E., and Polwart, A. (2002). Enhanced sandfly attraction to *Leishmania*-infected hosts. *Transactions of the Royal Society of Tropical Medicine and Hygiene* **96**, 117–118.

Otranto, D. and Traversa, D. (2002). A review of dicrocoeliosis of ruminants including recent advances in the diagnosis and treatment. *Veterinary Parasitology* **107**, 317–335.

Owen, S. F., Barber, I., and Hart, P. J.B. (1993). Low level infection by eye-fluke, *Diplostomum* spp., affects the vision of three-spined sticklebacks, *Gasterosteus aculeatus*. *Journal of Fish Biology* **42**, 803–806.

Pimentel, D., Zuniga, R., and Morrison, D. (2005). Update on the environmental and economic costs associated with alien-invasive species in the United States. *Ecological Economics* **52**, 273–288.

Poulin, R. (2007). *Evolutionary Ecology of Parasites*, 2nd edn. Princeton University Press, Princeton.

Poulin, R. and Thomas, F. (1999). Phenotypic variation induced by parasites: extent and evolutionary implications. *Parasitology Today* **15**, 28–32.

Prandota, J. (2010). Neuropathological changes and clinical features of autism spectrum disorder participants are similar to that reported in congenital and chronic cerebral toxoplasmosis in humans and mice. *Research in Autism Spectrum Disorders* **4**, 103–118.

Prenter, J., MacNeil, C., Dick, J. T. A., and Dunn, A. M. (2004). Roles of parasites in animal invasions. *Trends in Ecology and Evolution* **19**, 385–390.

Rau, M. E. and Caron, F. R. (1979). Parasite-induced susceptibility of moose to hunting. *Canadian Journal of Zoology* **57**, 2466–2468.

Ready, P. D. (2008). *Leishmania* manipulates sandfly feeding to enhance its transmission. *Trends in Parasitology* **24**, 151–153.

Rebollar-Tellez, E. A. (1999). Kairomone-mediated behaviour of members of the *Lutzomyia longipalpis* complex (Diptera; Psychodidae). PhD thesis, Keele University, Keele, UK.

Ribeiro, J. M. C., Rossignol, P. A., and Spielman, A. (1984). Role of mosquito saliva in blood vessel location. *Journal of Experimental Biology* **108**, 1–7.

Ricciardi, A. (2007). Are modern biological invasions an unprecedented form of global change? *Conservation Biology* **21**, 329–336.

Rogers, M. E. and Bates, P. A. (2007). *Leishmania* manipulation of sand fly feeding behavior results in enhanced transmission. *PLoS Pathogens* **3**, 818–825.

Rogers, M. E., Ilg, T., Nikolaev, A. V., Ferguson, M. J., and Bates, P. A. (2004). Transmission of cutaneous leishmaniasis by sand flies is enhanced by regurgitation of fPPG. *Nature* **430**, 463–467.

Rogers, M. E., Sizova, O. V., Ferguson, M. A. J., Nikolaev, A. V., and Bates, P. A. (2006). Synthetic glycovaccine protects against the bite of *Leishmania*-infected sand flies. *Journal of Infectious Diseases* **194**, 512–518.

Romig, T., Lucius, R., and Frank, W. (1980). Cerebral larvae in the second intermediate host of *Dicrocoelium dendriticum* (Rudolphi, 1819) and *Dicrocoelium hospes* Looss, 1907 (Trematoda, Dicrocoelidae). *Zeitschrift für Parasitenkunde* **63**, 277–286.

Rossignol, P. A., Ribeiro, J. M. C., and Spielman, A. (1986). Increased biting-rate and reduced fertility in sporozoite-infected mosquitoes. *American Journal of Tropical Medicine and Hygiene* **35**, 277–279.

Ruiz, G. M., Rawlings, T. K., Dobbs, F. C., Drake, L. A., Mullady, T., Huq, A., and Colwell, R. R. (2000). Global spread of microorganisms by ships. *Nature* **408**, 49–50.

Rupprecht, C., Hanlon, C., and Hemachudha, T. (2002). Rabies re-examined. *Lancet Infectious Diseases* **2**, 327–343.

Schaub, G. A. (2006). Parasitogenic alterations of vector behaviour. *International Journal of Medical Microbiology* **296**, 37–40.

Schwartz, A. and Koella, J. (2001). Trade-offs, conflicts of interest and manipulation in *Plasmodium*-mosquito interactions. *Trends in Parasitology* **17**, 189–194.

Schweiger, A., Ammann, R. W., Candinas, D., Clavien, P.-A., Eckert, J., Gottstein, B., Halkic, N., Muellhaupt, B., Prinz, B. M., Reichen, J., Tarr, P. E., Torgerson, P. R., and Deplazes, P. (2007). Human alveolar echinococcosis after fox population increase, Switzerland. *Emerging Infectious Diseases* **13**, 878–882.

Schwienbacher, I., Fendt, M., Richardson, R., and Schnitzler, H.-U. (2004). Temporary inactivation of the nucleus accumbens disrupts acquisition and expression of fear-potentiated startle in rats. *Brain Research* **1027**, 87–93.

Seppälä, O., Karvonen, A., and Valtonen, E. T. (2004). Parasite-induced change in host behaviour and susceptibility to predation in an eye fluke-fish interaction. *Animal Behaviour* **68**, 257–263.

Seppälä, O., Karvonen, A., and Valtonen, E. T. (2005). Impaired crypsis of fish infected with a trophically transmitted parasite. *Animal Behaviour* **70**, 895–900.

Seppälä, O., Karvonen, A., and Valtonen, E. T. (2006). Susceptibility of eye fluke-infected fish to predation by bird hosts. *Parasitology* **132**, 575–579.

Setian, N., Andrade, R. S. F., Kuperman, H., Della Manna, T., Dichtchekenian, V., and Damiani, D. (2002). Precocious puberty: An endocrine manifestation in congenital toxoplasmosis. *Journal of Pediatric Endocrinology & Metabolism* **15**, 1487–1490.

Sharif, M., Ziaei, H., Daryani, A., and Ajami, A. (2007). Seroepiderniological study of toxoplasmosis in intellectual disability children in rehabilitation centers of northern Iran. *Research in Developmental Disabilities* **28**, 219–224.

Skallova, A., Kodym, P., Frynta, D., and Flegr, J. (2006). The role of dopamine in Toxoplasma-induced behavioural alterations in mice: an ethological and ethopharmacological study. *Parasitology* **133**, 525–535.

Skallova, A., Novotna, M., Kolbekova, P., Gasova, Z., Vesely, V., and Flegr, J. (2005). Decreased level of novelty seeking in blood donors infected with toxoplasma. *Neuroendocrinology Letters* **26**, 480–486.

Stearns, S. C. and Koella, J. C. (2008). *Evolution in Health and Disease*, 2nd edn. Oxford University Press, Oxford.

Suzán, G. and Ceballos, G. (2005). The role of feral mammals on wildlife infectious disease prevalence in two nature reserves within Mexico City limits. *Journal of Zoo and Wildlife Medicine* **36**, 479–484.

Taraschewski, H. (2006). Hosts and parasites as aliens. *Journal of Helminthology* **80**, 99–128.

Tenter, A. M., Heckeroth, A. R., and Weiss, L. M. (2000). *Toxoplasma gondii*: from animals to humans. *International Journal for Parasitology* **30**, 1217–1258.

Thomas, F., Lafferty, K. D., Brodeur, J., Elguero, E., Gauthier-Clerc, M., and Missé, D. (2011). Incidence of

adult brain cancers is higher in countries where the protozoan parasite *Toxoplasma gondii* is common. *Biology
Letters*, in press. Published online before print July 27,
2011, doi: 10.1098/rsbl.2011.0588.

Thompson, R. C. A. (2008). The taxonomy, phylogeny and
transmission of *Echinococcus*. *Experimental Parasitology*
119, 439–446.

Torchin, M. E., Lafferty, K. D., Dobson, A. P., McKenzie,
V. J., and Kuris, A. M. (2003). Introduced species and
their missing parasites. *Nature* **421**, 628–630.

Torrey, E. F. and Yolken, R. H. (2003). *Toxoplasma gondii* and
schizophrenia. *Emerging Infectious Diseases* **9**, 1375–1380.

Torrey, E. F. and Yolken, R. H. (2007). Editors' introduction: Schizophrenia and toxoplasmosis. *Schizophrenia
Bulletin* **33**, 727–728.

Torrey, E. F., Bartko, J. J., Lun, Z. R., and Yolken, R. H.
(2007). Antibodies to *Toxoplasma gondii* in patients with
schizophrenia: A meta-analysis. *Schizophrenia Bulletin*
33, 729–736.

Valenzuela, J. G., Belkaid, Y., Garfield, M. K., Mendez, S.,
Kamhawi, S., Rowton, E. D., Sacks, D. L., and Ribeiro,
M. C. (2001). Toward a defined anti-*Leishmania* vaccine
targeting vector antigens: characterization of a protective salivary protein. *Journal of Experimental Medicine*
194, 331–342.

Valtonen, E. T. and Gibson, D. I. (1997). Aspects of the biology of diplostomid metacercarial (Digenea) populations
occurring in fishes in different localities of northern
Finland. *Annales Zoologici Fennici* **34**, 47–59.

Vlassoff, A., Leathwick, D. M., and Heath, A. C. G. (2001).
The epidemiology of nematode infections of sheep. *New
Zealand Veterinary Journal* **49**, 213–221.

Vyas, A. and Sapolsky, R. (2010). Manipulation of host
behaviour by *Toxoplasma gondii*: what is the minimum a
proposed proximate mechanism should explain? *Folia
Parasitologica* **57**, 88–94.

Vyas, A., Kim, S. K., and Sapolsky, R. M. (2007). The effects
of *Toxoplasma* infection on rodent behavior are dependent on dose of the stimulus. *Neuroscience* **148**, 342–348.

Webster, J. (2001). Rats, cats, people and parasites: the
impact of latent toxoplasmosis on behaviour. *Microbes
and Infection* **3**, 1037–1045.

Webster, J. and McConkey, G. A. (2010). *Toxoplasma gondii*-
altered host behaviour: clues as to mechanism of action.?
Folia Parasitologica **57**, 95–104.

Webster, J. P. (1994a). The effect of *Toxoplasma gondii* and
other parasites on activity levels in wild and hybrid
Rattus norvegicus. *Parasitology* **109**, 583–589.

Webster, J. P. (1994b). Prevalence and transmission of
Toxoplasma gondii in wild brown rats, *Rattus norvegicus*.
Parasitology **108**, 407–411.

Webster, J. P. (2007). The effect of *Toxoplasma gondii* on animal behavior: Playing cat and mouse. *Schizophrenia
Bulletin* **33**, 752–756.

Webster, J. P., Brunton, C. F. A., and Macdonald, D. W.
(1994). Effect of *Toxoplasma gondii* upon neophobic
behavior in wild brown rats, *Rattus norvegicus*.
Parasitology **109**, 37–43.

Webster, J. P., Lamberton, P. H. L., Donnelly, C. A., and
Torrey, E. F. (2006). Parasites as causative agents of
human affective disorders? The impact of anti-
psychotic, mood-stabilizer and anti-parasite medication on *Toxoplasma gondii*'s ability to alter host behaviour.
Proceedings of the Royal Society of London B **273**,
1023–1030.

White, N. J. (2003). Malaria. In: *Manson's Tropical Diseases*,
21st edn (eds Cook, G.C., and Zumla, A.), pp. 1205–
1295. Saunders, London.

Xiao, J., Buka, S. L., Cannon, T. D., Suzuki, Y., Viscidi, R. P.,
Torrey, E. F., and Yolken, R. H. (2009). Serological pattern consistent with infection with type I *Toxoplasma
gondii* in mothers and risk of psychosis among adult offspring. *Microbes and Infection* **11**, 1011–1018.

Xiao, Y., Yin, J., Jiang, N., Xiang, M., Hao, L., Sang, H., Liu,
X., Xu, H., Ankarklev, J., Lindh, J., and Chen, Q. (2010).
Seroepidemiology of human *Toxoplasma gondii* infection
in China. *BMC Infectious Diseases* **10**, 4.

Yagmur, F., Yazar, S., Temel, H. O., and Cavusoglu, M.
(2010). May *Toxoplasma gondii* increase suicide attempt?
Preliminary results in Turkish subjects. *Forensic Science
International* **199**, 15–17.

Yereli, K., Balcioglu, I. C., and Ozbilgin, A. (2006). Is
Toxoplasma gondii a potential risk for traffic accidents in
Turkey? *Forensic Science International* **163**, 34–37.

Yolken, R. E. and Torrey, E. F. (2006). Infectious agents
and gene-environmental interactions in the etiopatho-
genesis of schizophrenia. *Clinical Neuroscience Research*
6, 97–109.

Yolken, R. H., Bachmann, S., Rouslanova, I., Lillehoj, E.,
Ford, G., Torrey, E. F., and Schroeder, J. (2001). Antibodies
to *Toxoplasma gondii* in individuals with first-episode
schizophrenia. *Clinical Infectious Diseases* **32**, 842–884.

Yolken, R. H., Dickerson, F. B., and Torrey, E.F. (2009).
Toxoplasma and schizophrenia. *Parasite Immunology* **31**,
706–715.

Zhu, S. (2009). Psychosis may be associated with toxoplasmosis. *Medical Hypotheses* **73**, 799–801.

Zhu, S., Guo, M., Feng, Q., Fan, J., and Zhang, L. (2007).
Epidemiological evidences from China assume that
psychiatric-related diseases may be associated with
Toxoplasma gondii infection. *Neuroendocrinology Letters*
28, 115–120.

Afterword

Andrew Read and Victoria Braithwaite

Healthy body, unhealthy mind?

"The manipulation of host phenotype by parasites is a ubiquitous phenomenon...not limited to a few unusual species of purely academic interest" (Poulin and Levry 2012). Imagine Poulin, Levry, and others in this volume are right about that. That would mean that for at least 500 million years, parasites have been trying to make hosts do things they would not otherwise do. Such pressure must have left an indelible mark on host evolution. Here we speculate that this "ghost of manipulation past" is likely to have had a profound effect on the way hosts control their own behavior, with consequences for health and wealth as important as those listed by Poulin and Levry.

There are two ways hosts can protect themselves from behavior attack. One way is to kill or incapacitate the causal pathogen. The other way is to counter the manipulation itself, either by making behavior control systems less vulnerable to attack, or by recalibrating things to accommodate the manipulation. Immunologists study the first kind of defense; next to nothing is known about the other kind, but we contend anti-manipulation defenses must exist. We will be surprised if they do not become central to behavioral biology and neuroscience in the years to come.

The easiest way to see this is by analogy. The hygiene hypothesis in immunology comes in various forms but broadly speaking it posits that lack of exposure to infectious agents increases susceptibility to allergies, asthma, and other autoimmune problems. A growing body of literature supports the view that the developing immune system needs stimuli from infectious agents, symbiotic bacteria, and parasites in order to perform adequately. For example, in a large double-blind placebo study, infants of Ugandan mothers treated with anthelmintics were twice as likely to develop infantile eczema (Mpairwe 2011). Why do immune systems "need" pathogens to function properly? Immune systems are constantly being manipulated by pathogens. Helminths, for example, secrete substances that dampen mammalian immune responses. The evolutionary version of the hygiene hypothesis posits that immune systems evolved to cope with this down regulation. Remove that down regulation and you have excessively aggressive immune reactions. Increasingly it really looks like there is something in this. Give people with autoimmune disease worms and they get better (Broadhurst et al.).

Imagine that behavioral control systems are similarly adjusted by natural selection for the presence of manipulating parasites. The neural pathways and endocrine systems that underlie sensory and decision-making processes are prime targets for pathogens seeking to modify the behavior of their hosts to enhance transmission. How much of our neural complexity is a necessary defense against manipulative invaders? How much of the enormous redundancy is to provide system level functionality if part of the system is attacked? How much of the complex process of wiring a brain during development is to prevent pathogen re-wiring? Hormone and neurotransmitter concentrations can be key behavior modifiers; are base levels set to accommodate parasite-derived manipulation?

And what is the consequence of sudden removal of parasites and pathogens from hosts evolutionarily prepared for encounters with manipulative parasites? How many behaviors in a hygienic world are maladaptive, in the same sense that allergies are maladaptive immune responses? Behaviors associated with disease transmission, such as promiscuity, aggregation, activity levels, boldness, and predator avoidance, are likely to be the target of parasite manipulation. Is it possible that without parasites, behavior control systems perform in ways they were not designed for?

Consider for example the apparently profound impact of *Toxoplasma* infection on humans, ably summarized by Poulin and Levri. *Toxoplasma* infection is staggeringly common today, even in rich countries; in the conditions under which humans evolved, infection with *Toxoplasma* and many other animal-maintained parasites must have been at least as frequent. Humans should be well defended against *Toxoplasma*-induced behavioral alterations. We are a dead-end host for the parasite, so we should have won the evolutionary arms race with *Toxoplasma*: the other side was not partaking.

What are we to make of the variety of behavioral phenotypes induced in contemporary humans by *Toxoplasma* infection? At first glance, some look like pathology (men more likely to be jealous) but some look suspiciously useful (men more likely to be vigilant), and indeed it is relatively easy to imagine that in historical times, many *Toxoplasma*-induced traits conferred fitness advantages on human hosts (even jealousy). Did natural selection recalibrate human behavior control systems so that they were optimal in the face of manipulating parasites?

If so, we see intriguing dilemmas for applied biology. Again by analogy with the hygiene hypothesis, when might modern medicine and farm practice aimed at removing pathogens or preventing infections be maladaptive for a host that is evolutionarily prepared to encounter specific parasites and the suite of altered mechanisms that ensue? When do we need our parasites?

We are not arguing that patients should not be treated against infectious disease (!) or that we should abandon modern hygiene (!!). But if our logic is correct, it could be that solutions to what we nowadays view as behavioral pathologies lie in identifying the mechanisms by which parasites manipulate host behavior. Could administration of analogs of those mechanisms offer solutions to problems in mental health?

We are also not arguing that farm animals should be left unprotected from disease—pathogen burdens can be greatly elevated in intensive farming situations, with clear effects on yield. But are there conditions where farm animal welfare would be improved if it were possible to mimic the manipulative pressures that might come from more "natural" disease burdens or from parasite species present until recently? For instance, stereotypies, the repeated expression of sometimes maladaptive behavior brought about by the inability to perform more normal behavior, is a common feature of zoo animals. Is this because zoo animals have been deprived of the regulatory effects of behavioral manipulation by natural pathogens?

A different issue concerns animals reared in captivity for release as part of reintroduction programs for conservation. Normally, we strive to release animals with a full bill of health. The scrupulously clean living conditions and hygienic, plentiful supplies of food are free from the array of pathogens wild-living conspecifics will naturally be exposed to. Is such hygiene misplaced? We could be disadvantaging the long-term health of the animals if they fail in the pre-release phase to develop immunity to pathogens they will encounter in the wild. Could animals raised without stimulation from manipulative parasites have a poorly developed behavioral repertoire? They may be missing out on key behavioral adaptations that promote dispersal and recolonization. Parasites such as *Toxoplasma* have the capacity to alter temperament traits such as boldness and tendency to explore. Without this manipulation some individuals may fail to disperse and recolonize—behaviors that might deliver advantages in a conservation context.

We are only too well aware of the speculative nature of our arguments (but if you can't speculate in a commentary in a David Hughes book, when can you speculate?). If we are ball-park correct, then parasite manipulation of host behavior could become a key element in the nascent fields of evolutionary medicine and applied evolution. As always, it is a mistake to equate the natural or evolved state as that which maximizes societal good. But understanding selective forces helps generate hypotheses, structure thinking and unify otherwise disparate biological observations. Behavioral variation in our farm and companion animals, and indeed in us, is hugely important for human happiness and health. If much of this variation is due to defenses against parasites past, and the mechanisms involved can be understood, novel approaches to mental health, animal welfare and conservation should be possible.

Credit. These speculations were stimulated by Zuk (2007, chapter 4).

References

1. Poulin, R., Levri, E. P. (2012) Applied aspects of host manipulation by parasites. "In: *Host Manipulation by Parasites*. (eds Hughes, D. P., Brodeur, J., and Thomas, F.) pp. 172–94. Oxford University Press, Oxford."

2. Mpairwe, H., Webb, E. L., Muhangi, L., Ndibazza, J., Akishule, D., Nampijja, M., Ngom-wegi, S., Tumusime J., Jones, F. M., Fitzsimmons ,C., Dunne, D. W., Muwanga, M., Rodrigues. L. C., Elliott. A. M. (2011) Anthelminthic treatment during pregnancy is associated with increased risk of infantile eczema: randomised-controlled trial results. *Pediatric Allergy and Immunology* **22**: 305–312.

3. Broadhurst, M. J., Leung, J. M., Kashyap, V., McCune, J. M., Mahadevan, U., McKerrow, J. H., Loke, P. (2010) -22(+) CD4(+) T Cells Are Associated with Therapeutic Trichuris trichiura Infection in an Ulcerative Colitis Patient. Science Translational Medicine 2: 11.

4. Zuk, M. (2007) Riddled with Life. Friendly Worms, Ladybug Sex, and the Parasites That Make Us Who We Are. Orlando: Harcourt.

CHAPTER 11

Behavioral manipulation outside the world of parasites

Frank Cézilly and Frédéric Thomas

11.1 Introduction

The term "host manipulation" has been coined to cover a range of phenotypic, including behavioral, alterations observed in infected hosts and that appear to increase the fitness of parasites, most often at the expense of that of their hosts (Thomas et al. 2005, see however Fenton et al. 2010). As is often the case in evolutionary biology and behavioral ecology (Kennedy 1992), the term "manipulation" is obviously charged with some kind of emotional content as it may evoke in the human world a continuum of phenomena from some weak level of deceit or luring to sophisticated mental control and brainwashing (Biderman and Zimmer 1961; Taylor 2006). As a matter of fact, this probably partly explains why instances of host-manipulation by parasites have been popularized with so much success in the recent years (Zimmer 2001; Hertel and Cousteau 2008).

One of the pending questions regarding the origin and evolution of "host manipulation", however, is to what extent the phenomenon is peculiar to host–parasite systems or whether similar forms of interactions can take place between organisms belonging to the same species or to different ones outside the world of parasites. The aim of this chapter is precisely to review what can be regarded as cases of behavioral "manipulation" in the animal kingdom, and to examine the different mechanisms and strategies at work, in order to evaluate to what extent "host manipulation" by parasites is a unique phenomenon or not.

11.2 A categorization of manipulation

At an evolutionary level, behavioral manipulation occurs when an individual, the manipulator, increases its own fitness through altering or channelling the behavior of another individual, in ways that are detrimental to that of the manipulated individual. In that respect, several kinds of interactions between conspecific or heterospecific individuals, especially in the field of animal communication (Krebs and Dawkins 1984), can be grouped under the heading "behavioral manipulation." The concept is larger than that of "behavioral deception" (Semple and McComb 1996), as it is not restricted to signaling. Considering the ways by which some kind of behavioral control can be imposed by one individual over another may then help to establish a gradual categorization of cases of manipulation. We will consider here five categories in manipulation. The first corresponds to *deceit through sensory exploitation*, in which the manipulator simply exploits the sensory system of its host to lure it. This first step in manipulation does not alter in any durable way the behavior of the manipulated individual. Instead, natural selection will favor physical and/or characteristics of the manipulator species that will help to achieve instantaneous deception. The second category of manipulation lies in the *exploitation of compensatory mechanisms*. In this case, a manipulator induces a particular stress that, in turn, triggers some physiological mechanisms that modify the motivational state of the manipulated individual in ways that make it more likely to behave in a manner beneficial to the manipulator. The third category includes cases of *coercive exploitation*, in which a manipulated individual modifies its behavior to escape punishment by the manipulator. *Manipulation of information* constitutes the fourth category, and is supposed to imply higher cognitive abilities in the manipulator (Byrne and Corp 2004). The fifth and last category corresponds to *neuroendocrine manipulation*, that is an ability to manipulate the nervous system of the manipulated individuals, enabling the manipulator to control the behavior of manipulated individuals or negate their use of abilities.

11.2.1 Deceit through sensory exploitation

Whereas honesty in signaling is often assumed to play a crucial role in communication (Zahavi and Zahavi 1997), deceit is actually common, particularly in interspecific interactions (Maynard-Smith and Harper 2003). Indeed, natural history abounds with examples of predators luring their prey or flowers deceiving pollinators. A widespread form of deception is "appetitive mimicry" in which the manipulator produces a fake signal that fits the motivational needs of the manipulated individual and thus succeeds in modifying its behavior. Depending on the prey–predator system, deception can exploit different sensory systems, such as vision, olfaction, or audition.

In several ambush predator species, parts of the body, or their use, have been modified during the course of evolution to function as visual lures that attract prey close enough to seize. For instance, in order to attract larval anurans and juvenile salmonids, juveniles of the Pacific Coast aquatic garter snakes (*Thamophilis atratus*) remain in an ambush position while extending and holding their tongue out straight, with the tongue tips shimmering on the water surface (Welsh and Lind 2000). Through mimicking insects that way, juvenile snakes attract young fish and tadpoles within striking range. The same behavior has recently been observed in another snake species, the Mangrove Saltmarsh Snake (*Nerodia Clarkii Compressicauda*) feeding on fish (Hansknecht 2008). Other famous examples include the vermiform appendage situated on the

floor of the oral cavity of alligator snapping turtles *Macroclemys temminckii* (Drummond and Gordon 1979), the modified tail-tip of several snake species (Chiszar et al. 1990; Hagman et al. 2008; Reiserer and Schuett 2008), toes of frog and toads (Murphy 1976; Radcliffe et al. 1986; Grafe 2008; Hagman and Shine 2008). A particular case of visual lure is that of anglerfishes (order Lophiiformes), named from their characteristic method of predation. In anglerfish the first spine of the anterior dorsal fin is located out of the tip of the snout and modified into a long filament, the illicium, that protrudes above the fish eye and terminates into a fleshy appendage, the esca (see Shimazaki and Nakaya 2004 and references therein for details). The esca can be moved as to look like a living animal and is used to lure prey (Laurenson et al. 2004). Some species of anglerfish live on the Continental Shelf, whereas others live in the deep sea. In most species of anglerfish living in deep sea, the esca includes a light gland that contains bioluminescent symbiotic bacteria (Munk 1999). The light gland appears to reinforce the luring ability of the esca in the dark environment characteristic of the deep sea.

Although visual cues are predominant in prey luring, olfaction can also be exploited. For instance, the tropical forest rove beetle *Leistotrophus versicoloris* is known to use chemical luring to attract flies that ordinarily exploit dung or carrions (Forsyth and Alcock 1990). A similar phenomenon has recently be observed (Di Giusto et al. 2010) in carnivorous plants that produce volatiles that mimic those emitted by flowers to attract insect pollinators.

In some cases, however, the manipulator does not attract its prey through exploiting its appetite for food, but its appetite for sex, as observed in the North and South American genera *Photuris* and *Bicellychonia* which are nocturnal specialist predators of other fireflies (Lloyd and Ballantyne 2003). Fireflies are famous for relying mainly on bioluminescent signals during courtship (Lewis and Cratsley 2008). Lloyd (1965) showed that *Photuris* "femmes fatales" attract male fireflies of the genus *Photina* by responding to their courtship signals with flashes mimicking the response delays and flash patterns of females of the same species. In addition, female *Photuris versicolor* possess signal repertoires that allow them to modulate their response, depending on which prey species they are attempting to attract (Lloyd 1975). Similarly, bolas spiders, *Mastophora hatchinsoni*, attract males of several moth species through mimicking their specific sex pheromones (Haynes et al. 2002). Sexual mimicry is also observed in plants, as in the case of the orchid *Ophrys sphegodes* that produces the same mix of compounds as that found in the sex pheromone of its pollinator species, the solitary bee *Andrena nigroaenea* (Schiestl et al. 1999).

Predators may also exploit the social behavior of their prey to lure them. For instance, Atkinson (1997) observed Northern shrikes, *Lanius excubitor*, using mimetic singing to attract and capture their passerine prey, whereas Callea et al. (2009) reported the observation of a margay, *Leopardus wiedii*, producing calls similar to those emitted by pied tamarin pups, *Saguinus bicolour*, to lure adults. One spectacular case of a manipulator exploiting the social behavior of its prey is that of the ant-eating spiders *Zodarion germanicum* and *Z. rubidum* that resemble their prey (the large dark ant, *Formica cinerea*, and the red ant, *Myrmica sabuleti*, respectively) in size, color, and setosity, and even move their body and antennae in a similar fashion (Pekár and Král 2002). This perfect mimicry is further enhanced by the behavior of the spider. When moving through the ant colony while carrying one captured ant, the spider deceives approaching ants by exploiting their system of recognition between nest mates. Using its front legs (which are remarkably almost devoid of macrosetae, much like the antennae of ants, while its other limbs are covered with flattened setae that mimic the dense setosity of ants' limbs), the spider first taps the antennae of the approaching ant, thus mimicking a tactile cue, before presenting its prey (the ant corpse), thus exploiting an olfactory cue.

From an evolutionary point of view, the deception is stable as long as the frequency of false signals in the environment remains low. Most of the time, the signal is not corrupted and failure to respond to it would result in lost opportunities of feeding or mating. Furthermore, the ability of manipulated species to resist luring is probably limited by the cost associated with developing more efficient discriminating abilities. First, a more efficient discriminative ability might come with a risk of an increased number of false positives (or type 1 errors), such that actual food items might sometimes be confounded with potential predators. Second, in many systems, the manipulated and the manipulator species are probably engaged in a coevolutionary arms race, where any improvement in the ability of the manipulated species to resist luring acts as a selection pressure on manipulators to improve their deceptive abilities.

11.2.2 Exploitation of compensatory mechanisms

A different category of manipulation includes cases in which the manipulator induces an alteration in the environment that elicits in the manipulated an automatic compensatory response that can then be exploited by the manipulator. A classical example is infanticide by immigrating males famously observed in lions (Packer and Pusey 1983), and also in primates (Cords and Fuller 2010), brown bears, *Ursus arctos* (Bellemain et al. 2006), or rodents (Labov 1980). When male lions overtake a pride, several females are usually lactating and therefore unavailable for mating. The suppression of ovulation during lactation is presumably adaptive for females as it prevents them from becoming pregnant and spreading their energy resources too thinly among different offspring. However, following the death of the dependent young, ovulation resumes and the female becomes fertile again. This quick return into estrus is regarded as an adaptive automatic compensatory response by which females increase their reproductive success. In this sense, infanticide by males can be seen as a form of exploitation of compensatory mechanism.

An automatic compensatory response is a biological inevitability and must be obligatory. It is typically a physiological consequence of some kind of stress. In contrast, a volitional compensatory response tends to be deliberate and behavioral, which the individual intentionally performs. Although female mammals have no direct control on the suppression or reactivation of estrus, they appear to have developed a large repertoire of counter-strategies to male infanticide, such as aggression, female coalitions, female choice of dominant males, and promiscuity to confuse paternity (reviewed in Agrell et al. 1998). One important point then is to what extent females can decide between resorting to any of these strategies or remaining passive when males attempt to kill dependent young, depending on the costs of infanticide and the benefits of rapidly mating with a male of superior quality. If such a strategic flexibility exists, then the manipulative dimension of infanticide by males would be questionable.

The exploitation of compensatory mechanisms has also been invoked in relation to sexual conflict. In several insect species, such as the bruchid seed beetle, *Callosobruchus* spp. (Crudgington and Siva-Jothy 2000; Rönn et al. 2007), the dung fly, *Sepsis cynipsea* (Blanckenhorn et al. 2002), and several species of drosophila (Kamimura 2007, 2010) males inflict wounds on female genitalia during copulation. Wounding behavior by males has been seen as a form of adaptive manipulation because it reduces the probability that a female will remate (Johnstone and Keller 2000) or leads female to increase their investment into current reproduction (e.g., Lessells 1999). However, wounding behaviors may not directly enhance male post-copulatory fitness but may arise as a side-effect of the male mating strategy in another context (Morrow et al. 2003), such that the manipulative dimension of damaging mating behavior remains elusive (e.g., Teuschl et al. 2007; Kamimura 2008). A different situation occurs in the green-veined white butterfly, *Pieris napi* (Wedell et al. 2009). In this species, males produce both normal-fertilizing sperm (eupyrene sperm) and a large quantity of non-fertile sperm (apyrene sperm). The function of the latter is to fill the female's sperm storage organ, with the effect of switching off receptivity and thus reducing the propensity of the female to remate. By preventing the female from mating again, the male protects his paternity and ensures that only his paternal genes appear in progeny, despite direct benefits from polyandry to females (Wedell et al. 2002). Thus, sexual conflict over female remating rate acts to favor manipulation in this species, with males using their

apyrene sperm to exploit a female physiological function designed to monitor sperm in storage.

11.2.3 Coercive exploitation

Coercive exploitation occurs in principle when manipulators enforce modified behavior on other individuals through punishment in the absence of compliance. According to this view, manipulated individuals may "accept" modifying their behavior to escape retaliation by manipulators. The possibility was originally suggested by Zahavi (1979) and later developed by Soler et al. (1995, 1998) in the context of interspecific brood parasitism by great spotted cuckoos, *Clamator glandarius*. Host species may *a priori* protect themselves from brood parasitism by recognizing cuckoo's eggs and rejecting them. However, the cuckoo might retaliate by coming back to the nest of rejecters to destroy or prey upon host eggs or nestlings. On a short time scale, cuckoos are supposed to benefit from destroying host offspring by inducing some hosts to re-nest and subsequently accept a cuckoo egg. On an evolutionary time, reiterated retaliation by cuckoos may ultimately result in preventing the diffusion of the genes underlying the rejection behavior. However, theoretical accounts suggest that the evolution of coercive manipulation in the case of brood parasitism is contingent on particular conditions that may not be often met in the wild. First, it can only work if the host species can raise at least one young while accepting taking charge of the young cuckoo, or if non-ejectors have lower rates of parasitism in later clutches compared to ejectors (Pagel et al. 1998). Second, assuming a cost for the host species in discriminating between its own eggs and that of the brood parasite, and a cost for the brood parasite for monitoring nests, the coevolution between retaliators and ejectors remains unstable, with populations of both hosts and brood parasites cycling indefinitely (Robert et al. 1999). Not surprisingly then, empirical confirmation of the existence of mafia-like behavior in brood parasites is very limited, even though a recent study (Hoover and Robinson 2007) provided some evidence of coercive manipulation in another brood parasite, the brown-headed cowbird, *Molothrus ater* parasitizing the broods of the prothonotary warbler (*Protonotaria citrea*). However other studies on the same brood parasite species, but with different host species (Lorenzana and Sealy 2001; Smith et al. 2003; Rasmussen et al. 2009; Peer and Rothstein 2010) provide a different picture, thus suggesting that coercive manipulation plays only a minor role in the complex dynamics of the interaction between brood parasites and their hosts.

Examples of coercive exploitation outside of brood parasites are scarce. We can think however of a few possible cases. The first is the observation that rhesus monkeys that withhold information about food discovery receive significantly more aggression if detected with food by other group members than vocal discoverers (Hauser 1992), potentially explaining why deception through withholding information is rare in the wild. The second corresponds to the down regulation of the reproductive endocrine axis in subordinate females observed in many animal societies. Although the phenomenon may result from a chronic "stress" aggressively imposed by dominant females, it has been suggested that it could actually correspond to some kind of physiological restraint self-imposed by surbordinates in response to the threat of infanticide by dominants (Young et al. 2008). The third is the recent observation of intimidation of females by male water striders (Han and Jablonski 2010). In this species of aquatic insect, the female vulvar opening is protected by a shield that prevents mounted males from coercive intromission. Copulation can therefore only be achieved if the female exposes her genitalia. However, during mounting, males emit vibration signals that attract insect and fish predators, which attack females rather than males. According to Han and Jablonski (2010), attraction of predators by males is a coercive strategy to force reluctant females into mating. Indeed, in response to increased predation risk, males may even increase their rate of signaling, whereas females permit copulation sooner, prompting males to stop signaling more quickly.

Overall, only a few cases of coercive manipulation have been evidenced in nature. One pending question is whether they are rare or only rarely observed. Careful examinations of conflicts of inter-

ests and costs of deception in long term associations, such as brood parasites, sibling rivalry, and sexual conflict may thus help to assess the general importance of this particular mode of manipulation.

11.2.4. Manipulation of information

One rather sophisticated form of manipulation is the production of false information in order to deceive other individuals. A classical example is the use of distractive (or diversionary) displays, under the form of injury feigning, observed in some bird and fish species (Amstrong 1949; Wiens 1966; Foster 1994; Mallory et al. 1998), in reaction to predators approaching nest or progeny. One spectacular example is the broken-wing display famously used by several shorebird (Deane 1944; Heckenlively 1972; Bergstrom 1988; Byrkjedal 1989) and columbid species (Wiley 1991; Baskett et al. 1993). Typically a bird on a nest reacts to some approaching danger by walking on the ground with one wing hanging low and dragging on the ground, thus looking like easy prey for a predator. However, once the predator has been driven away from the nest, the bird resumes normal posture and takes flight. Several interpretations have been proposed for this behavior in the framework of classical ethology (reviewed by Ristau 1991). One particular point of debate is to decide whether the broken-wing display can be reduced to some form of pre-programmed or reflexive behavior, arises as a ritualized outcome of a motivational conflict, or can be regarded as intentional. Although flexibility in the use of the broken-wing display depending on the nature of the approaching danger, together with the fact that birds appear to monitor the behavior of the predator while driving him away from the nest and young, has been regarded as evidence for purposeful behavior (Ristau 1991), no study so far has provided unambiguous evidence of intentionality in injury feigning. As suggested by Premack (2007), one criteria by which to judge whether the use of the broken-wing display is intentional or not could be to establish whether a bird is able to cease performing the broken-wing display when it is inefficient in driving predators away, as intentional acts are supposed to extinguish when they fail to realize their goal. However, to our knowledge, no experimental investigation of the broken-wing has been realized so far.

Another way of manipulating information is to perform an act from the regular repertoire of the species in the absence of the stimulus that would normally trigger it, with the effect of confusing other individuals (referred to as *tactical deception*, Byrne and Whitten 1988). A much discussed example is the production of false alarm calls by various vertebrate species to drive potential competitors away from food sources (Munn 1986; Wheeler 2009; Flower 2011), or to secure matings (Møller 1990; Bro-Jørgensen and Pangle 2010). In an influential paper, Munn (1986) suggested that through giving alarm calls in the absence of predators, birds in mixed-species flocks can induce other individuals to drop food that they then consume. Although similar observations have been published in two other bird species (Matsuoka 1980; Møller 1988; but see Haftorn 2000; Evans et al. 2004) and in a monkey, the wild tufted capuchin, *Cebus apella nigritus* (Wheeler 2009), the manipulative dimension of false alarm calls remains an issue. In this respect, recent studies on drongos are of particular interest. Drongos are kleptoparasite bird species that typically forage in mixed-species flocks. It has been recently reported (Flower 2011) that the production by fork-tailed drongos, *Dicrurus adsimilis*, of both drongo-specific and mimicked false alarm calls induced target species, such as meerkats, *Suricata suricatta*, and pied babblers, *Turdoides bicolour*, to flee for cover and abandon the food they were handling. However, according to another study on the same species (Radford et al. 2010), vocal signals broadcast by drongos would actually serve to advertise their presence to their victims, that would in turn reduce their individual levels of vigilance and thus increase their foraging efficiency. The higher food discovery rate would in turn benefit the drongos as they would be faced with more opportunities for stealing food from the babblers. To the extent that the supposed "victims" appear to benefit from the sentinel calls given by the drongos, Radford et al. (2010) suggest that the relationship between drongos and babblers is more akin to mutualism than to manipulation, although such an interaction could also be seen to a certain extent as a case of exploitation of compensatory responses. However, based on a detailed acoustical analysis, another recent study (Harsha et al. 2010) on a closely-related species, the Greater Racket-tailed drongo, *Dicrurus paradiseus*, suggests that supposedly "false alarm" may actually signal aggressive intent, thus casting doubt on their "deceptive" nature.

The production of false alarm calls to retain sexual partners has been so far documented in one bird, the barn swallow, *Hirundo rustica* (Møller 1990), and two mammal species, the Formosan squirrel, *Callosciurus erythraeus thaiwanensis* (Tamura 1995) and the Topi, *Damaliscus lunatus* (Bro-Jørgensen and Pangle 2010). Bro-Jørgensen and Pangle (2010) argued that sexual conflict over mating frequency could easily lead to the falsification of alarm calls, particularly if the benefits are important for males while the costs are limited for females. However, these observations are open to alternative interpretations. First, data based on observations by sight in natural environments are not always reliable, as it is quite difficult to ascertain that the behavior of animals is not altered by the presence of the observer. For instance, Bro-Jørgensen and Pangle (2010) mentioned that their observations on topis were conducted from a car "to which the animals were habituated" (page E34), but provide no evidence for the degree of habituation. Second, the identification of false alarm calls by human observers is dependent on their own performance in locating predators that are most often very precautious in their approaches. However, mammals and birds may surpass human beings in their sensory ability to detect olfactory (Gilad et al. 2003), auditory (Nelson and Suthers 2004), or visual cues (Heesy and Hall 2010), and may then outperform them in the ability to detect long-distance evidence of predator presence. Third, in several species, females have been shown to benefit from male vigilance against predators (Bull and Pamula 1998; Forslund 1993; Squires et al. 2007) and to prefer more vigilant males (Dahlgren 1990; Pizzari 2002), such that the production of alarm calls by males, even in reaction to the least cue of danger, might be positively selected.

Finally, interpreting false alarm calls as evidence for deception might not be parsimonious. Indeed, theoretical evidence suggests that an optimal system for regulating all-or-none defences against the uncertain presence of serious threats (such as predator attacks can be considered to be) is bound to express false alarms, a phenomenon known as the "smoke detector principle" (Nesse 2005). In addition, the mere presence of sexually receptive females may arouse males, with the effect of reducing their ability to distinguish stimuli arising from real danger from other noise stimuli. More simply, the need to divide attention between visually monitoring the behavior of females and detecting cues from approaching predators early enough may decrease the efficiency of the later and result in an increase in the production of false alarms, if, for instance, visual attention reduces auditory or olfactory sensitivity and, hence, discriminative abilities (see Delano et al. 2007). The performance of certain behavioral acts in a misleading context is, however, not limited to the production of false alarm calls. It has for instance been recently reported in a totally different context: food hoarding. Several species of birds and mammals store food that they later retrieve from caches. For the hoarding strategy to be evolutionary stable, it is generally considered that food hoarders must have a higher probability of recovering food from their own caches than other foragers (Andersson and Krebs 1978), or that all hoarders must also be pilferers (Vander Wall and Jenkins 2003). Scatter-hoarding, that is to space caches far apart, is enough to reduce the risk of pilferage of hoarded food (Jenkins et al. 1995), but does not suppress it. It has therefore been suggested that deception could be used by food hoarding species to protect their caches. The evidence in favor of such a phenomenon remains limited, however. Heinrich (1999) reported that one hand-reared raven, *Corvus corax*, cached inedible items presumably to deceive conspecifics, while Bugnyar and Kortschal (2004) observed one individual of the same species leading conspecifics away from food caches provisioned by experimenters. Similarly, Dally et al. (2006) observed rooks, *Corvus frugilegus*, caching inedible items or covering empty caches. Recently, Steele et al. (2008) reported that tree squirrels, *Sciurus carolinensis*, were more likely to cover additional empty sites where nothing had been cached in the close

presence (> 20 m away) of conspecifics. However experimentally simulating cache pilferage had no effect on the intensity of false food caching by tree squirrels.

Overall, apart from monkeys and possibly corvids, evidence for tactical deception and, hence, intentionality in supposedly deceptive behaviors is thin, and corresponds mainly to cases of witholding of information in primates, as for instance when a knowledgeable subordinate monkey refrains from exploiting a food source in the presence of a naïve dominant individual (Amici et al. 2009). One reason for that might be that intentional deception requires highly developed cognitive skills (Byrne 2010). In particular, intentional manipulation of information in a social context appears to involve rapid learning and extensive social knowledge, two capacities which are constrained by the size of the neocortex.

Indeed, comparative anatomical studies indicate that the use of deception among primate species is well predicted by the neocortical volume, but not by the size of the rest of the brain (Byrne and Corp 2004).

11.2.5 Neuroendocrine manipulation

This last category includes cases where the manipulator is able to exert some control on the behavior of the manipulated individual through direct action on its neuroendocrine system and thus its behavior. In social insects, queen pheromones have a deep influence on the behavioral development of workers, thus regulating developmental polyethism (Vergoz et al. 2007, Holman 2010). More specifically, the queen mandibular pheromone (QMP) in honey bees, *Apis melifera*, directly influences the brain chemistry of workers in an age-dependent manner, through acting on levels of dopamine in the brain (Vergoz et al. 2007). More specifically, young bees exposed to QMP fail to learn to respond to noxious stimuli though extending their sting. Young bees under the influence of the QMP remain in the hive, are less aggressive, and less active. This apparently prevents young bees from developing aversive memories against odors, including that of the queen. The extent to which the production of QMP can be regarded as evidence for neuroendocrine manipulation depends on whether it plays a significant part in regulating the conflict between the workers and the queen (Wenseleers et al. 2003; Gardner and Grafen 2009; Nowak et al. 2010).

Another form of conflict, potentially open to manipulation, is sexual conflict (Arnqvist and Rowe 2005). The ability of males to alter the reproductive physiology and behavior of females in ways which are beneficial for males and detrimental for females has recently gained evidence. For instance, the seminal fluid of *Drosophila melanogaster* contains a sex peptide, Acp 70a, that causes in females both a reduction in sexual receptivity and an increase in egg production (Chapman et al. 2003), with detrimental effects on the longevity of females (Wigby and Chapman 2005). Although the precise mechanism by which Acp 70a alters female behavior is unknown, Yapici et al. (2008) have recently identified a receptor for Acp70A, expressed in the female's reproductive tract and central nervous system. Females without the sex peptide receptor CG16752 do not respond to Acp 70a and continue to show sexual receptivity after mating. Acp 70a might act in part by modulating the activity the subset of neurons that also express the *fruitless* gene, a key determinant of sex-specific reproductive behavior (Kvitsiani and Dickson 2006; Yapici et al. 2008).

11.3 A brief critique of the "manipulation" concept

The above examples suggest that several behavioral traits commonly observed in natural environments can be considered as evidence for deception and, possibly, manipulation. However the anecdotal nature of a certain number of studies clearly limits their importance (Ristau 1991; Semple and McComb 1996; but see Bates and Byrne 2007), such that the generality of the phenomenon is difficult to assess. Potential cases of deception and manipulation might be rare in nature or difficult to observe, partly explaining the lack of careful and detailed analysis. On the other hand, claimed cases of manipulation might be prone to false identification and misinterpretation. One major problem is the recurrent absence of alternative explanations based on more parsimonious interpretations. Parsimony

here should be sought at two levels. The first concerns the complexity of the cognitive mechanisms which are explicitly or implicitly invoked in the interpretation of deceptive behaviors (see Mitchell 1986). The second corresponds to the adaptive nature of manipulation. From an evolutionary point of view, establishing that manipulative behavioral traits represent true adaptations, that is that they evolved as a direct consequence of the benefit they confer, remains a crucial issue that deserves further consideration. In some of the above cases, for instance, manipulation is supposedly achieved through either exploiting some sensory bias or matching search images or recognition system of victims, although direct evidence for this remains limited. One noticeable exception, however, is the recent study of Nelson et al. (2010) on caudal luring in the death adder, *Acanthophis antarcticus*, that suggests that the pre-existing biases in the nervous system of one of its common prey, the jacky dragon, *Amphibolorus muricatus*, originally selected for prey recognition, might have shaped the design of the caudal luring signal. Detailed investigations of the origin and evolution of other seemingly deceptive behaviors, such as for instance the broken-wing display observed in several bird species, would be particularly valuable. In this respect, the use of the comparative method (Harvey and Pagel 1991) could be particularly rewarding to identify morphological, behavioral, or ecological characteristics that are common to species resorting to the broken-wing display when their progeny is under threat. More generally, interpretations of manipulative behavior, whether implying parasitic or non-parasitic organisms, should always be put forward with caution, and attempts should be made to consider alternative evolutionary scenarios, rather than endorsing adaptive explanations too quickly (see Cézilly et al. 2010).

11.4 Manipulation inside and outside the world of parasites: convergence and divergence

As a scientific field, host manipulation by parasites has, until recently, developed in relative isolation from behavioral ecology approaches. This probably explains why, despite the important effort invested in studying manipulative parasites, no study has, until now, established parallels between manipulative strategies inside and outside the world of parasites. In light of the previous sections, it seems clear that behavioral manipulation is far from being a strategy restricted to host–parasite systems; instead it appears as a widespread phenomenon in the animal kingdom. Not only is it common outside the world of parasites, it also occurs in quite a large variety of interactive contexts, both within and between species. Should we conclude that behavioral manipulations are more common and/or diversified among free-living organisms than in host–parasite systems? This question cannot be properly answered at the moment and is inherently complex for several reasons. A first problem is linked to the definitions. To date, host manipulation by parasites has (by definition) exclusively been examined in the context of parasitic transmission (Poulin 1995; Moore 2002; Thomas et al. 2005; Poulin 2010). In this chapter, which is centered on free-living species, we considered the manipulation concept in a broader perspective, bearing in mind that behavioral manipulations occur solely when an individual (the manipulator) increases its own fitness through altering or channeling the behavior of another individual, in ways that are detrimental to that of the manipulated individual. Unless host–parasite systems are also investigated for manipulation in this perspective, it is rather difficult to make direct comparisons. For instance, the greatly enhanced appetite of crickets *Nemobius sylvestris* harbouring juvenile hairworms *Paragordius tricuspidatus* (Thomas, unpublished observations) is not traditionally considered a case of manipulation, despite the fact that it is probably a behavioral change induced by the worm to boost its growth/fecundity. Further research would clearly be needed at the moment to decipher the diversity of manipulative changes that are parasite-induced outside the context of transmission. This step is essential before appropriate comparisons and generalizations can be made.

Despite this limitation, several preliminary conclusions can be derived from this chapter. First, it appears that the main modes used by parasites to achieve behavioral manipulations (see Chapter 2) are all present in the world of free-living species. There are no parasite exclusivities. Aside from the broad similarities concerning manipulative mechanisms, several differences can be put forward between host–parasite and free-living systems. For instance, unlike the world of parasites, that of the free-living includes numerous vertebrate species with highly developed brains and advanced cognitive abilities. In this respect, we do not expect the same kind of manipulation to occur, at least at the same frequencies. Manipulative interactions among vertebrate species are likely to involve protagonists that are aware of their status (manipulator or manipulated) during the interaction. Therefore, manipulations based on punishments in absence of compliance should have, everything else being equal, more opportunities to evolve in these systems than in those involving two invertebrates, or an invertebrate parasitizing a vertebrate. This prediction is difficult to test at the moment because only a few cases of coercive manipulation have been documented in nature. However, in accordance with the above statement, the only well documented example of manipulation based on punishment in a host–parasite system concerns an exception among host–parasite systems since both the parasite and the host are vertebrates (cuckoo and other birds). A second important aspect to consider is that parasitic manipulators, unlike free-living ones, generally live on, or inside, manipulated individuals. This difference also influences the opportunities for the different kinds of manipulation to evolve. It is for instance easier for a manipulative individual to directly interfere with the physiology of another individual and/or with its morphology/color by being inside rather than outside the host. Neuroendocrine manipulations would then have more opportunities to evolve in host–parasite systems than in free-living ones.

A major difference between host-manipulation by parasites and manipulation outside the world of parasites is the duration of the interaction. Although in both cases the issue might be drastic for the manipulated individual (e.g., castration and/or death), the time frame over which the interaction unfolds may vary from a few seconds in the case of a predator luring prey to several weeks or months in some host–parasite systems. The length of the manipulative period exerted by parasites can, however, be short as well, especially if the manipulation is efficient (the goal is achieved rapidly), and if parasites start to manipulate only once they are mature and/or infective to the next host.

Another important difference concerns the fitness costs for individuals that are victims of manipulation. With few exceptions, host manipulation by parasites severely impairs the host fitness, as the latter dies or is castrated. Even in the few cases where host manipulation is reversible (e.g., Ponton et al. 2011; Maure et al. 2011), dramatic fitness consequences remain for the host (Biron et al. 2005). Manipulations among free-living organisms also have detrimental, sometimes major consequences for the manipulated individual (e.g., manipulation by predators), but in many cases the effect of manipulation is not too severe or of short duration. For instance, the fitness consequences of being momentarily confused following the production of a false alarm call by a conspecific are probably minor and easy to compensate for in the short term. Reversibility and recovery are thus more frequent in manipulations exerted by free-living species compared to those involving a parasite. Although further theoretical explorations would be necessary to explore the evolutionary consequences of this phenomenon, common sense suggests that it cannot be neutral. For instance, resistance against manipulation, or avoidance strategies towards manipulators, should be high in host–parasite systems, and everything else being equal, less systematic in free-living species. Lower rates of resistance to manipulative attempts should in return favor the diversification of those strategies in free-living species. Along the same vein, examples presented in this chapter indicated that behavioral manipulations in free-living species often provide only a punctual benefit to the manipulator (i.e., one meal or one copulation). This is probably another major difference with host–parasite systems, since for parasites the host manipulation is crucial to complete the life cycle and produce offspring (but see Poulin and Cribb 2002). Using the analogy of the life–dinner principle traditionally applied to predator–prey interactions (Dawkins and Krebs 1979), there should be stronger selection pressure on parasites to manipulate than on free-living species (this can however be influenced by the definition problem mentioned above).

Both parasitic and non-parasitic species may have a major impact on the growth, reproductive capacity, and survival of manipulated species. Intensive manipulation is therefore expected to favor adaptations in manipulated species that reduce either their risk of being manipulated or the negative effects of manipulation (Bauer et al. 2009). Such adaptations could consist of avoidance strategies, including escape in time as well as chemical and mechanical defenses, and tolerance strategies related to compensation after having been manipulated. However, very little is known about the ability of manipulated species to oppose manipulation, or at least to decrease its negative effects, such that it is difficult to know if the phenomenon is common or rare. One possibility is that the various deception devices used by non-parasitic manipulating species function in a frequency-dependent manner, thus preventing adaptive reaction by the manipulated. For instance, evolving a discrimination against false alarm calls might come at the risk of ignoring true alarms, with immediate and drastic consequences. As long as false alarm calls remain at low frequency, the cost of evolving an ability to discriminate between true and false alarm calls might be higher than that of being occasionally deceived, thus leading to a simple rule of thumb consisting in trusting any alarm call. On the other hand, there is some evidence for discrimination against unreliable signalers in both primates (Cheney and Seyfarth 1988) and rodents (Hare and Atkins 2001), through associating an individual's identity with that individual's past behavior. However, reliability assessment requires that false alarm calls convey sufficient information to allow individual recognition of the manipulative signaler. In addition, manipulated individuals should possess sufficient sensory acuity, cognitive ability, and memory capacity to discriminate between true and false alarm calls and remember the identity of manipulative signalers, such that reliability assessment should evolve more easily in stable social groups or between close kin (see Pollard 2011). To what extent the same phenomenon can apply to other sorts of luring devices remains to be documented. Another possibility is that of a coevolutionary arm race between manipulators and manipulated species, eventually leading to more refined manipulation devices or more intense levels of manipulation. Recent studies of sexual conflict suggest, for instance, that females can evolve to resist manipulation by males, leading in turn to more harmful males (Wigby and Chapman 2004; Gay et al. 2011). The relevance of antagonistic coevolution in the evolution of manipulation in both parasitic and non-parasitic organisms clearly deserves further consideration.

Finally, a remarkable aspect of behavioral manipulations in free-living species is the possibility of role alternation, with a manipulator in one episode becoming the manipulated in another episode. For instance, individuals able to produce false alarm calls can themselves be usurped by conspecifics on other occasions. To our knowledge, role alternation is not documented in host–parasite systems, probably because host–parasite systems are heterospecific associations, while role alternation is more likely to occur within intraspecific contexts.

References

Agrell, J., Wolff, J. O., and Ylonen, H. (1998). Counter-strategies to infanticide in mammals: costs and consequences. *Oikos* **83**, 507–517.

Amici, F., Call, J., and Aureli, F. (2009). Variation in withholding of information in three monkey species. *Proceedings of the Royal Society Series B* **276**, 3311–3318.

Amstrong, E. A. (1949). Diversionary display. *Ibis* **91**, 179–188.

Andersson, M. and Krebs, J. R. (1978). On the evolution of hoarding behaviour. *Animal Behaviour* **26**, 707–711.

Arnqvist, G. and Rowe, L. (2005). *Sexual Conflict*. Princeton University Press, Princeton.

Atkinson, E. C. (1997). Singing for your supper: acoustical luring of avian prey by northern shrikes. *Condor* **99**, 203–206.

Baskett, T. S., Sayre, M. W., Tomlinson, R. E., and Mirarchi, R. E. (1993). *Ecology and Management of the Mourning Dove*. Stackpole Books, Harrisburg, P. A.

Bates, L. A. and Byrne, R. W. (2007). Creative or created: Using anecdotes to investigate animal cognition. *Methods* **42**, 12–21.

Bauer, S., Witte, V, Böhm, M., and Foitzik, S. (2009). Fight or flight? A geographic mosaic in host reaction and potency of a chemical weapon in the social parasite *Harpagoxenus sublaevis*. *Behavioral Ecology and Sociobiology* **64**, 45–56.

Bellemain, E., Swenson, J. E., and Taberlet, P. (2006). Mating strategies in relation to sexually selected infanticide in a non-social carnivore: The brown bear. *Ethology* **112**, 238–246.

Bergstrom, P. W. (1988). Breeding displays and vocalizations of Wilson's plovers. *Wilson Bulletin* **100**, 36–49.

Biderman, A. D. and Zimmer, H. (eds) (1961). *The Manipulation of Human Behavior*. John Wiley, New York.

Biron, D. G., Ponton, F., Joly, C., Menigoz, A., Hanelt, B., and Thomas, F. (2005). Water seeking behavior in insects harboring hairworms: should the host collaborate? *Behavioral Ecology* **16**, 656–660.

Blanckenhorn, W. U., Hosken, D. J., Martin, O. Y., Reim, C., Teuschl, Y., and Ward, P. I. (2002). The costs of copulating in the dung fly *Sepsis cynipsea*. *Behavioral Ecology* **13**, 353–358.

Bro-Jørgensen, J. and Pangle, W. M. (2010). Male topi antelopes alarm snort deceptively to retain females for mating. *American Naturalist* **176**, E33–E39.

Bugnyar, T. and Kortschal, K. (2004). Leading a conspecific away from food in ravens (*Corvus corax*)? *Animal Cognition* **7**, 69–76.

Bull, C. M. and Pamula, Y. (1998). Enhanced vigilance in monogamous pairs of the lizard, *Tiliqua rugosa*. *Behavioral Ecology* **9**, 452–455.

Byrkjedal, I. (1989). Nest defense behavior of lesser golden-plovers. The *Wilson Bulletin* **101**, 579–590.

Byrne, R. W. (2010). Deception: competition by misleading behavior. In: *Encyclopedia of Animal Behavior Vol. 1* (eds Breed, M. D., and Moore, J.) pp. 461–465, Academic Press, New York.

Byrne, R. W. and Corp, N. (2004). Neocortex predicts deception rate in primates. *Proceedings of the Royal Society London Series B* **271**, 1693–1699.

Byrne R. W. and Whitten, A. (1988). Tactical deception in primates. *Behavioural and brain sciences* **11**, 267–271.

Callea, F. O., Rohe, F., and Gordo, M. (2009). Hunting strategy of the margay (*Leopardos wiedii*) to attract the wild pied tamarin (*Saguinus bicolor*). *Neotropical Primates* **16**, 32–34.

Cézilly, F., Thomas, F., Médoc, V., and Perrot-Minnot, M. J. (2010). Host manipulation by parasites with complex life-cycles: adaptive or not? *Trends in Parasitology* **26**, 311–317.

Chapman, T., Bangham, J., Vinti, G., Seifried, B., Lung, O., Wolfner, M. F., Smith, H. K., and Partridge, L. (2003).

The sex peptide of *Drosophila melanogaster*: Female post-mating responses analyzed by using RNA interference. *Proceedings of the National Academy of Sciences USA* **100**, 9923–9928.

Cheney, D. L. and Seyfarth, R. M. (1988). Assessment of meaning and the detection of unreliable signals by vervet monkeys. *Animal Behaviour* **36**, 477–486.

Chiszar, D. D., Boyer, D., Lee, R., Murphy, J. B., and Radcliffe, C. W. (1990). Caudal luring in the southern death adder, *Acanthophis antarcticus. Journal of Herpetology* **24**, 253–260.

Cords, M. and Fuller, J. L. (2010). Infanticide in *Cercopithecus mitis stuhlmanni* in the Kakamega forest, Kenya: variation in the occurrence of an adaptive behavior. *International Journal of Primatology* **31**, 409–431.

Crudgington, H. S. and Siva-Jothy, M. T. (2000). Genital damage, kicking and early death. *Nature* **407**, 855–856.

Dahlgren, J. (1990). Females choose vigilant males: an experiment with the monogamous grey partridge, *Perdrix perdrix. Animal Behaviour* **39**, 646–651.

Dally, J. M., Clayton, N. S., and Emery, N. J. (2006). The behaviour and evolution of cache protection and pilferage. *Animal behaviour* **72**, 13–23.

Dawkins, R., Krebs, J. R. (1979). Arms races between and within species. *Proceedings of the Royal Society of London, B* **20**, 489–511.

Deane, C. D. (1944). The broken-wing behavior of the Killdeer. *Auk* **61**, 243–247.

Delano, P. H., Elgueda, D., Hamame, C. M., and Robles, L. (2007). Selective attention to visual stimuli reduces cochlear sensitivity in chinchillas. *Journal of Neurosciences* **27**, 4146–4153.

Di Giusto, B, Bessières, J-M., Guéroult, M., Lim, L. B., Marshall, D. J., Hossaert-McKey, M., and Gaume, L. (2010). Flower-scent mimicry masks a deadly trap in the carnivorous plant *Nepehntes rafflesiana. Journal of Ecology* **98**, 845–856.

Drummond, H. and Gordon, E. R. (1979). Luring in the neonate Alligator Snapping Turtle (*Macroclemys temminckii*): description and experimental analysis. *Zeitschrift fur Tierpsychologie* **50**, 136–152.

Evans, D. M., Ruxton, G. D., and Ruxton, D. A. (2004). Do false alarm anti-predatory flushes provide a foraging benefit to subdominant species? *Biologia Bratislva* **59**, 675–678.

Fenton, A., Magoolagan, L., Kennedy, Z., and Spencer, K. A. (2010). Parasite-induced warning coloration: a novel form of host manipulation. *Animal Behaviour* **81**, 417–422

Flower, T. (2011). Fork-tailed drongos use deceptive mimicked alarm calls to steal food. *Proceedings of the Royal Society B* **278**, 1548–1555.

Forslund, P. (1993). Vigilance in relation to brood size and predator abundance in the barnacle goose *Branta leucopsis. Animal Behaviour* **45**, 1199–1217.

Forsyth, A. and Alcock, J. (1990). Ambushing and prey-luring as alternative foraging tactics of the fly-catching rove beetle *Leistotrophus versicolor* (Coleoptera: Staphylinidae). *Journal of Insect Behavior* **3**, 703–718.

Foster, S. A. (1994). Inference of evolutionary pattern: diversionary displays of three-spined sticklebacks. *Behavioral Ecology* **5**, 114–121.

Gardner, A. and Grafen, A. (2009). Capturing the superorganism: a formal theory of group adaptation. *Journal of Evolutionary Biology* **22**, 659–671.

Gay, L., Hosken, D. J., Eady, P., Vasudev, R., and Tregenza, T. (2011). The evolution of harm-effect of sexual conflicts and population size. *Evolution* **65**, 725–737.

Gilad, Y., Man, O., Pääbo, S., and Lancet, D. (2003). Human specific loss of olfactory receptor genes. *Proceedings of the National Academy of sciences USA* **100**, 3324–3327.

Grafe, T. U. (2008). Toe waving in the Brown Marsh Frog *Rana baramica*: pedal luring to attract prey? *Scientia Bruneiana* **9**, 3–5.

Haftorn, S. (2000). Contexts and possible functions of alarm calling in the willow tit, *Parus montanus*; the principle of "better safe than sorry." *Behaviour* **137**, 437–449.

Hagman, M., Phillips, B. L., and Shine, R. (2008). Tails of enticement: caudal luring by an ambush foraging snake (*Acanthophis praelongus*, Elapidae). *Functional Ecology* **22**, 1134–1139.

Hagman, M. and Shine, R. (2008). Deceptive digits: the functional significance of toe waving by cannibalistic cane toads, *Chaunus marinus. Animal Behaviour* **75**, 123–131.

Han, C. S. and Jablonski, P. G. (2010). Male water striders attract predators to intimidate females into copulation. *Nature Communications* **1**, 52.

Hansknecht, K. A. (2008). Lingual luring by Mangrove Saltmarsh Snakes (*Nerodia Clarkii Compressicauda*). *Journal of Herpetology* **42**, 9–15.

Hare, J. F. and Atkins, B. A. (2001). The squirrel that cried wolf: reliability detection by juvenile Richardson's ground squirrels (*Spermophilus richardsonii*). *Behavioral Ecology and Sociobiology* **51**, 108–112.

Harsha, S., Satischandra, K., Kodituwakku, S. W., and Goodale, E. (2010). Assessing "false" alarm calls by a drongo (*Dicrurus paradiseus*) in mixed-species bird flocks. *Behavioral Ecology* **21**, 396–403.

Harvey, P. H., and Pagel, M. D. (1991). *The comparative method in evolutionary biology.* Oxford University Press, Oxford.

Hauser, M. D. (1992). Costs of deception-cheaters are punished in Rhesus-monkeys (*Macaca-Mulatta*). *Proceedings of the National Academy of Sciences* **89**, 12137–12139.

Haynes, K. F., Gemeno, C., Yeargan, K. V., Millar, J. G., and Johnson, K. M. (2002). Aggressive chemical mimicry of moth pheromones by a bolas spider: how does this specialist predator attract more than one species of prey? *Chemoecology* **12**, 99–105.

Heckenlively, D. B. (1972). Responses of adult killdeers to a downy young distress call. *Condor* **74**, 107–108.

Heesy, C. P. and Hall, M. I. (2010). The nocturnal bottleneck and the evolution of mammalian vision. *Brain, Behavior, and Evolution* **75**, 195–203.

Heinrich, B. (1999). *Mind of the Raven*. Harper Collins, New York.

Hertel, O. and Cousteau, C. (2008). *La Malédiction du Cloporte*. Tallandier, Paris.

Holman, L. (2010). Queen pheromones. The chemical crown governing insect social life. *Communicative and Integrative Biology* **3**, 1–3.

Hoover, J. P. and Robinson, S. K. (2007). Retaliatory mafia behavior by a parasitic cowbird favors host acceptance of parasitic eggs. *Proceedings of the National Academy of Sciences USA* **104**, 4479–4483.

Jenkins, S. H., Rothstein, A., and Green, W. C. H. (1995). Food-hoarding by kangaroo rats: a test of alternative hypotheses. *Ecology* **76**, 2470–2481.

Johnstone, R. A. and Keller, L. (2000). How males can gain by harming their mates: sexual conflict, seminal toxins, and the cost of mating. *American Naturalist* **156**, 368–377.

Kamimura, Y. (2007). Twin intromittent organs of *Drosophila* for traumatic insemination. *Biology Letters* **3**, 401–404.

Kamimura, Y. (2008). Copulatory wounds in the monandrous ant species *Formica japonica* (Hymenoptera: Formicidae). *Insectes Sociaux* **55**, 51–53.

Kamimura, Y. (2010). Copulation anatomy of *Drosophila melanogaster* (Diptera: Drosophilidae): wound-making organs and their possible roles. *Zoomorphology* **129**, 163–174.

Kennedy, J. S. (1992). *The New Anthropomorphism*. Cambridge University Press, Cambridge.

Krebs, J. R. and Dawkins, R. (1984). Animal signals: mind reading and manipulation. In: *Behavioural Ecology: An Evolutionary Approach*, 2nd edn (eds Krebs, J.R., and Davies, N.B.,) pp. 380–402. Blackwell, Oxford.

Kvitsiani, D. and Dickson, B. J. (2006). Shared neural circuitry for female and male sexual behaviours in Drosophila. *Current Biology* **16**, R355–R356.

Labov, J. B. (1980). Factors influencing infanticidal behavior in wild male house mice (*Mus musculus*). *Behavioral Ecology and Sociobiology* **6**, 297–303.

Laurenson, CH, Hudson, I. R., Jones, D. O. B., and Priede, I. G. (2004). Deep water observations of *Lophius piscato-rius* in the north-eastern Atlantic Ocean by means of a remotely operated vehicle. *Journal of Fish Biology* **65**, 947–960.

Lessells, C. M. (1999). Sexual conflict in animals. In: *Levels of Selection in Evolution*. (ed. Keller, L.) pp. 75–99, Princeton University Press, Princeton.

Lewis, S. M. and Cratsley, C. K. (2008). Flash signal evolution, matechoice, and predation in fireflies. *Annual Review of Entomology* **53**, 293–321.

Lloyd J. E. (1965). Aggressive mimicry in *Photuris*: firefly femmes fatales. *Science* **149**, 653–654.

Lloyd J. E. (1975). Aggressive mimicry in *Photuris* fireflies: signal repertoires by femmes fatales. *Science* **187**, 452–453.

Lloyd, James E. and Ballantyne, L. A., (2003). Taxonomy and behavior of *Photuris trivittata* sp. n. (Coleoptera: Lampyridae: Photurinae); redescription of *Aspisoma trilineata* (Say) comb. n. (Coleoptera: Lampyridae: Lampyrinae: Cratomorphini). *Florida Entomologist* **86**, 464–473.

Lorenzana, J. C. and Sealy, S. G. (2001). Fitness costs and benefits of cowbird egg ejection by gray catbirds. *Behavioral Ecology* **12**, 325–329.

Mallory, M. L., McNicol, D. K., and Walton, R. A. (1998). Risk-taking by incubating common goldeneyes and hooded mergansers. *Condor* **100**, 694–701.

Matsuoka, S. (1980). Pseudo alarm calling in titmice. *Tori* **29**, 87–90.

Maure, F., Brodeur, J., Ponlet, N., Doyon, J., Firlej, A., Elguero, E., and Thomas, F. (2011). The cost of a bodyguard. *Biology Letters*. **7**, 843–846.

Maynard-Smith, J. and Harper, D. (2003). *Animal Signals*. Oxford University Press, Oxford.

Mitchell, R. W. 1986. A framework for discussing deception. In: *Deception, Perspectives on Human and Nonhuman Deceit* (eds Mitchell, R. W., and Thompson, N.S.,) pp. 3–40. State University of New York Press, Albany.

Møller, A. P. (1988). False alarm calls as a means of resource usurpation in the great tit, *Parus major*. *Ethology* **79**, 25–30.

Møller, A. P. (1990). Parasites and sexual selection: current status of the Hamilton and Zuk hypothesis. *Journal of Evolutionary Biology* **3**, 319–328.

Moore, J. (2002). *Parasites and the Behavior of Animals*. Oxford Series in Ecology and Evolution. Oxford University Press. USA.

Morrow, E. H., Arnqvist, G., and Pitnick, S. (2003). Adaptation versus pleiotropy: why do males harm their mates. *Behavioral Ecology* **14**, 802–806.

Munk, O. (1999). The escal photophore of ceratioids (Pisces; Ceratioidei)—a review of structure and function. *Acta Zoologica* **80**, 265–284.

Munn, C. A. (1986). Birds that "cry wolf." *Nature* **391**, 143–145.

Murphy, J. B. (1976). Pedal luring in the Ptedodactylid frog, *Ceratophrys calcarata* Boulenger. *Herpetologica* **32**, 339–341.

Nelson, B. and Suthers R. A. (2004). Sound localization in a small passerine bird: discrimination of azimuths as a function of head orientation and sound frequency. *Journal of Experimental Biology* **207**, 4121–4133.

Nelson, X., Garnett, D. T., and Evans, C. S. (2010). Receiver psychology and the design of the deceptive caudal luring signal of the death adder. *Animal Behaviour* **79**, 555–561.

Nesse, R. M. (2005). Natural selection and the regulation of defenses. A signal detection analysis of the smoke detector principle. *Evolution and Human Behavior* **26**, 88–105.

Nowak, M. A., Tarnita, C. E., and Wilson, E. O. (2010). The evolution of eusociality. *Nature* **466**, 1057–1062.

Packer C. and Pusey, A. E. (1983). Adaptations of female lions to infanticides by incoming males. *American Naturalist* **121**, 716–728.

Pagel, M., Moller, A. P., Pomiankowski, A. (1998). Reduced parasitism by retaliatory cuckoos selects for hosts that rear cuckoo nestlings. *Behavioral Ecology* **9**, 566–572

Peer, B. D. and Rothstein, S. I. (2010). Phenotypic plasticity in common grackles (*Quiscalus quiscalus*) in response to repeated brood parasitism. *Auk* **127**, 293–299.

Pekár, S. and Křál, J. (2002). Mimicry complex in two central European zodariid spiders (Araneae: Zoodariidae): how *Zodarion* deceive ants. *Biological Journal of the Linnean Society* **75**, 517–532.

Pizzari, T. (2002). Food, vigilance, and sperm: the role of male direct benefits in the evolution of female preference in a polygamous bird. *Behavioral Ecology* **14**, 593–601.

Pollard, K. A. (2011). Making the most of alarm signals: the adaptive value of individual discrimination in an alarm context. *Behavioral Ecology* **22**, 93–100.

Ponton, F., Otálora-Luna, F., Lefèvre, T., Guerin, P., Lebarbenchon, C., Duneau, D., Biron, D. G., and Thomas, F. (2011). Water-seeking behavior in worm-infected crickets and reversibility of parasitic manipulation. *Behavioral Ecology* **22**, 392–400.

Poulin, R., (1995). "Adaptive" change in the behaviour of parasitized animals: a critical review. *International Journal of Parasitology* **25**, 1371–1383.

Poulin, R. (2010). Parasite manipulation of host behavior: an update and frequently asked questions. *Advances in the Study of Behavior* **41**, 151–186.

Poulin, R. and Cribb, TH. (2002). Trematode life cycles: short is sweet ? *Trends in Parasitology* **18**, 176–183.

Premack, D. (2007). Human and animal cognition: Continuity and discontinuity. *Proceedings of the National Academy of Sciences USA* **104**, 13861–13867.

Radcliffe, C. W., Chiszar, D., Estep, K. Murphy J. B., and Smith H. M. (1986). Observations of pedal luring and pedal movements in Leptodactylid frogs. *Journal of Herpetology* **20**, 300–306.

Radford, A. N., Bell, M. B. V., Hollén, L. I., and Ridley, A. R. (2010). Singing for your supper: sentinel calling by kleptoparasites can mitigate the cost to victims. *Evolution* **65**, 900–906.

Rasmussen, J. L., Sealy, S. G., and Underwood, T. J. (2009). Video recordings reveal the method of ejection of brown-headed cowbird eggs and no cost in American robins and gray catbirds. *Condor* **111**, 570–574.

Reiserer, R. S. and Schuett, G. W. (2008). Aggressive mimicry in neonates of the sidewinder rattlesnake, *Crotalus cerastes* (Sepentes: Viperidae): stimulus control and visual perception of prey luring. *Biological Journal of the Linnean Society* **95**, 81–91.

Ristau, C. A. (1991). Aspects of the cognitive ethology of an injury-feigning bird, the Piping Plover. In: *Cognitive Ethology: The Minds of Other Animals* (ed.Ristau, C. A.,) pp. 79–89. Lawrence Erlbaum Associates, Hillsdale.

Robert, M., Sorci, G., Møller, A. P., Hochberg, M. E., Pomiankowski, A., and Pagel, M. (1999). Retaliatory cuckoos and the evolution of host resistance to brood parasites. *Animal Behaviour* **58**, 817–824.

Rönn, J., Katvala, M., and Arnqvist, G. (2007). Coevolution between harmful male genitalia and female resistance in seed beetles. *Proceedings of the National Academy of Sciences of the USA* **104**, 10921–10925.

Schiestl, F. P., Ayasse M., Paulus, H. F., Lofstedt, C., Hansson, B. S., Ibarra, F., and Francke, W. (1999). Orchid pollination by sexual swindle. *Nature* **399**, 421–422.

Semple, S. and McComb, K. (1996). Behavioural deception. *Trends in Ecology and Evolution* **11**, 434–437.

Shimazaki, M. and Nakaya, K. (2004). Functional anatomy of the luring apparatus of the deep-sea ceratioid anglerfish *Cryptopsaras couesii* (Lophiiformes: Ceratiidae). *Ichtyological Research* **51**, 33–37.

Smith, J. N. M., Taitt, M. J., Zanette, L., and Myers-Smith, J. H. (2003). How do Brown-headed Cowbirds (*Molothrus ater*) cause nest failures in Song Sparrows (*Melospiza melodia*)? A removal experiment. *Auk* **120**, 772–783.

Soler, J. J., Møller, A. P., and Soler, M. (1998). Mafia behaviour and the evolution of facultative virulence. *Journal of theoretical Biology* **191**, 267–277.

Soler, M., Soler, J., Martinez, J. J., and Møller, A. P. (1995). Magpie host manipulation by great spotted cuckoos— Evidence for an avian mafia. *Evolution* **49**, 770–775.

Squires, K. A., Martin, K., and Goudie, R. I. (2007). Vigilance behaviour in the Harlequin Duck (*Histrionicus histrionicus*) during the preincubation period in Labrador: Are males vigilant for self or social partner? *Auk* **124**, 241–252.

Steele, M. A., Halkin, S. L., Smallwood, P. D., McKenna, T. J., Mitsopoulos, K., and Beam, M. (2008). Cache protection strategies of a scatter-hoarding rodent: do tree squirrels engage in behavioural deception? *Animal Behaviour* **75**, 705–714.

Tamura, N. (1995). Postcopulatory mate guarding by vocalization in the Formosan squirrel. *Behavioral Ecology and Sociobiology* **36**, 377–386.

Taylor, K. (2006). *Brainwashing: The Science of Thought Control.* Oxford University Press, New York.

Teuschl, Y., Hosken, D. J., and Blanckenhorn, W. U. (2007). Is reduced female survival after mating a by-product of male-male competition in the dung fly *Sepsis cynipsea*? *BMC Evolutionary Biology* **7**, 94.

Thomas, F., Adamo, S., and Moore, J. (2005). Parasitic manipulation: where are we and where should we go? *Behavioural Processes* **68**, 185–199.

Vander Wall, S. B. and Jenkins, S. H. (2003). Reciprocal pilferage and the evolution of food-hoarding behavior. *Behavioral Ecology* **14**, 656–667.

Vergoz, V., Schreurs, H. A., and Mercer, A. R. (2007). Queen pheromone blocks aversive learning in young worker bees. *Science* **317**, 384–386.

Wedell, N., Wiklund, C., and Bergström, J. (2009). Coevolution of non-fertile sperm and female receptivity in a butterfly. *Biology Letters* **5**, 678–681.

Wedell, N., Wiklund, C., and Cook, P. A. (2002). Monandry and polyandry as alternative lifestyles in a butterfly. *Behavioral Ecology* **13**, 450–455.

Welsh, H. H. and Lind, A. J. (2000). Evidence of lingual-luring by an aquatic snake. *Journal of Herpetology* **34**, 67–74.

Wenseleers, T., Ratnieks, F. L. W., and Billen, J. (2003). Caste fate conflict in swarm-founding social Hymenoptera: an inclusive fitness analysis. *Journal of Evolutionary Biology* **16**, 647–658.

Wheeler, B. C. (2009). Monkeys crying wolf? Tufted capuchin monkeys use anti-predator calls to usurp resources from conspecifics. *Proceedings of the Royal Society B* **276**, 3013–3018.

Wiens, J. A. (1966). Notes on the distraction display of the Virginia rail. *Wilson Bulletin* **78**, 229–231.

Wigby, S. and Chapman, T. (2004). Female resistance to male harm evolves in response to manipulation of sexual conflict. *Evolution* **58**, 1028–1037.

Wigby, S. and Chapman, T. (2005). Sex peptide causes mating costs in female *Drosophila melanogaster*. *Current Biology* **15**, 316–321.

Wiley, J. W. (1991) Ecology and behavior of the Zenaida Dove. *Ornitologia Neotropical* **2**, 49–75.

Yapici, N., Kim, Y. J., Ribeiro, C., and Dickson, B. J. (2008). A receptor that mediates the post-mating switch in *Drosophila* reproductive behaviour. *Nature* **451**, 33–37.

Young, A. J., Monfort, S. L., and Clutton-Brock, T. H. (2008). The causes of physiological suppression among female meerkats: A role for subordinate restraint due to the threat of infanticide? *Hormones and Behavior* **53**, 131–139.

Zahavi, A. (1979). Parasitism and nest predation in parasitic cuckoos. *American Naturalist* **113**, 157–159.

Zahavi, A. and Zahavi, A. (1997). *The Handicap Principle: a Missing Piece of Darwin's Puzzle.* Oxford University Press, Oxford.

Zimmer, C. (2001). *Parasite Rex.* Free Press, London.

Afterword

Alex Kacelnik

Everyday language has many words to address ways in which individuals manipulate the behavior or beliefs of others. We talk about inducing, tempting, persuading, seducing, and so on, whenever one organism causes another to behave in its interest through sensory stimulation rather than force. The different terms recognize subtle variations. For instance, to seduce (from the Latin *se*, meaning "away," or "astray" and *ducere*—"to lead") hints that the action elicited is not to the benefit of the receiver, while to induce (from the Latin *in*, roughly "in" and again *ducere*) emphasizes the actor's causing of the receiver's behavior, with no implication of the effect on the receiver. Because humans often assume that their behavior is caused by beliefs, some of these terms refer to the formation of beliefs rather than behavioral change. For instance, persuasion (from the Latin *per*, meaning "thoroughly" and *suadeo* meaning "I advise") puts the weight on changing the receiver's beliefs rather than directly causing a change in behavior, and again does not imply that the receiver will lose by being persuaded. It seems appropriate to use the term manipulation for the broad category of interactions that are performed by an actor to modify others' behavior, rather than restricting the notion only to cases where the receiver is led astray.

The interactions between non-human organisms are rich in such manipulatory effects, and this becomes more obvious when there is sharp conflict between the interests of actor and receiver. In this sense, all communication is manipulation: evolution can only favor signalers that produce stimuli that increase the probability of receivers acting in the actor's benefit, however indirect this may be. The song of male birds or crickets, the begging calls of nestlings, the pheromone release by female moths, the deimatic display of a preying mantis flashing away predators, are all cases when an actor's behavior results from an adaptive history of influencing others' behavior through the receivers' sensory input. The immediate question is highly predictable: why are receivers manipulable? And so is the answer: on balance, for a receiver's response to be maintained, the expected outcome of responding must be positive over the range of natural conditions in which a species evolves. If some stimuli have a distribution of positive and negative outcomes, then the frequency and magnitude of each kind will exert overall control of the selective pressures. The conceptual approach of Signal Detection Theory handles this. Basically, the consequences of responding or not to any stimulus can always be seen as 2 x 2 matrix, with the entries in the rows being whether to act or not, and those in the columns showing a binary classification of the stimulus ("true" or "false," "good" or "bad," "food" or "predator," "own egg" or "parasite's egg," etc.). A small bird flying away from an owl-looking stimulus may, with complementary probabilities, miss a juicy displaying mantis or successfully avoid becoming breakfast to an owl, while a conspecific that attacks the same stimulus instead of flying away may benefit by a meal or become the meal of an owl, also with complementary probabilities. The play between probabilities and fitness consequences determine the net selective advantage of responding to noisy signals. Applied to these instances when an actor influences a receiver to change its behavior, both manipulator and manipulated respond adaptively in the face of uncertain information and probabilistic outcomes.

Parasites also can induce host's behaviors that favor them, but, in contrast to the examples discussed so far, they can take shortcuts. When a fungus of the genus *Ophiocordyceps* (= *Cordyceps*) takes control of the behavior of its host ant, it does not need to act through the ant's sensory processes. The fungus can affect efferent pathways in the ants' nervous system to cause locomotion and other actions that lead to the ant's demise in favor of the fungus' reproduction. The ant is not "persuaded" by stimuli it has evolved to attend to, but "forced" to do it by the fungus using the ant's motor system as a puppeteer uses strings attached to the limbs of its puppets. The right framework is more distantly related to signal detection theory.

Cézilly and Thomas review, examine, and classify many examples of behavioral manipulation among non-parasites, asking whether the concept of host manipulation, often used to address parasite–host interactions, is equivalent to manipulation between free-living interacting organisms. Quoting, they say:

"The aim of this chapter is precisely to review what can be regarded as cases of behavioral 'manipulation' in the animal kingdom, and to examine the different mechanisms and strategies at work, in order to evaluate to what extent 'host manipulation' by parasites is a unique phenomenon or not."

Since non-parasites by definition do not have hosts, host manipulation is indeed exclusive to parasites, but this is not what is meant. Further, since virtually all communication is manipulation, it is hardly controversial that animals (or plants) manipulate the behavior of other animals through their senses to their benefit. What interests Cézilly and Thomas is whether, as a category of adaptations, manipulatory behaviors in parasites and non-parasites share sufficient features to be productively lumped into one common category, so that theoretical interpretations can fertilize thinking across fields. It is not news that some scholars are natural splitters and others are lumpers, nor is it that there is a role for both attitudes in the clarification of arguments. Cézilly and Thomas seem to lean in the direction of lumping, and this commentator would perhaps lean towards splitting. In any case, Cézilly and Thomas

provide an insightful classification of manipulatory behaviors, including five categories: deceit through sensory exploitation, exploitation of compensatory mechanisms, coercive exploitation, manipulation of information, and neuroendocrine manipulation. The typical host-manipulation scenario, exemplified above with *Ophiocordyceps* control of ant behavior, would fall in the very last category.

Their classification is an excellent platform for thinking about these issues, but I see a substantial difference between the last category, common in endoparasites, and the rest. Only endoparasites have direct access to the efferent control paths of their host, and thus they do not need to act through behavioral adaptations of the receiver. Notice that this is not even true for all parasites. When offspring of avian brood parasites use begging calls to control the provisioning behavior of their host, the system is immediately suitable for a signal detection approach that assumes that both parties behave adaptively (viz. Davies et al. 1996). I feel that manipulations by endoparasites do differ from the other four categories at this important level, and while it of course makes sense not to neglect manipulations by free-living animals (little work for behavioral ecologists would be left if one did), host manipulations do appeal to a different theoretical approach.

It is possible to ask why, in the case of manipulations such as those by *Ophiocordyceps*, don't ants' nervous systems evolve internal pathways by which downstream motor units "disobey" central commands when the probability that the central system has been hijacked by a fungus is high. This would align the theoretical perspective of host manipulation closer to other forms of behavioural control.

Cézilly and Thomas's review is an important contribution, and their classification of manipulatory interactions should hopefully induce, seduce, and persuade Behavioral ecologists dealing with social interactions in their broadest sense.

References

Davies, N. B., Brooke, M. L., and Kacelnik, A., (1996). Recognition errors and probability of parasitism determine whether reed warblers should accept or reject mimetic cuckoo eggs. *Proceeding of the Royal Society of London B* **263**, 925–931.

Index

Page numbers in *italic* refer to figures and tables, and those in **bold** to boxes.